ASTRONOMY AND ASTROPHYSICS LIBRARY

Series Editors: G. Börner, Garching, Germany
A. Burkert, München, Germany
W. B. Burton, Charlottesville, VA, USA and
 Leiden, The Netherlands
M. A. Dopita, Canberra, Australia
A. Eckart, Köln, Germany
T. Encrenaz, Meudon, France
B. Leibundgut, Garching, Germany
J. Lequeux, Paris, France
A. Maeder, Sauverny, Switzerland
V. Trimble, College Park, MD, and Irvine, CA, USA

J.-L. Starck   F. Murtagh

Astronomical Image and Data Analysis

Second Edition

With 119 Figures

 Springer

Jean-Luc Starck
Service d'Astrophysique CEA/Saclay
Orme des Merisiers, Bat 709
91191 Gif-sur-Yvette Cedex, France

Fionn Murtagh
Dept. Computer Science
Royal Holloway
University of London
Egham, Surrey TW20 0EX, UK

Cover picture: The cover image to this 2nd edition is from the Deep Impact project. It was taken approximately 8 minutes after impact on 4 July 2005 with the CLEAR6 filter and deconvolved using the Richardson-Lucy method. We thank Don Lindler, Ivo Busko, Mike A'Hearn and the Deep Impact team for the processing of this image and for providing it to us.

Library of Congress Control Number: 2006930922

ISSN 0941-7834
ISBN-10 3-540-33024-0 2nd Edition Springer Berlin Heidelberg New York
ISBN-13 978-3-540-33024-0 2nd Edition Springer Berlin Heidelberg New York
ISBN 3-540-42885-2 1st Edition Springer Berlin Heidelberg New York

This work is subject to copyright. All rights are reserved, whether the whole or part of the material is concerned, specifically the rights of translation, reprinting, reuse of illustrations, recitation, broadcasting, reproduction on microfilm or in any other way, and storage in data banks. Duplication of this publication or parts thereof is permitted only under the provisions of the German Copyright Law of September 9, 1965, in its current version, and permission for use must always be obtained from Springer. Violations are liable to prosecution under the German Copyright Law.

Springer is a part of Springer Science+Business Media

springer.com

Springer-Verlag Berlin Heidelberg 2006

The use of general descriptive names, registered names, trademarks, etc. in this publication does not imply, even in the absence of a specific statement, that such names are exempt from the relevant protective laws and regulations and therefore free for general use.

Typesetting: by the authors
Final layout: Data conversion and production by LE-TEX Jelonek, Schmidt & VöcklerGbR, Leipzig, Germany
Cover design: *design & production* GmbH, Heidelberg

Printed on acid-free paper   SPIN: 11595496   55/3141 - 5 4 3 2 1 0

Preface to the Second Edition

This book presents material which is more algorithmically oriented than most alternatives. It also deals with topics that are at or beyond the state of the art. Examples include *practical* and *applicable* wavelet and other multiresolution transform analysis. New areas are broached like the ridgelet and curvelet transforms. The reader will find in this book an engineering approach to the interpretation of scientific data.

Compared to the 1st Edition, various additions have been made throughout, and the topics covered have been updated. The background or environment of this book's topics include continuing interest in e-science and the virtual observatory, which are based on web based and increasingly web service based science and engineering.

Additional colleagues whom we would like to acknowledge in this 2nd edition include: Bedros Afeyan, Nabila Aghanim, Emmanuel Candès, David Donoho, Jalal Fadili, and Sandrine Pires, We would like to particularly acknowledge Olivier Forni who contributed to the discussion on compression of hyperspectral data, Yassir Moudden on multiwavelength data analysis and Vicent Martínez on the genus function.

The cover image to this 2nd edition is from the Deep Impact project. It was taken approximately 8 minutes after impact on 4 July 2005 with the CLEAR6 filter and deconvolved using the Richardson-Lucy method. We thank Don Lindler, Ivo Busko, Mike A'Hearn and the Deep Impact team for the processing of this image and for providing it to us.

Paris, London
June, 2006

Jean-Luc Starck
Fionn Murtagh

Preface to the First Edition

When we consider the ever increasing amount of astronomical data available to us, we can well say that the needs of modern astronomy are growing by the day. Ever better observing facilities are in operation. The fusion of information leading to the coordination of observations is of central importance.

The methods described in this book can provide effective and efficient ripostes to many of these issues. Much progress has been made in recent years on the methodology front, in line with the rapid pace of evolution of our technological infrastructures.

The central themes of this book are *information* and *scale*. The approach is astronomy-driven, starting with real problems and issues to be addressed. We then proceed to comprehensive theory, and implementations of demonstrated efficacy.

The field is developing rapidly. There is little doubt that further important papers, and books, will follow in the future.

Colleagues we would like to acknowledge include: Alexandre Aussem, Albert Bijaoui, François Bonnarel, Jonathan G. Campbell, Ghada Jammal, René Gastaud, Pierre-François Honoré, Bruno Lopez, Mireille Louys, Clive Page, Eric Pantin, Philippe Querre, Victor Racine, Jérôme Rodriguez, and Ivan Valtchanov.

The cover image is from Jean-Charles Cuillandre. It shows a five minute exposure (5 60-s dithered and stacked images), R filter, taken with CFH12K wide field camera (100 million pixels) at the primary focus of the CFHT in July 2000. The image is from an extremely rich zone of our Galaxy, containing star formation regions, dark nebulae (molecular clouds and dust regions), emission nebulae (H_α), and evolved stars which are scattered throughout the field in their two-dimensional projection effect. This zone is in the constellation of Saggitarius.

Paris, Belfast   *Jean-Luc Starck*
June, 2002   *Fionn Murtagh*

Table of Contents

1. **Introduction to Applications and Methods** 1
   1.1 Introduction 1
   1.2 Transformation and Data Representation 3
       1.2.1 Fourier Analysis 5
       1.2.2 Time-Frequency Representation 6
       1.2.3 Time-Scale Representation: The Wavelet Transform .. 9
       1.2.4 The Radon Transform 12
       1.2.5 The Ridgelet Transform 12
       1.2.6 The Curvelet Transform 14
   1.3 Mathematical Morphology 15
   1.4 Edge Detection 18
       1.4.1 First Order Derivative Edge Detection 18
       1.4.2 Second Order Derivative Edge Detection 20
   1.5 Segmentation 23
   1.6 Pattern Recognition 24
   1.7 Chapter Summary 27

2. **Filtering** 29
   2.1 Introduction 29
   2.2 Multiscale Transforms 31
       2.2.1 The A Trous Isotropic Wavelet Transform 31
       2.2.2 Multiscale Transforms Compared
             to Other Data Transforms 33
       2.2.3 Choice of Multiscale Transform 36
       2.2.4 The Multiresolution Support 37
   2.3 Significant Wavelet Coefficients 38
       2.3.1 Definition 38
       2.3.2 Noise Modeling 39
       2.3.3 Automatic Estimation of Gaussian Noise 40
       2.3.4 Detection Level Using the FDR 48
   2.4 Filtering and Wavelet Coefficient Thresholding 50
       2.4.1 Thresholding 50
       2.4.2 Iterative Filtering 51
       2.4.3 Other Wavelet Denoising Methods 52

         2.4.4 Experiments 54
         2.4.5 Iterative Filtering with a Smoothness Constraint 56
    2.5 Filtering from the Curvelet Transform 57
         2.5.1 Contrast Enhancement 57
         2.5.2 Curvelet Denoising 59
         2.5.3 The Combined Filtering Method 61
    2.6 Haar Wavelet Transform and Poisson Noise 63
         2.6.1 Haar Wavelet Transform 63
         2.6.2 Poisson Noise and Haar Wavelet Coefficients 64
         2.6.3 Experiments 67
    2.7 Chapter Summary 70

3. **Deconvolution** 71
    3.1 Introduction 71
    3.2 The Deconvolution Problem 74
    3.3 Linear Regularized Methods 75
         3.3.1 Least Squares Solution 75
         3.3.2 Tikhonov Regularization 75
         3.3.3 Generalization 76
    3.4 CLEAN 78
    3.5 Bayesian Methodology 78
         3.5.1 Definition 78
         3.5.2 Maximum Likelihood with Gaussian Noise 79
         3.5.3 Gaussian Bayes Model 79
         3.5.4 Maximum Likelihood with Poisson Noise 80
         3.5.5 Poisson Bayes Model 81
         3.5.6 Maximum Entropy Method 81
         3.5.7 Other Regularization Models 82
    3.6 Iterative Regularized Methods 84
         3.6.1 Constraints 84
         3.6.2 Jansson-Van Cittert Method 85
         3.6.3 Other Iterative Methods 85
    3.7 Wavelet-Based Deconvolution 86
         3.7.1 Introduction 86
         3.7.2 Wavelet-Vaguelette Decomposition 87
         3.7.3 Regularization from the Multiresolution Support 90
         3.7.4 Wavelet CLEAN 93
         3.7.5 The Wavelet Constraint 98
    3.8 Deconvolution and Resolution 104
    3.9 Super-Resolution 105
         3.9.1 Definition 105
         3.9.2 Gerchberg-Saxon Papoulis Method 106
         3.9.3 Deconvolution with Interpolation 107
         3.9.4 Undersampled Point Spread Function 107
    3.10 Conclusions and Chapter Summary 109

4. Detection 111
4.1 Introduction 111
4.2 From Images to Catalogs 112
4.3 Multiscale Vision Model 116
4.3.1 Introduction 116
4.3.2 Multiscale Vision Model Definition 117
4.3.3 From Wavelet Coefficients to Object Identification 117
4.3.4 Partial Reconstruction 120
4.3.5 Examples 122
4.3.6 Application to ISOCAM Data Calibration 122
4.4 Detection and Deconvolution 126
4.5 Detection in the Cosmological Microwave Background 130
4.5.1 Introduction 130
4.5.2 Point Sources on a Gaussian Background 132
4.5.3 Non-Gaussianity 132
4.6 Conclusion 135
4.7 Chapter Summary 135

5. Image Compression 137
5.1 Introduction 137
5.2 Lossy Image Compression Methods 139
5.2.1 The Principle 139
5.2.2 Compression with Pyramidal Median Transform 140
5.2.3 PMT and Image Compression 142
5.2.4 Compression Packages 145
5.2.5 Remarks on these Methods 146
5.2.6 Other Lossy Compression Methods 148
5.3 Comparison 149
5.3.1 Quality Assessment 149
5.3.2 Visual Quality 150
5.3.3 First Aladin Project Study 151
5.3.4 Second Aladin Project Study 155
5.3.5 Computation Time 159
5.3.6 Conclusion 160
5.4 Lossless Image Compression 161
5.4.1 Introduction 161
5.4.2 The Lifting Scheme 161
5.4.3 Comparison 166
5.5 Large Images: Compression and Visualization 167
5.5.1 Large Image Visualization Environment: LIVE 167
5.5.2 Decompression by Scale and by Region 168
5.5.3 The SAO-DS9 LIVE Implementation 169
5.6 Hyperspectral Compression for Planetary Space Missions 170
5.7 Chapter Summary 173

6. Multichannel Data ... 175
- 6.1 Introduction ... 175
- 6.2 The Wavelet-Karhunen-Loève Transform ... 176
  - 6.2.1 Definition ... 176
  - 6.2.2 Correlation Matrix and Noise Modeling ... 178
  - 6.2.3 Scale and Karhunen-Loève Transform ... 179
  - 6.2.4 The WT-KLT Transform ... 179
  - 6.2.5 The WT-KLT Reconstruction Algorithm ... 180
- 6.3 Noise Modeling in the WT-KLT Space ... 180
- 6.4 Multichannel Data Filtering ... 181
  - 6.4.1 Introduction ... 181
  - 6.4.2 Reconstruction from a Subset of Eigenvectors ... 181
  - 6.4.3 WT-KLT Coefficient Thresholding ... 183
  - 6.4.4 Example: Astronomical Source Detection ... 183
- 6.5 The Haar-Multichannel Transform ... 183
- 6.6 Independent Component Analysis ... 184
  - 6.6.1 Definition ... 184
  - 6.6.2 JADE ... 185
  - 6.6.3 FastICA ... 186
- 6.7 CMB Data and the SMICA ICA Method ... 189
  - 6.7.1 The CMB Mixture Problem ... 189
  - 6.7.2 SMICA ... 190
- 6.8 ICA and Wavelets ... 193
  - 6.8.1 WJADE ... 193
  - 6.8.2 Covariance Matching in Wavelet Space: WSMICA ... 194
  - 6.8.3 Numerical Experiments ... 195
- 6.9 Chapter Summary ... 198

7. An Entropic Tour of Astronomical Data Analysis ... 201
- 7.1 Introduction ... 201
- 7.2 The Concept of Entropy ... 204
- 7.3 Multiscale Entropy ... 210
  - 7.3.1 Definition ... 210
  - 7.3.2 Signal and Noise Information ... 212
- 7.4 Multiscale Entropy Filtering ... 215
  - 7.4.1 Filtering ... 215
  - 7.4.2 The Regularization Parameter ... 215
  - 7.4.3 Use of a Model ... 217
  - 7.4.4 The Multiscale Entropy Filtering Algorithm ... 218
  - 7.4.5 Optimization ... 219
  - 7.4.6 Examples ... 220
- 7.5 Deconvolution ... 220
  - 7.5.1 The Principle ... 220
  - 7.5.2 The Parameters ... 224
  - 7.5.3 Examples ... 225

	7.6	Multichannel Data Filtering 225
	7.7	Relevant Information in an Image 228
	7.8	Multiscale Entropy and Optimal Compressibility 230
	7.9	Conclusions and Chapter Summary 231

8. Astronomical Catalog Analysis 233
8.1 Introduction 233
8.2 Two-Point Correlation Function 234
    8.2.1 Introduction 234
    8.2.2 Determining the 2-Point Correlation Function 235
    8.2.3 Error Analysis 236
    8.2.4 Correlation Length Determination 237
    8.2.5 Creation of Random Catalogs 237
    8.2.6 Examples 238
    8.2.7 Limitation of the Two-Point Correlation Function:
          Toward Higher Moments 242
8.3 The Genus Curve 245
8.4 Minkowski Functionals 247
8.5 Fractal Analysis 249
    8.5.1 Introduction 249
    8.5.2 The Hausdorff and Minkowski Measures 250
    8.5.3 The Hausdorff and Minkowski Dimensions 251
    8.5.4 Multifractality 251
    8.5.5 Generalized Fractal Dimension 253
    8.5.6 Wavelets and Multifractality 253
8.6 Spanning Trees and Graph Clustering 257
8.7 Voronoi Tessellation and Percolation 259
8.8 Model-Based Clustering 260
    8.8.1 Modeling of Signal and Noise 260
    8.8.2 Application to Thresholding 262
8.9 Wavelet Analysis 263
8.10 Nearest Neighbor Clutter Removal 265
8.11 Chapter Summary 266

9. Multiple Resolution in Data Storage and Retrieval 267
9.1 Introduction 267
9.2 Wavelets in Database Management 267
9.3 Fast Cluster Analysis 269
9.4 Nearest Neighbor Finding on Graphs 271
9.5 Cluster-Based User Interfaces 272
9.6 Images from Data 273
    9.6.1 Matrix Sequencing 273
    9.6.2 Filtering Hypertext 277
    9.6.3 Clustering Document-Term Data 278
9.7 Chapter Summary 282

10. Towards the Virtual Observatory 285
  10.1 Data and Information 285
  10.2 The Information Handling Challenges Facing Us............ 287

Appendix

A. A Trous Wavelet Transform 291

B. Picard Iteration 297

C. Wavelet Transform Using the Fourier Transform.......... 299

D. Derivative Needed for the Minimization 303

E. Generalization of the Derivative
   Needed for the Minimization 307

F. Software and Related Developments 309

Bibliography... 311

Index 331

1. Introduction to Applications and Methods

1.1 Introduction

"May you live in interesting times!" ran the old Chinese wish. The early years of the third millennium are interesting times for astronomy, as a result of the tremendous advances in our computing and information processing environment and infrastructure. The advances in signal and image processing methods described in this book are of great topicality as a consequence. Let us look at some of the overriding needs of contemporary observational astronomical.

Unlike in Earth observation or meteorology, astronomers do not want to interpret data and, having done so, delete it. Variable objects (supernovae, comets, etc.) bear witness to the need for astronomical data to be available indefinitely. The unavoidable problem is the sheer overwhelming quantity of data which is now collected. The only basis for selective choice for what must be kept long-term is to associate more closely the data capture with the information extraction and knowledge discovery processes. We have got to understand our scientific knowledge discovery mechanisms better in order to make the correct selection of data to keep long-term, including the appropriate resolution and refinement levels.

The vast quantities of visual data collected now and in the future present us with new problems and opportunities. Critical needs in our software systems include compression and progressive transmission, support for differential detail and user navigation in data spaces, and "thinwire" transmission and visualization. The technological infrastructure is one side of the picture.

Another side of this same picture, however, is that our human ability to interpret vast quantities of data is limited. A study by D. Williams, CERN, has quantified the maximum possible volume of data which can conceivably be interpreted at CERN. This points to another more fundamental justification for addressing the critical technical needs indicated above. This is that selective and prioritized transmission, which we will term intelligent streaming, is increasingly becoming a key factor in human understanding of the real world, as mediated through our computing and networking base. We need to receive condensed, summarized data first, and we can be aided in our understanding of the data by having more detail added progressively. A hyperlinked and networked world makes this need for summarization more and more acute. We need to take resolution scale into account in our information and knowledge spaces. This is a key aspect of an intelligent streaming system.

A further area of importance for scientific data interpretation is that of storage and display. Long-term storage of astronomical data, we have already noted, is part and parcel of our society's memory (a formulation due to Michael Kurtz, Center for Astrophysics, Smithsonian Institute). With the rapid obsolescence of storage devices, considerable efforts must be undertaken to combat social amnesia. The positive implication is the ever-increasing complementarity of professional observational astronomy with education and public outreach.

Astronomy's data centers and image and catalog archives play an important role in our society's collective memory. For example, the SIMBAD database of astronomical objects at Strasbourg Observatory contains data on 3 million objects, based on 7.5 million object identifiers. Constant updating of SIMBAD is a collective cross-institutional effort. The MegaCam camera at the Canada-France-Hawaii Telescope (CFHT), Hawaii, is producing images of dimensions 16000×16000, 32-bits per pixel. The European Southern Observatory's VLT (Very Large Telescope) is beginning to produce vast quantities of very large images. Increasingly, images of size 1 GB or 2 GB, for a single image, are not exceptional. CCD detectors on other telescopes, or automatic plate scanning machines digitizing photographic sky surveys, produce lots more data. Resolution and scale are of key importance, and so also is region of interest. In multiwavelength astronomy, the fusion of information and data is aimed at, and this can be helped by the use of resolution similar to our human cognitive processes. Processing (calibration, storage and transmission formats and approaches) and access have not been coupled as closely as they could be. Knowledge discovery is the ultimate driver.

Many ongoing initiatives and projects are very relevant to the work described in later chapters.

Image and Signal Processing. The major areas of application of image and signal processing include the following.

- **Visualization:** Seeing our data and signals in a different light is very often a revealing and fruitful thing to do. Examples of this will be presented throughout this book.
- **Filtering:** A signal in the physical sciences rarely exists independently of noise, and noise removal is therefore a useful preliminary to data interpretation. More generally, data cleaning is needed, to bypass instrumental measurement artifacts, and even the inherent complexity of the data. Image and signal filtering will be presented in Chapter 2.
- **Deconvolution:** Signal "deblurring" is used for reasons similar to filtering, as a preliminary to signal interpretation. Motion deblurring is rarely important in astronomy, but removing the effects of atmospheric blurring, or quality of seeing, certainly is of importance. There will be a wide-ranging

discussion of the state of the art in deconvolution in astronomy in Chapter 3.
- **Compression:** Consider three different facts. Long-term storage of astronomical data is important. A current trend is towards detectors accommodating ever-larger image sizes. Research in astronomy is a cohesive but geographically distributed activity. All three facts point to the importance of effective and efficient compression technology. In Chapter 5, the state of the art in astronomical image compression will be surveyed.
- **Mathematical morphology:** Combinations of dilation and erosion operators, giving rise to opening and closing operations, in boolean images and in greyscale images, allow for a truly very esthetic and immediately practical processing framework. The median function plays its role too in the context of these order and rank functions. Multiple scale mathematical morphology is an immediate generalization. There is further discussion on mathematical morphology below in this chapter.
- **Edge detection:** Gradient information is not often of central importance in astronomical image analysis. There are always exceptions of course.
- **Segmentation and pattern recognition:** These are discussed in Chapter 4, dealing with object detection. In areas outside astronomy, the term feature selection is more normal than object detection.
- **Multidimensional pattern recognition:** General multidimensional spaces are analyzed by clustering methods, and by dimensionality mapping methods. Multiband images can be taken as a particular case. Such methods are pivotal in Chapter 6 on multichannel data, 8 on catalog analysis, and 9 on data storage and retrieval.
- **Hough and Radon transforms, leading to 3D tomography and other applications:** Detection of alignments and curves is necessary for many classes of segmentation and feature analysis, and for the building of 3D representations of data. Gravitational lensing presents one area of potential application in astronomy imaging, although the problem of faint signal and strong noise is usually the most critical one. Ridgelet and curvelet transforms (discussed below in this chapter) offer powerful generalizations of current state of the art ways of addressing problems in these fields.

A number of outstanding general texts on image and signal processing are available. These include Gonzalez and Woods (1992), Jain (1990), Pratt (1991), Parker (1996), Castleman (1995), Petrou and Bosdogianni (1999), Bovik (2000). A text of ours on image processing and pattern recognition is available on-line (Campbell and Murtagh, 2001). Data analysis texts of importance include Bishop (1995), and Ripley (1995).

1.2 Transformation and Data Representation

Many different transforms are used in data processing, – Haar, Radon, Hadamard, etc. The Fourier transform is perhaps the most widely used. The goal of these transformations is to obtain a *sparse* representation of the data, and to pack most information into a small number of samples. For example, a sine signal $f(t) = \sin(2\pi\nu t)$, defined on N pixels, requires only two samples (at frequencies $-\nu$ and ν) in the Fourier domain for an exact representation. Wavelets and related multiscale representations pervade all areas of signal processing. The recent inclusion of wavelet algorithms in JPEG 2000 – the new still-picture compression standard – testifies to this lasting and significant impact. The reason for the success of wavelets is due to the fact that wavelet bases represent well a large class of signals. Therefore this allows us to detect roughly isotropic elements occurring at all spatial scales and locations. Since noise in the physical sciences is often not Gaussian, modeling in wavelet space of many kind of noise – Poisson noise, combination of Gaussian and Poisson noise components, non-stationary noise, and so on – has been a key motivation for the use of wavelets in scientific, medical, or industrial applications. The wavelet transform has also been extensively used in astronomical data analysis during the last ten years. A quick search with ADS (NASA Astrophysics Data System, adswww.harvard.edu) shows that around 500 papers contain the keyword "wavelet" in their abstract, and this holds for all astrophysical domains, from study of the sun through to CMB (Cosmic Microwave Background) analysis:

- Sun: active region oscillations (Ireland et al., 1999; Blanco et al., 1999), determination of solar cycle length variations (Fligge et al., 1999), feature extraction from solar images (Irbah et al., 1999), velocity fluctuations (Lawrence et al., 1999).
- Solar system: asteroidal resonant motion (Michtchenko and Nesvorny, 1996), classification of asteroids (Bendjoya, 1993), Saturn and Uranus ring analysis (Bendjoya et al., 1993; Petit and Bendjoya, 1996).
- Star studies: Ca II feature detection in magnetically active stars (Soon et al., 1999), variable star research (Szatmary et al., 1996).
- Interstellar medium: large-scale extinction maps of giant molecular clouds using optical star counts (Cambrésy, 1999), fractal structure analysis in molecular clouds (Andersson and Andersson, 1993).
- Planetary nebula detection: confirmation of the detection of a faint planetary nebula around IN Com (Brosch and Hoffman, 1999), evidence for extended high energy gamma-ray emission from the Rosette/Monoceros Region (Jaffe et al., 1997).
- Galaxy: evidence for a Galactic gamma-ray halo (Dixon et al., 1998).
- QSO: QSO brightness fluctuations (Schild, 1999), detecting the non-Gaussian spectrum of QSO Ly_α absorption line distribution (Pando and Fang, 1998).
- Gamma-ray burst: GRB detection (Kolaczyk, 1997; Norris et al., 1994) and GRB analysis (Greene et al., 1997; Walker et al., 2000).
- Black hole: periodic oscillation detection (Steiman-Cameron et al., 1997; Scargle, 1997)

1.2 Transformation and Data Representation

- Galaxies: starburst detection (Hecquet et al., 1995), galaxy counts (Aussel et al., 1999; Damiani et al., 1998), morphology of galaxies (Weistrop et al., 1996; Kriessler et al., 1998), multifractal character of the galaxy distribution (Martínez et al., 1993a).
- Galaxy cluster: sub-structure detection (Pierre and Starck, 1998; Krywult et al., 1999; Arnaud et al., 2000), hierarchical clustering (Pando et al., 1998a), distribution of superclusters of galaxies (Kalinkov et al., 1998).
- Cosmic Microwave Background: evidence for scale-scale correlations in the Cosmic Microwave Background radiation in COBE data (Pando et al., 1998b), large-scale CMB non-Gaussian statistics (Popa, 1998; Aghanim et al., 2001), massive CMB data set analysis (Gorski, 1998).
- Cosmology: comparing simulated cosmological scenarios with observations (Lega et al., 1996), cosmic velocity field analysis (Rauzy et al., 1993).

This broad success of the wavelet transform is due to the fact that astronomical data generally gives rise to complex hierarchical structures, often described as fractals. Using multiscale approaches such as the wavelet transform, an image can be decomposed into components at different scales, and the wavelet transform is therefore well-adapted to the study of astronomical data.

This section reviews briefly some of the existing transforms.

1.2.1 Fourier Analysis

The Fast Fourier Transform. The Fourier transform of a continuous function $f(t)$ is defined by:

$$\hat{f}(\nu) = \int_{-\infty}^{+\infty} f(t) e^{-i 2\pi \nu t} dt \tag{1.1}$$

and the inverse Fourier transform is:

$$f(t) = \int_{-\infty}^{+\infty} \hat{f}(\nu) e^{i 2\pi \nu t} du \tag{1.2}$$

The discrete Fourier transform is given by:

$$\hat{f}(u) = \frac{1}{N} \sum_{k=-\infty}^{+\infty} f(k) e^{-i 2\pi \frac{uk}{N}} \tag{1.3}$$

and the inverse discrete Fourier transform is:

$$\hat{f}(k) = \sum_{u=-\infty}^{+\infty} f(u) e^{i 2\pi \frac{uk}{N}} \tag{1.4}$$

In the case of images (two variables), this is:

$$\hat{f}(u,v) = \frac{1}{MN}\sum_{l=-\infty}^{+\infty}\sum_{k=-\infty}^{+\infty} f(k,l)e^{-2i\pi(\frac{uk}{M}+\frac{vl}{N})}$$

$$f(k,l) = \sum_{u=-\infty}^{+\infty}\sum_{v=-\infty}^{+\infty} \hat{f}(u,v)e^{2i\pi(\frac{uk}{M}+\frac{vl}{N})} \qquad (1.5)$$

Since $\hat{f}(u,v)$ is generally complex, this can be written using its real and imaginary parts:

$$\hat{f}(u,v) = Re[\hat{f}(u,v)] + iIm[\hat{f}(u,v)] \qquad (1.6)$$

with:

$$Re[\hat{f}(u,v)] = \frac{1}{MN}\sum_{l=-\infty}^{+\infty}\sum_{k=-\infty}^{+\infty} f(k,l)\cos(2\pi\left(\frac{uk}{M}+\frac{vl}{N}\right))$$

$$Im[\hat{f}(u,v)] = -\frac{1}{MN}\sum_{l=-\infty}^{+\infty}\sum_{k=-\infty}^{+\infty} f(k,l)\sin(2\pi\left(\frac{uk}{M}+\frac{vl}{N}\right)) \qquad (1.7)$$

It can also be written using its modulus and argument:

$$\hat{f}(u,v) = \mid \hat{f}(u,v) \mid e^{i\arg\hat{f}(u,v)} \qquad (1.8)$$

$\mid \hat{f}(u,v) \mid^2$ is called the power spectrum, and $\Theta(u,v) = \arg\hat{f}(u,v)$ the phase.

Two other related transforms are the cosine and the sine transforms. The discrete cosine transform is defined by:

$$DCT(u,v) = \frac{1}{\sqrt{2N}}c(u)c(v)\sum_{k=0}^{N-1}\sum_{l=0}^{N-1} f(k,l)$$
$$\cos\left(\frac{(2k+1)u\pi}{2N}\right)\cos\left(\frac{(2l+1)v\pi}{2N}\right)$$

$$IDCT(k,l) = \frac{1}{\sqrt{2N}}\sum_{u=0}^{N-1}\sum_{v=0}^{N-1} c(u)c(v)DCT(u,v)$$
$$\cos\left(\frac{(2k+1)u\pi}{2N}\right)\cos\left(\frac{(2l+1)v\pi}{2N}\right)$$

with $c(i) = \frac{1}{\sqrt{2}}$ when $i=0$ and 1 otherwise.

1.2.2 Time-Frequency Representation

The Wigner-Ville Transform. The Wigner-Ville distribution (Wigner, 1932; Ville, 1948) of a signal $s(t)$ is

$$W(t,\nu) = \frac{1}{2\pi}\int s^*(t-\frac{1}{2}\tau)s(t+\frac{1}{2}\tau)e^{-i\tau 2\pi\nu}d\tau \qquad (1.9)$$

where s^* is the conjugate of s. The Wigner-Ville transform is always real (even for a complex signal). In practice, its use is limited by the existence of interference terms, even if they can be attenuated using specific averaging approaches. More details can be found in (Cohen, 1995; Mallat, 1998).

The Short-Term Fourier Transform. The Short-Term Fourier Transform of a 1D signal f is defined by:

$$STFT(t,\nu) = \int_{-\infty}^{+\infty} e^{-j2\pi\nu\tau} f(\tau)g(\tau-t)d\tau \qquad (1.10)$$

If g is the Gaussian window, this corresponds to the Gabor transform. The energy density function, called the *spectrogram*, is given by:

$$SPEC(t,\nu) = |STFT(t,\nu)|^2 = |\int_{-\infty}^{+\infty} e^{-j2\pi\nu\tau} f(\tau)g(\tau-t)d\tau|^2 \qquad (1.11)$$

Fig. 1.1 shows a quadratic chirp $s(t) = \sin(\frac{\pi t^3}{3N^2})$, N being the number of pixels and $t \in \{1,..,N\}$, and its spectrogram.

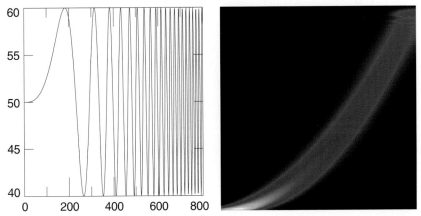

Fig. 1.1. *Left:* a quadratic chirp and *right:* its spectrogram. The y-axis in the spectrogram represents the frequency axis, and the x-axis the time. In this example, the instantaneous frequency of the signal increases with the time.

The inverse transform is obtained by:

$$f(t) = \int_{-\infty}^{+\infty} g(t-\tau) \int_{-\infty}^{+\infty} e^{j2\pi\nu\tau} STFT(\tau,\nu) d\nu d\tau \qquad (1.12)$$

Example: QPO Analysis. Fig. 1.2, top, shows an X-ray light curve from a galactic binary system, formed from two stars of which one has collapsed to a compact object, very probably a black hole of a few solar masses. Gas from the companion star is attracted to the black hole and forms an accretion disk around it. Turbulence occurs in this disk, which causes the gas to accrete

1. Introduction to Applications and Methods

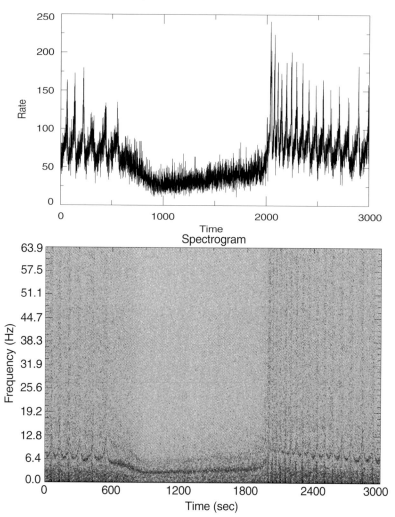

Fig. 1.2. *Top:* QPO X-ray light curve, and *bottom:* its spectrogram.

slowly to the black hole. The X-rays we see come from the disk and its corona, heated by the energy released as the gas falls deeper into the potential well of the black hole. The data were obtained by RXTE, an X-ray satellite dedicated to the observation of this kind of source, and in particular their fast variability which gives us information on the processes in the disk. In particular they show sometimes a QPO (quasi-periodic oscillation) at a varying frequency of the order of 1 to 10 Hz (see Fig. 1.2, bottom), which probably corresponds to a standing feature rotating in the disk.

1.2.3 Time-Scale Representation: The Wavelet Transform

The Morlet-Grossmann definition (Grossmann et al., 1989) of the continuous wavelet transform for a 1-dimensional signal $f(x) \in L^2(R)$, the space of all square integrable functions, is:

$$W(a,b) = \frac{1}{\sqrt{a}} \int_{-\infty}^{+\infty} f(x) \psi^* \left(\frac{x-b}{a} \right) dx \quad (1.13)$$

where:

- $W(a,b)$ is the wavelet coefficient of the function $f(x)$
- $\psi(x)$ is the analyzing wavelet
- $a\ (>0)$ is the scale parameter
- b is the position parameter

The inverse transform is obtained by:

$$f(x) = \frac{1}{C_\chi} \int_0^{+\infty} \int_{-\infty}^{+\infty} \frac{1}{\sqrt{a}} W(a,b) \psi \left(\frac{x-b}{a} \right) \frac{da\ db}{a^2} \quad (1.14)$$

where:

$$C_\psi = \int_0^{+\infty} \frac{\hat{\psi}^*(\nu)\hat{\psi}(\nu)}{\nu} d\nu = \int_{-\infty}^0 \frac{\hat{\psi}^*(\nu)\hat{\psi}(\nu)}{\nu} d\nu \quad (1.15)$$

Reconstruction is only possible if C_ψ is defined (admissibility condition) which implies that $\hat{\psi}(0) = 0$, i.e. the mean of the wavelet function is 0.

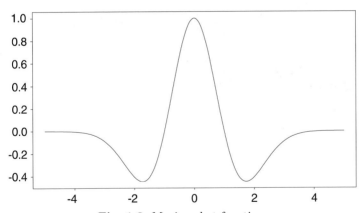

Fig. 1.3. Mexican hat function.

Fig. 1.3 shows the Mexican hat wavelet function, which is defined by:

$$g(x) = (1-x^2)e^{-x^2/2} \quad (1.16)$$

This is the second derivative of a Gaussian. Fig. 1.4 shows the continuous wavelet transform of a 1D signal computed with the Mexican Hat wavelet. This diagram is called a *scalogram*. The y-axis represents the scale.

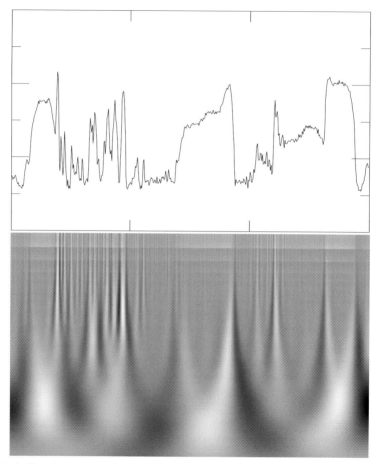

Fig. 1.4. Continuous wavelet transform of a 1D signal computed with the Mexican Hat wavelet.

The Orthogonal Wavelet Transform. Many discrete wavelet transform algorithms have been developed (Mallat, 1998; Starck et al., 1998a). The most widely-known one is certainly the orthogonal transform, proposed by Mallat (1989) and its bi-orthogonal version (Daubechies, 1992). Using the orthogonal wavelet transform, a signal s can be decomposed as follows:

$$s(l) = \sum_k c_{J,k} \phi_{J,l}(k) + \sum_k \sum_{j=1}^{J} \psi_{j,l}(k) w_{j,k} \qquad (1.17)$$

with $\phi_{j,l}(x) = 2^{-j}\phi(2^{-j}x - l)$ and $\psi_{j,l}(x) = 2^{-j}\psi(2^{-j}x - l)$, where ϕ and ψ are respectively the scaling function and the wavelet function. J is the number of resolutions used in the decomposition, w_j the wavelet (or detail) coefficients at scale j, and c_J is a coarse or smooth version of the original signal s. Thus, the algorithm outputs $J+1$ subband arrays. The indexing is such that, here, $j = 1$ corresponds to the finest scale (high frequencies). Coefficients $c_{j,k}$ and $w_{j,k}$ are obtained by means of the filters h and g:

$$\begin{aligned} c_{j+1,l} &= \sum_k h(k-2l)c_{j,k} \\ w_{j+1,l} &= \sum_k g(k-2l)c_{j,k} \end{aligned} \tag{1.18}$$

where h and g verify:

$$\begin{aligned} \frac{1}{2}\phi(\frac{x}{2}) &= \sum_k h(k)\phi(x-k) \\ \frac{1}{2}\psi(\frac{x}{2}) &= \sum_k g(k)\phi(x-k) \end{aligned} \tag{1.19}$$

and the reconstruction of the signal is performed with:

$$c_{j,l} = 2\sum_k [\tilde{h}(k+2l)c_{j+1,k} + \tilde{g}(k+2l)w_{j+1,k}] \tag{1.20}$$

where the filters $\tilde{h}$ and $\tilde{g}$ must verify the conditions of dealiasing and exact reconstruction:

$$\begin{aligned} \hat{h}\left(\nu+\frac{1}{2}\right)\hat{\tilde{h}}(\nu) + \hat{g}\left(\nu+\frac{1}{2}\right)\hat{\tilde{g}}(\nu) &= 0 \\ \hat{h}(\nu)\hat{\tilde{h}}(\nu) + \hat{g}(\nu)\hat{\tilde{g}}(\nu) &= 1 \end{aligned} \tag{1.21}$$

The two-dimensional algorithm is based on separate variables leading to prioritizing of horizontal, vertical and diagonal directions. The scaling function is defined by $\phi(x,y) = \phi(x)\phi(y)$, and the passage from one resolution to the next is achieved by:

$$c_{j+1}(k_x,k_y) = \sum_{l_x=-\infty}^{+\infty}\sum_{l_y=-\infty}^{+\infty} h(l_x-2k_x)h(l_y-2k_y)f_j(l_x,l_y) \tag{1.22}$$

The detail signal is obtained from three wavelets:

- vertical wavelet : $\psi^1(x,y) = \phi(x)\psi(y)$
- horizontal wavelet: $\psi^2(x,y) = \psi(x)\phi(y)$
- diagonal wavelet: $\psi^3(x,y) = \psi(x)\psi(y)$

which leads to three wavelet subimages at each resolution level. For three dimensional data, seven wavelet subcubes are created at each resolution level, corresponding to an analysis in seven directions. Other discrete wavelet transforms exist. The à trous wavelet transform which is very well-suited for astronomical data is discussed in the next chapter, and described in detail in Appendix A.

1.2.4 The Radon Transform

The Radon transform of an object f is the collection of line integrals indexed by $(\theta, t) \in [0, 2\pi) \times \mathbf{R}$ given by

$$Rf(\theta, t) = \int f(x_1, x_2) \delta(x_1 \cos\theta + x_2 \sin\theta - t) \, dx_1 dx_2, \qquad (1.23)$$

where δ is the Dirac distribution. The two-dimensional Radon transform maps the spatial domain (x, y) to the Radon domain (θ, t), and each point in the Radon domain corresponds to a line in the spatial domain. The transformed image is called a *sinogram* (Liang and Lauterbur, 2000).

A fundamental fact about the Radon transform is the projection-slice formula (Deans, 1983):

$$\hat{f}(\lambda \cos\theta, \lambda \sin\theta) = \int Rf(t, \theta) e^{-i\lambda t} dt.$$

This says that the Radon transform can be obtained by applying the one-dimensional inverse Fourier transform to the two-dimensional Fourier transform restricted to radial lines going through the origin.

This of course suggests that approximate Radon transforms for digital data can be based on discrete fast Fourier transforms. This is a widely used approach, in the literature of medical imaging and synthetic aperture radar imaging, for which the key approximation errors and artifacts have been widely discussed. See (Toft, 1996; Averbuch et al., 2001) for more details on the different Radon transform and inverse transform algorithms. Fig. 1.5 shows an image containing two lines and its Radon transform. In astronomy, the Radon transform has been proposed for the reconstruction of images obtained with a rotating Slit Aperture Telescope (Touma, 2000), for the BATSE experiment of the Compton Gamma Ray Observatory (Zhang et al., 1993), and for robust detection of satellite tracks (Vandame, 2001). The Hough transform, which is closely related to the Radon transform, has been used by Ballester (1994) for automated arc line identification, by Llebaria (1999) for analyzing the temporal evolution of radial structures on the solar corona, and by Ragazzoni and Barbieri (1994) for the study of astronomical light curve time series.

1.2.5 The Ridgelet Transform

The two-dimensional continuous ridgelet transform in $\mathbf{R}^2$ can be defined as follows (Candès and Donoho, 1999). We pick a smooth univariate function $\psi : \mathbf{R} \to \mathbf{R}$ with sufficient decay and satisfying the admissibility condition

$$\int |\hat{\psi}(\xi)|^2 / |\xi|^2 \, d\xi < \infty, \qquad (1.24)$$

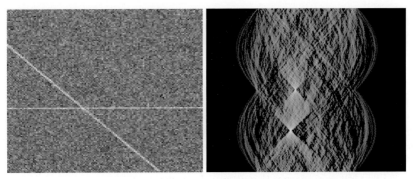

Fig. 1.5. *Left:* image with two lines and Gaussian noise. *Right:* its Radon transform.

which holds if, say, ψ has a vanishing mean $\int \psi(t)dt = 0$. We will suppose that ψ is normalized so that $\int |\hat{\psi}(\xi)|^2 \xi^{-2} d\xi = 1$.

For each $a > 0$, each $b \in \mathbf{R}$ and each $\theta \in [0, 2\pi]$, we define the bivariate *ridgelet* $\psi_{a,b,\theta} : \mathbf{R}^2 \to \mathbf{R}$ by

$$\psi_{a,b,\theta}(\mathbf{x}) = a^{-1/2} \cdot \psi((x_1 \cos\theta + x_2 \sin\theta - b)/a). \tag{1.25}$$

Given an integrable bivariate function $f(\mathbf{x})$, we define its ridgelet coefficients by:

$$\mathcal{R}_f(a, b, \theta) = \int \overline{\psi_{a,b,\theta}(\mathbf{x})} f(\mathbf{x}) d\mathbf{x}.$$

We have the exact reconstruction formula

$$f(\mathbf{x}) = \int_0^{2\pi} \int_{-\infty}^{\infty} \int_0^{\infty} \mathcal{R}_f(a, b, \theta) \psi_{a,b,\theta}(\mathbf{x}) \frac{da}{a^3} db \frac{d\theta}{4\pi} \tag{1.26}$$

valid for functions which are both integrable and square integrable.

It has been shown (Candès and Donoho, 1999) that the ridgelet transform is precisely the application of a 1-dimensional wavelet transform to the slices of the Radon transform. Fig. 1.6 (left) shows an example ridgelet function. This function is constant along lines $x_1 \cos\theta + x_2 \sin\theta = const$. Transverse to these ridges it is a wavelet: Fig. 1.6 (right).

Local Ridgelet Transform

The ridgelet transform is optimal for finding only lines of the size of the image. To detect line segments, a partitioning must be introduced. The image is decomposed into smoothly overlapping blocks of side-length B pixels in such a way that the overlap between two vertically adjacent blocks is a rectangular array of size $B \times B/2$; we use an overlap to avoid blocking artifacts. For an

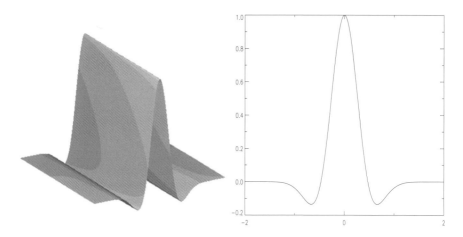

Fig. 1.6. Example of 2D ridgelet function.

$n \times n$ image, we count $2n/B$ such blocks in each direction. The partitioning introduces redundancy, since a pixel belongs to 4 neighboring blocks.

More details on the implementation of the digital ridgelet transform can be found in Starck et al. (2002; 2003a). The ridgelet transform is therefore optimal for detecting lines of a given size, equal to the block size.

1.2.6 The Curvelet Transform

The curvelet transform (Donoho and Duncan, 2000; Candès and Donoho, 2000a; Starck et al., 2003a) opens the possibility to analyze an image with different block sizes, but with a single transform. The idea is to first decompose the image into a set of wavelet bands, and to analyze each band with a local ridgelet transform. The block size can be changed at each scale level. Roughly speaking, different levels of the multi-scale ridgelet pyramid are used to represent different sub-bands of a filter bank output.

The side-length of the localizing windows is doubled *at every other* dyadic sub-band, hence maintaining the fundamental property of the curvelet transform, that elements of length about $2^{-j/2}$ serve for the analysis and synthesis of the jth subband $[2^j, 2^{j+1}]$. Note also that the coarse description of the image c_J is not processed. In our implementation, we used the default block size value $B_{min} = 16$ pixels. This implementation of the curvelet transform is also redundant. The redundancy factor is equal to $16J + 1$ whenever J scales are employed. A given curvelet band is therefore defined by the resolution level j ($j = 1 \ldots J$) related to the wavelet transform, and by the ridgelet scale r. This method is optimal for detecting anisotropic structures of different lengths.

A sketch of the discrete curvelet transform algorithm is:

1. apply the à trous wavelet transform algorithm (Appendix A) with J scales,
2. set $B_1 = B_{min}$,
3. for $j = 1, \ldots, J$ do,
   - partition the subband w_j with a block size B_j and apply the digital ridgelet transform to each block,
   - if j modulo $2 = 1$ then $B_{j+1} = 2B_j$,
   - else $B_{j+1} = B_j$.

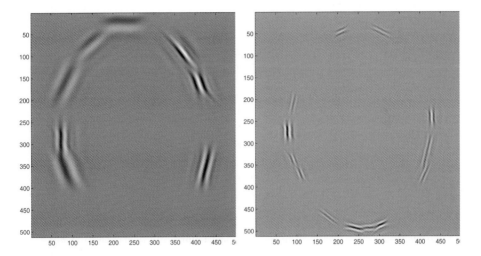

Fig. 1.7. A few curvelets.

Fig. 1.7 shows a few curvelets at different scales, orientations and locations. A fast curvelet transform algorithm has also recently been published by Candes et al. (2005).

In Starck et al. (2004), it has been shown that the curvelet transform could be useful for the detection and the discrimination of non-Gaussianity in CMB (Cosmic Microwave Background) data.

1.3 Mathematical Morphology

Mathematical morphology is used for nonlinear filtering. Originally developed by Matheron (1967; 1975) and Serra (1982), mathematical morphology is based on two operators: the *infimum* (denoted ∧) and the *supremum* (denoted ∨). The infimum of a set of images is defined as the greatest lower bound while the *supremum* is defined as the least upper bound. The basic morphological transformations are erosion, dilation, opening and closing. For grey-level images, they can be defined in the following way:

- *Dilation* consists of replacing each pixel of an image by the maximum of its neighbors.

$$\delta_B(f) = \bigvee_{b \in B} f_b$$

where f stands for the image, and B denotes the structuring element, typically a small convex set such as a square or disk.

The dilation is commonly known as "fill", "expand", or "grow." It can be used to fill "holes" of a size equal to or smaller than the structuring element. Used with binary images, where each pixel is either 1 or 0, dilation is similar to convolution. At each pixel of the image, the origin of the structuring element is overlaid. If the image pixel is nonzero, each pixel of the structuring element is added to the result using the "or" logical operator.

- *Erosion* consists of replacing each pixel of an image by the minimum of its neighbors:

$$\epsilon_B(f) = \bigwedge_{b \in B} f_{-b}$$

where f stands for the image, and B denotes the structuring element.

Erosion is the dual of dilation. It does to the background what dilation does to the foreground. This operator is commonly known as "shrink" or "reduce". It can be used to remove islands smaller than the structuring element. At each pixel of the image, the origin of the structuring element is overlaid. If each nonzero element of the structuring element is contained in the image, the output pixel is set to one.

- *Opening* consists of doing an erosion followed by a dilation.

$$\alpha_B = \delta_B \epsilon_B \text{ and } \alpha_B(f) = f \circ B$$

- *Closing* consists of doing a dilation followed by an erosion.

$$\beta_B = \epsilon_B \delta_B \text{ and } \beta_B(f) = f \bullet B$$

In a more general way, *opening* and *closing* refer to morphological filters which respect some specific properties (Breen et al., 2000). Such morphological filters were used for removing "cirrus-like" emission from far-infrared extragalactic IRAS fields (Appleton et al., 1993), and for astronomical image compression (Huang and Bijaoui, 1991).

The skeleton of an object in an image is a set of lines that reflect the shape of the object. The set of skeletal pixels can be considered to be the medial axis of the object. More details can be found in (Breen et al., 2000; Soille, 2003). Fig. 1.8 shows an example of the application of the morphological operators with a square binary structuring element.

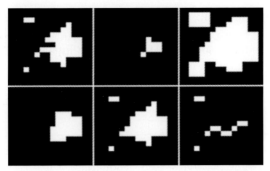

Fig. 1.8. Application of the morphological operators with a square binary structuring element. *Top, from left to right:* original image and images obtained by erosion and dilation. Bottom, images obtained respectively by the opening, closing and skeleton operators.

Undecimated Multiscale Morphological Transform. Mathematical morphology has been up to now considered as another way to analyze data, in competition with linear methods. But from a multiscale point of view (Starck et al., 1998a; Goutsias and Heijmans, 2000; Heijmans and Goutsias, 2000), mathematical morphology or linear methods are just filters allowing us to go from a given resolution to a coarser one, and the multiscale coefficients are then analyzed in the same way.

By choosing a set of structuring elements B_j having a size increasing with j, we can define an undecimated morphological multiscale transform by

$$\begin{aligned} c_{j+1,l} &= \mathcal{M}_j(c_j)(l) \\ w_{j+1,l} &= c_{j,l} - c_{j+1,l} \end{aligned} \quad (1.27)$$

where $\mathcal{M}_j$ is a morphological filter (erosion, opening, etc.) using the structuring element B_j. An example of B_j is a box of size $(2^j + 1) \times (2^j + 1)$. Since the detail signal w_{j+1} is obtained by calculating a simple difference between the c_j and c_{j+1}, the reconstruction is straightforward, and is identical to the reconstruction relative to the "à trous" wavelet transform (see Appendix A). An exact reconstruction of the image c_0 is obtained by:

$$c_{0,l} = c_{J,l} + \sum_{j=1}^{J} w_{j,l} \quad (1.28)$$

where J is the number of scales used in the decomposition. Each scale has the same number N of samples as the original data. The total number of pixels in the transformation is $(J+1)N$.

1.4 Edge Detection

An edge is defined as a local variation of image intensity. Edges can be detected by the computation of a local derivative operator.

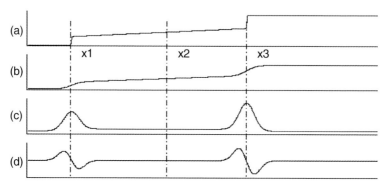

Fig. 1.9. First and second derivative of $G_\sigma * f$. (**a**) Original signal, (**b**) signal convolved by a Gaussian, (**c**) first derivative of (**b**), (**d**) second derivative of (**b**).

Fig. 1.9 shows how the inflection point of a signal can be found from its first and second derivative. Two methods can be used for generating first order derivative edge gradients.

1.4.1 First Order Derivative Edge Detection

Gradient. The gradient of an image f at location (x, y), along the line normal to the edge slope, is the vector (Pratt, 1991; Gonzalez and Woods, 1992; Jain, 1990):

$$\nabla f = \begin{bmatrix} f_x \\ f_y \end{bmatrix} = \begin{bmatrix} \frac{\partial f}{\partial x} \\ \frac{\partial f}{\partial y} \end{bmatrix} \qquad (1.29)$$

The spatial gradient amplitude is given by:

$$G(x, y) = \sqrt{f_x^2 + f_y^2} \qquad (1.30)$$

and the gradient direction with respect to the row axis is

$$\Theta(x, y) = \arctan \frac{f_y}{f_x} \qquad (1.31)$$

The first order derivative edge detection can be carried out either by using two orthogonal directions in an image or by using a set of directional derivatives.

1.4 Edge Detection

Gradient Mask Operators. Gradient estimates can be obtained by using gradient operators of the form:

$$f_x = f * H_x \tag{1.32}$$
$$f_y = f * H_y \tag{1.33}$$

where $*$ denotes convolution, and H_x and H_y are 3×3 row and column operators, called gradient masks. Table 1.1 shows the main gradient masks proposed in the literature. Pixel difference is the simplest one, which consists just of forming the difference of pixels along rows and columns of the image:

$$\begin{aligned} f_x(x_m, y_n) &= f(x_m, y_n) - f(x_m - 1, y_n) \\ f_y(x_m, y_n) &= f(x_m, y_n) - f(x_m, y_n - 1) \end{aligned} \tag{1.34}$$

The Roberts gradient masks (Roberts, 1965) are more sensitive to diagonal edges. Using these masks, the orientation must be calculated by

$$\Theta(x_m, y_n) = \frac{\pi}{4} + \arctan\left[\frac{f_y(x_m, y_n)}{f(x_m, y_n)}\right] \tag{1.35}$$

Prewitt (1970), Sobel, and Frei-Chen (1977) produce better results than the pixel difference, separated pixel difference and Roberts operator, because the mask is larger, and provides averaging of small luminance fluctuations. The Prewitt operator is more sensitive to horizontal and vertical edges than diagonal edges, and the reverse is true for the Sobel operator. The Frei-Chen edge detector has the same sensitivity for diagonal, vertical, and horizontal edges.

Compass Operators. Compass operators measure gradients in a selected number of directions. The directions are $\Theta_k = k\frac{\pi}{4}$, $k = 0, \ldots, 7$. The edge template gradient is defined as:

$$G(x_m, y_n) = \max_{k=0}^{7} | f(x_m, y_n) * H_k(x_m, y_n) | \tag{1.36}$$

Table 1.2 shows the principal template gradient operators.

Derivative of Gaussian. The previous methods are relatively sensitive to the noise. A solution could be to extend the window size of the gradient mask operators. Another approach is to use the derivative of the convolution of the image by a Gaussian. The derivative of a Gaussian (DroG) operator is

$$\begin{aligned} \nabla(g * f) &= \frac{\partial(g * f)}{\partial x} + \frac{\partial(g * f)}{\partial y} \\ &= f_x + f_y \end{aligned} \tag{1.37}$$

with $g = e^{-\frac{x^2+y^2}{2\sigma^2}}$. Partial derivatives of the Gaussian function are

$$\begin{aligned} g_x(x, y) &= \frac{\partial g}{\partial x} = -\frac{x}{\sigma^2} e^{-\frac{x^2+y^2}{2\sigma^2}} \\ g_y(x, y) &= \frac{\partial g}{\partial y} = -\frac{y}{\sigma^2} e^{-\frac{x^2+y^2}{2\sigma^2}} \end{aligned} \tag{1.38}$$

20   1. Introduction to Applications and Methods

The filters are separable so we have
$$g_x(x,y) = g_x(x) * g(y)$$
$$g_y(x,y) = g_y(y) * g(x) \tag{1.39}$$
Then
$$f_x = g_x(x) * g(y) * f$$
$$f_y = g_y(y) * g(x) * f \tag{1.40}$$

Thinning the Contour. From the gradient map, we may want to consider only pixels which belong to the contour. This can be done by looking for each pixel in the direction of gradient. For each point P0 in the gradient map, we determine the two adjacent pixels P1,P2 in the direction orthogonal to the gradient. If P0 is not a maximum in this direction (i.e. P0 < P1, or P0 < P2), then we threshold P0 to zero. Fig. 1.10 shows the Saturn image and the detected edges by the DroG method.

1.4.2 Second Order Derivative Edge Detection

Second derivative operators allow us to accentuate the edges. The most frequently used operator is the Laplacian operator, defined by
$$\nabla^2 f = \frac{\partial^2 f}{\partial x^2} + \frac{\partial^2 f}{\partial y^2} \tag{1.41}$$

Table 1.1. Gradient edge detector masks.

Operator	H_x	H_y	Scale factor
Pixel difference	$\begin{bmatrix} 0 & 0 & 0 \\ 0 & 1 & -1 \\ 0 & 0 & 0 \end{bmatrix}$	$\begin{bmatrix} 0 & -1 & 0 \\ 0 & 1 & 0 \\ 0 & 0 & 0 \end{bmatrix}$	1
Separated pixel difference	$\begin{bmatrix} 0 & 0 & 0 \\ 1 & 0 & -1 \\ 0 & 0 & 0 \end{bmatrix}$	$\begin{bmatrix} 0 & -1 & 0 \\ 0 & 0 & 0 \\ 0 & 1 & 0 \end{bmatrix}$	1
Roberts	$\begin{bmatrix} 0 & 0 & -1 \\ 0 & 1 & 0 \\ 0 & 0 & 0 \end{bmatrix}$	$\begin{bmatrix} -1 & 0 & 0 \\ 0 & 1 & 0 \\ 0 & 0 & 0 \end{bmatrix}$	1
Prewitt	$\begin{bmatrix} 1 & 0 & -1 \\ 1 & 0 & -1 \\ 1 & 0 & -1 \end{bmatrix}$	$\begin{bmatrix} -1 & -1 & -1 \\ 0 & 0 & 0 \\ 1 & 1 & 1 \end{bmatrix}$	1
Sobel	$\begin{bmatrix} 1 & 0 & -1 \\ 2 & 0 & -2 \\ 1 & 0 & -1 \end{bmatrix}$	$\begin{bmatrix} -1 & -2 & -1 \\ 0 & 0 & 0 \\ 1 & 2 & 1 \end{bmatrix}$	$\frac{1}{4}$
Fei-Chen	$\begin{bmatrix} 1 & 0 & -1 \\ \sqrt{2} & 0 & \sqrt{2} \\ 1 & 0 & -1 \end{bmatrix}$	$\begin{bmatrix} -1 & -\sqrt{2} & -1 \\ 0 & 0 & 0 \\ 1 & \sqrt{2} & 1 \end{bmatrix}$	$\frac{1}{2+\sqrt{2}}$

1.4 Edge Detection   21

Table 1.2. Template gradients.

Gradient direction	Prewitt compass gradient	Kirsch	Robinson 3-level	Robinson 5-level
East	$\begin{bmatrix} 1 & 1 & -1 \\ 1 & -2 & -1 \\ 1 & 1 & -1 \end{bmatrix}$	$\begin{bmatrix} 5 & -3 & -3 \\ 5 & 0 & -3 \\ 5 & -3 & -3 \end{bmatrix}$	$\begin{bmatrix} 1 & 0 & -1 \\ 1 & 0 & -1 \\ 1 & 0 & -1 \end{bmatrix}$	$\begin{bmatrix} 1 & 0 & -1 \\ 2 & 0 & -2 \\ 1 & 0 & -1 \end{bmatrix}$
Northeast	$\begin{bmatrix} 1 & 1 & -1 \\ 1 & -2 & -1 \\ 1 & 1 & 1 \end{bmatrix}$	$\begin{bmatrix} -3 & -3 & -3 \\ 5 & 0 & -3 \\ 5 & 5 & -3 \end{bmatrix}$	$\begin{bmatrix} 0 & -1 & -1 \\ 1 & 0 & -1 \\ 1 & 1 & 0 \end{bmatrix}$	$\begin{bmatrix} 0 & -1 & -2 \\ 1 & 0 & -1 \\ 2 & 1 & 0 \end{bmatrix}$
North	$\begin{bmatrix} -1 & -1 & -1 \\ 1 & -2 & 1 \\ 1 & 1 & 1 \end{bmatrix}$	$\begin{bmatrix} -3 & -3 & -3 \\ -3 & 0 & -3 \\ 5 & 5 & 5 \end{bmatrix}$	$\begin{bmatrix} -1 & -1 & -1 \\ 0 & 0 & 0 \\ 1 & 1 & 1 \end{bmatrix}$	$\begin{bmatrix} -1 & -2 & -1 \\ 0 & 0 & 0 \\ 1 & 2 & 1 \end{bmatrix}$
Northwest	$\begin{bmatrix} -1 & -1 & -1 \\ -1 & -2 & 1 \\ 1 & 1 & 1 \end{bmatrix}$	$\begin{bmatrix} -3 & -3 & -3 \\ -3 & 0 & 5 \\ -3 & 5 & 5 \end{bmatrix}$	$\begin{bmatrix} -1 & -1 & 0 \\ -1 & 0 & 1 \\ 0 & 1 & 1 \end{bmatrix}$	$\begin{bmatrix} -2 & -1 & 0 \\ -1 & 0 & 1 \\ 0 & 1 & 2 \end{bmatrix}$
West	$\begin{bmatrix} -1 & 1 & 1 \\ -1 & -2 & 1 \\ -1 & 1 & 1 \end{bmatrix}$	$\begin{bmatrix} -3 & -3 & 5 \\ -3 & 0 & 5 \\ -3 & -3 & 5 \end{bmatrix}$	$\begin{bmatrix} -1 & 0 & 1 \\ -1 & 0 & 1 \\ -1 & 0 & 1 \end{bmatrix}$	$\begin{bmatrix} -1 & 0 & 1 \\ -2 & 0 & 2 \\ -1 & 0 & 1 \end{bmatrix}$
Southwest	$\begin{bmatrix} -1 & 1 & 1 \\ -1 & -2 & 1 \\ -1 & -1 & 1 \end{bmatrix}$	$\begin{bmatrix} -3 & -3 & -3 \\ -3 & 0 & 5 \\ -3 & 5 & 5 \end{bmatrix}$	$\begin{bmatrix} -1 & 0 & 1 \\ -1 & 0 & 1 \\ 0 & 1 & 1 \end{bmatrix}$	$\begin{bmatrix} 0 & 1 & 2 \\ -1 & 0 & 1 \\ -2 & -1 & 0 \end{bmatrix}$
South	$\begin{bmatrix} 1 & 1 & 1 \\ 1 & -2 & 1 \\ -1 & -1 & -1 \end{bmatrix}$	$\begin{bmatrix} 5 & 5 & 5 \\ -3 & 0 & -3 \\ -3 & -3 & -3 \end{bmatrix}$	$\begin{bmatrix} 1 & 1 & 1 \\ 0 & 0 & 0 \\ -1 & -1 & -1 \end{bmatrix}$	$\begin{bmatrix} 1 & 2 & 1 \\ 0 & 0 & 0 \\ -1 & -2 & -1 \end{bmatrix}$
Southeast	$\begin{bmatrix} 1 & 1 & 1 \\ 1 & -2 & -1 \\ 1 & -1 & -1 \end{bmatrix}$	$\begin{bmatrix} 5 & 5 & -3 \\ 5 & 0 & -3 \\ -3 & -3 & -3 \end{bmatrix}$	$\begin{bmatrix} 1 & 1 & 0 \\ 1 & 0 & -1 \\ 0 & -1 & -1 \end{bmatrix}$	$\begin{bmatrix} 2 & 1 & 0 \\ 1 & 0 & -1 \\ 0 & -1 & -2 \end{bmatrix}$
Scale factor	$\frac{1}{5}$	$\frac{1}{15}$	$\frac{1}{3}$	$\frac{1}{4}$

22   1. Introduction to Applications and Methods

Fig. 1.10. Saturn image (left) and DroG detected edges.

Table 1.3 gives three discrete approximations of this operator.

Table 1.3. Laplacian operators.

Laplacian 1	Laplacian 2	Laplacian 3
$\frac{1}{4}\begin{bmatrix} 0 & -1 & 0 \\ -1 & 4 & -1 \\ 0 & -1 & 0 \end{bmatrix}$	$\frac{1}{8}\begin{bmatrix} -1 & -1 & -1 \\ -1 & 8 & -1 \\ -1 & -1 & -1 \end{bmatrix}$	$\frac{1}{8}\begin{bmatrix} -1 & -2 & -1 \\ -2 & 4 & -2 \\ -1 & -2 & -1 \end{bmatrix}$

Marr and Hildreth (1980) have proposed the Laplacian of Gaussian (LoG) edge detector operator. It is defined as

$$L(x,y) = \frac{1}{\pi s^4}\left[1 - \frac{x^2+y^2}{2s^2}\right]e^{-\frac{x^2+y^2}{2s^2}} \tag{1.42}$$

where σ controls the width of the Gaussian kernel.

Zero-crossings of a given image f convolved with L give its edge locations.

A simple algorithm for zero-crossings is:

1. For all pixels i,j do
2. ZeroCross(i,j) = 0
3. P0 = G(i,j); P1 = G(i,j-1); P2 = G(i-1,j); P3 = G(i-1,j-1)
4. If (P0*P1 < 0) or (P0*P2 < 0) or (P0*P3 < 0) then ZeroCross(i,j) = 1

1.5 Segmentation

Image segmentation is a process which partitions an image into regions (or segments) based upon similarities within regions – and differences between regions. An image represents a scene in which there are different objects or, more generally, regions. Although humans have little difficulty in separating the scene into regions, this process can be difficult to automate.

Segmentation takes stage 2 into stage 3 in the following information flow:

1. Raw image: pixel values are intensities, noise-corrupted.
2. Preprocessed image: pixels represent physical attributes, e.g. thickness of absorber, greyness of scene.
3. Segmented or symbolic image: each pixel labeled, e.g. into object and background.
4. Extracted features or relational structure.
5. Image analysis model.

Taking stage 3 into stage 4 is feature extraction, such as line detection, or use of moments. Taking stage 4 into stage 5 is shape detection or matching, identifying and locating object position. In this schema we start off with raw data (an array of grey-levels) and we end up with information – the identification and position of an object. As we progress, the data and processing move from low-level to high-level.

Haralick and Shapiro (1985) give the following wish-list for segmentation: "What should a good image segmentation be? Regions of an image segmentation should be uniform and homogeneous with respect to some characteristic (property) such as grey tone or texture. Region interiors should be simple and without many small holes. Adjacent regions of a segmentation should have significantly different values with respect to the characteristic on which they (the regions themselves) are uniform. Boundaries of each segment should be simple, not ragged, and must be spatially accurate".

Three general approaches to image segmentation are: single pixel classification, boundary-based methods, and region growing methods. There are other methods – many of them. Segmentation is one of the areas of image processing where there is certainly no agreed theory, nor agreed set of methods.

Broadly speaking, single pixel classification methods label pixels on the basis of the pixel value alone, i.e. the process is concerned only with the position of the pixel in grey-level space, or color space in the case of multi-valued images. The term *classification* is used because the different regions are considered to be populated by pixels of different *classes*.

Boundary-based methods detect boundaries of regions; subsequently pixels enclosed by a boundary can be labeled accordingly.

Finally, region growing methods are based on the identification of spatially connected groups of similarly valued pixels; often the grouping procedure is applied iteratively – in which case the term relaxation is used.

1.6 Pattern Recognition

Pattern recognition encompasses a broad area of study to do with automatic decision making. Typically, we have a collection of data about a situation; completely generally, we can assume that these data come as a set of p values, $\{x_1, x_2, \ldots x_p\}$. Usually, they will be arranged as a tuple or vector, $x = (x_1, x_2, \ldots x_p)^T$. An example is the decision whether a burgular alarm state is {intruder, no intruder}, based on a set of radar, acoustic, and electrical measurements. A pattern recognition system may be defined as taking an input data vector, $x = (x_1, x_2, \ldots x_p)^T$, and outputting a class label, w, taken from a set of possible labels $\{w_1, w_2, \ldots, w_C\}$.

Because it is deciding/selecting to which of a number of classes the vector x belongs, a pattern recognition system is often called a *classifier* – or a pattern classification system. For the purposes of most pattern recognition theory, a *pattern* is merely an ordered collection of numbers. This abstraction is a powerful one and is widely applicable.

Our p input numbers could be simply raw measurements, e.g. pixels in an area surrounding an object under investigation, or from the burgular alarm sensor referred to above. Quite often it is useful to apply some problem-dependent processing to the raw data before submitting them to the decision mechanism. In fact, what we try to do is to derive some data (another vector) that are sufficient to discriminate (classify) patterns, but eliminate all superfluous and irrelevant details (e.g. noise). This process is called *feature extraction*.

The components of a pattern vector are commonly called features, thus the term feature vector introduced above. Other terms are attribute, characteristic. Often all patterns are called feature vectors, despite the literal unsuitability of the term if it is composed of raw data.

It can be useful to classify feature extractors according to whether they are high- or low-level.

A typical low-level feature extractor is a transformation $\mathbb{R}^{p'} \longrightarrow \mathbb{R}^p$ which, presumably, either enhances the separability of the classes, or, at least, reduces the dimensionality of the data ($p < p'$) to the extent that the recognition task more computationally tractable, or simply to compress the data. Many data compression schemes are used as feature extractors, and vice-versa.

Examples of low-level feature extractors are:

- Fourier power spectrum of a signal – appropriate if frequency content is a good discriminator and, additionally, it has the property of shift invariance.
- Karhunen-Loève transform – transforms the data to a space in which the features are ordered according to information content based on variance.

At a higher-level, for example in image shape recognition, we could have a vector composed of: length, width, circumference. Such features are more in keeping with the everyday usage of the term feature.

1.6 Pattern Recognition

As an example of features, we will take two-dimensional invariant moments for planar shape recognition (Gonzalez and Woods, 1992). Assume we have isolated the object in the image. Two-dimensional moments are given by:

$$m_{pq} = \sum_x \sum_y x^p y^q f(x,y)$$

for $p, q = 0, 1, 2, \ldots$.

These are not invariant to anything, yet.

$$\tilde{x} = m_{10}/m_{00}$$

gives the x-center of gravity of the object, and

$$\tilde{y} = m_{01}/m_{00}$$

gives the y-center of gravity.

Now we can obtain shift invariant features by referring all coordinates to the center of gravity $(\tilde{x}, \tilde{y})$. These are the central moments:

$$m'_{pq} = \sum_x \sum_y (x - \tilde{x})^p (y - \tilde{y})^q f(x,y)$$

The first few m' can be interpreted as follows:
$m'_{00} = m_{00}$ = sum of the grey-levels in the object,
$m'_{10} = m'_{01} = 0$, always, i.e. center of gravity is (0,0) with respect to itself.
m'_{20} = measure of width along x-axis
m'_{02} = measure of width along y-axis.
From the m'_{pq} can be derived a set of normalized moments:

$$\mu_{pq} = m'_{pq}/((m'_{00})^g)$$

where $g = (p+q)/2 + 1$

Finally, a set of seven fully shift, rotation, and scale invariant moments can be defined:

$$p_1 = n_{20} + n_{02}$$
$$p_2 = (n_{20} - n_{02})^2 + 4n_{11}^2$$

etc.

The crucial principles behind feature extraction are:

1. Descriptive and discriminating feature(s).
2. As few as possible of them, leading to a simpler classifier.

An important practical subdivision of classifiers is between *supervised* and *unsupervised* classifiers. In the case of supervised classification, a training set is used to define the classifier parameter values. Clustering or segmentation are examples of (usually) unsupervised classification, because we approach these tasks with no prior knowledge of the problem.

A supervised classifier involves:

Training: gathering and storing example feature vectors – or some summary of them,

Operation: extracting features, and classifying, i.e. by computing similarity measures, and either finding the maximum, or applying some sort of thresholding.

When developing a classifier, we distinguish between *training* data, and *test* data:

- training data are used to train the classifier, i.e. set its parameters,
- test data are used to check if the trained classifier works, i.e. if it can generalize to new and unseen data.

Statistical classifiers use maximum likelihood (probability) as a criterion. In a wide range of cases, likelihood corresponds to closeness to the class cluster, i.e. closeness to the center or mean, or closeness to individual points. Hence, *distance* is an important criterion or metric. Consider a decision choice between class i and class j. Then, considering probabilities, if $p(i) > p(j)$ we decide in favor of class i. This is a maximum probability, or maximum likelihood, rule. It is the basis of all statistical pattern recognition. Training the classifier simply involves histogram estimation. Histograms though are hard to measure well, and usually we use *parametric* representations of probability density.

Assume two classes, w_0, w_1. Assume we have the two probability densities $p_0(x)$, $p_1(x)$. These may be denoted by

$$p(x \mid w_0), \ p(x \mid w_1)$$

the class conditional probability densities of x. Another piece of information is vital: what is the relative probability of occurrence of w_0, and w_1? These are the *prior* probabilities, P_0, P_1 – upper-case Ps represent priors. In this case the "knowledge" of the classifier is represented by the $p(x \mid w_j)$, P_j; $j = 0, 1$.

Now if we receive a feature vector x, we want to know what is the probability (likelihood) of each class. In other words, what is the probability of w_j given x ? – the *posterior* probability.

Bayes' law gives a method of computing the posterior probabilities:

$$p(w_j \mid x) = P_j p(x \mid w_j) / (\sum_{j=0} P_j p(x \mid w_j))$$

Each of the quantities on the right-hand side of this equation is known – through training.

In Bayes' equation the denominator of the right hand side is merely a normalizing factor, to ensure that $p(w_j \mid x)$ is a proper probability, and so can be neglected in cases where we just want maximum probability.

Now, classification becomes a matter of computing Bayes' equation, and choosing the class, j, with maximum $p(w_j \mid x)$.

The Bayes classifier is optimal based on an objective criterion: the class chosen is the most probable, with the consequence that the Bayes rule is also a minimum error classifier, i.e. in the long run it will make fewer errors than any other classifier.

Neural network classifiers, and in particular the multilayer perceptron, are a class of non-parametric, trainable classifiers, which produce a nonlinear mapping between inputs (vectors, x), and outputs (labels, w). Like all trainable classifiers, neural networks need good training data which covers the entire feature space quite well. The latter is a requirement which becomes increasingly harder to accomplish as the dimensionality of the feature space becomes larger.

Examples of application of neural net classifiers or neural nets as nonlinear regression methods (implying, respectively, categorical or quantitative outputs) include the following.

- Gamma-ray bursts (Balastegui et al., 2001).
- Stellar spectral classification (Snider et al., 2001).
- Solar atmospheric model analysis (Carroll and Staude, 2001).
- Star-galaxy discrimination (Cortiglioni et al., 2001).
- Geophysical disturbance prediction (Gleisner and Lundstedt, 2001).
- Galaxy morphology classification (Lahav et al., 1996; Bazell and Aha, 2001).
- Studies of the Cosmic Microwave Background (Baccigalupi et al., 2000a).

Many more applications can be found in the literature. A special issue of the journal *Neural Networks* on "Analysis of Complex Scientific Data – Astronomy and Geology" in 2003 (Tagliaferri et al., 2003) testifies to the continuing work in both theory and application with neural network methods.

1.7 Chapter Summary

In this chapter, we have surveyed key elements of the state of the art in image and signal processing. Fourier, wavelet and Radon transforms were introduced. Edge detection algorithms were specified. Signal segmentation was discussed. Finally, pattern recognition in multidimensional feature space was overviewed.

Subsequent chapters will take these topics in many different directions, motivated by a wide range of scientific problems.

2. Filtering

2.1 Introduction

Data in the physical sciences are characterized by the all-pervasive presence of noise, and often knowledge is available of the detector's and data's noise properties, at least approximately.

It is usual to distinguish between the *signal*, of substantive value to the analyst, and *noise* or clutter. The data signal can be a 2D image, a 1D time-series or spectrum, a 3D data cube, and variants of these.

Signal is what we term the scientifically interesting part of the data. Signal is often very compressible, whereas noise by definition is not compressible. Effective separation of signal and noise is evidently of great importance in the physical sciences.

Noise is a necessary evil in astronomical image processing. If we can reliably estimate noise, through knowledge of instrument properties or otherwise, subsequent analyses would be very much better behaved. In fact, major problems would disappear if this were the case – e.g. image restoration or sharpening based on solving inverse equations could become simpler.

One perspective on the theme of this chapter is that we present a coherent and integrated algorithmic framework for a wide range of methods which may well have been developed elsewhere on pragmatic and heuristic grounds. We put such algorithms on a firm footing, through explicit noise modeling followed by computational strategies which benefit from knowledge of the data. The advantages are clear: they include objectivity of treatment; better quality data analysis due to far greater thoroughness; and possibilities for automation of otherwise manual or interactive procedures.

Noise is often taken as additive Poisson (related to arrival of photons) and/or Gaussian. Commonly used electronic CCD (charge-coupled device) detectors have a range of Poisson noise components, together with Gaussian readout noise (Snyder et al., 1993). Digitized photographic images were found by Tekalp and Pavlović (1991) to be also additive Poisson and Gaussian (and subject to nonlinear distortions which we will not discuss here).

The noise associated with a particular detector may be known in advance. In practice rule-of-thumb calculation of noise is often carried out. For instance, limited convex regions of what is considered as background are sampled, and the noise is determined in these regions. For common noise distributions, noise is specified by its variance.

There are different ways to more formally estimate the standard deviation of Gaussian noise in an image. Olsen (1993) carried out an evaluation of six methods and showed that the best was the average method, which is the simplest also. This method consists of filtering the data I with the average filter (filtering with a simple box function) and subtracting the filtered image from I. Then a measure of the noise at each pixel is computed. To keep image edges from contributing to the estimate, the noise measure is disregarded if the magnitude of the intensity gradient is larger than some threshold, T.

Other approaches to automatic estimation of noise, which improve on the methods described by Olsen, are given in this chapter. Included here are methods which use multiscale transforms and the multiresolution support data structure.

As has been pointed out, our initial focus is on accurate determination of the noise. Other types of signal modeling, e.g. distribution mixture modeling or density estimation, are more easily carried out subsequently. Noise modeling is a desirable, and in many cases necessary, preliminary to such signal modeling.

In Chapter 1, we introduced the wavelet transform, which furnishes a multi-faceted approach for describing and modeling data. There are many 2D wavelet transform algorithms (Chui, 1992; Mallat, 1998; Burrus et al., 1998; Starck et al., 1998a; Cohen, 2003). The most widely-used is perhaps the biorthogonal wavelet transform (Mallat, 1989; Cohen et al., 1992). This method is based on the principle of reducing the redundancy of the information in the transformed data. Other wavelet transform algorithms exist, such as the Feauveau algorithm (Feauveau, 1990) (which is an orthogonal transform, but using an isotropic wavelet), or the à trous algorithm which is non-orthogonal and furnishes a redundant dataset (Holschneider et al., 1989). The à trous algorithm presents the following advantages:

- The computational requirement is reasonable.
- The reconstruction algorithm is trivial.
- The transform is known at each pixel, allowing position detection without any error, and without interpolation.
- We can follow the evolution of the transform from one scale to the next.
- Invariance under translation is completely verified.
- The transform is isotropic.

The last point is important if the image or the cube contains isotropic features. This is the case for most astronomical data sets, and this explains why the à trous algorithm has been so successful in astronomical data processing.

Section 2.2 describes the à trous algorithm and discusses the choice of this wavelet transform in the astronomical data processing framework. Section 2.3 introduces noise modeling relative to the wavelet coefficients. Section 2.4

presents how to filter a data set once the noise has been modeled, and some experiments are presented in section 2.4.4. Recent papers have argued for the use the Haar wavelet transform when the data contain Poisson noise. This approach is discussed in section 2.6, and we compare it to the à trous algorithm based filtering method.

2.2 Multiscale Transforms

2.2.1 The A Trous Isotropic Wavelet Transform

The wavelet transform of a signal produces, at each scale j, a set of zero-mean coefficient values $\{w_j\}$. Using an algorithm such as the à trous method (Holschneider et al., 1989; Shensa, 1992), this set $\{w_j\}$ has the same number of pixels as the signal and thus this wavelet transform is a redundant one. Furthermore, using a wavelet defined as the difference between the scaling functions of two successive scales ($\frac{1}{2}\psi(\frac{x}{2}) = \phi(x) - \phi(\frac{x}{2})$), the original signal c_0, with a pixel at position k, can be expressed as the sum of all the wavelet scales and the smoothed array c_J

$$c_{0,k} = c_{J,k} + \sum_{j=1}^{J} w_{j,k} \tag{2.1}$$

A summary of the à trous wavelet transform algorithm is as follows.

1. Initialize j to 0, starting with a signal $c_{j,k}$. Index k ranges over all pixels.
2. Carry out a discrete convolution of the data $c_{j,k}$ using a filter h (see Appendix A), yielding $c_{j+1,k}$. The convolution is an interlaced one, where the filter's pixel values have a gap (growing with level, j) between them of 2^j pixels, giving rise to the name à trous ("with holes"). "Mirroring" is used at the data extremes.
3. From this smoothing we obtain the discrete wavelet transform, $w_{j+1,k} = c_{j,k} - c_{j+1,k}$.
4. If j is less than the number J of resolution levels wanted, then increment j and return to step 2.

The set $w = \{w_1, w_2, ..., w_J, c_J\}$, where c_J is a last smooth array, represents the wavelet transform of the data. We denote as $\mathcal{W}$ the wavelet transform operator. If the input data set s has N pixels, then its transform w ($w = \mathcal{W}s$) has $(J+1)N$ pixels. The redundancy factor is $J+1$ whenever J scales are employed.

The discrete filter h is derived from the scaling function $\phi(x)$ (see Appendix A). In our calculations, $\phi(x)$ is a spline of degree 3, which leads (in one dimension) to the filter $h = (\frac{1}{16}, \frac{1}{4}, \frac{3}{8}, \frac{1}{4}, \frac{1}{16})$. A 2D or a 3D implementation can be based on two 1D sets of (separable) convolutions.

The associated wavelet function is of mean zero, of compact support, with a central bump and two negative side-lobes. Of interest for us is that, like the

32   2. Filtering

scaling function, it is isotropic (point symmetric). More details can be found in Appendix A.

Fig. 2.1. Galaxy NGC 2997.

Fig. 2.2 shows the à trous transform of the galaxy NGC 2997 displayed in Fig. 2.1 . Five wavelet scales are shown and the final smoothed plane (lower right). The original image is given exactly by the sum of these six images.

Fig. 2.3 shows each scale as a perspective plot.

Example: Dynamic Range Compression Using the à Trous Algorithm. Since some features in an image may be hard to detect by the human eye due to low contrast, we often process the image before visualization. Histogram equalization is certainly one the most well-known methods for contrast enhancement. Images with a high dynamic range are also difficult to analyze. For example, astronomers generally visualize their images using a logarithmic look-up-table conversion.

Wavelets can be used to compress the dynamic range at all scales, and therefore allow us to clearly see some very faint features. For instance, the wavelet-log representation consists of replacing $w_{j,k,l}$ by $\text{sgn}(w_{j,k,l}) \log(|w_{j,k,l}|)$, leading to the alternative image

$$I_{k,l} = \log(c_{J,k,l}) + \sum_{j=1}^{J} \text{sgn}(w_{j,k,l}) \log(|w_{j,k,l}| + \epsilon) \qquad (2.2)$$

where ϵ is a small number (for example $\epsilon = 10^{-3}$). Fig. 2.4 shows a Hale-Bopp Comet image (logarithmic representation) (top left), its histogram equaliza-

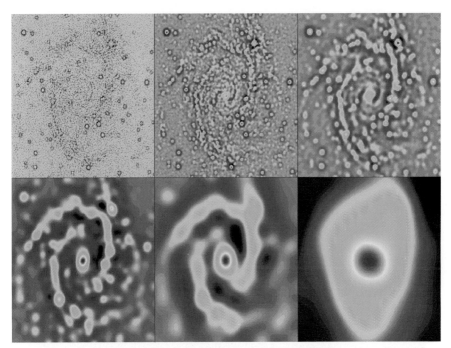

Fig. 2.2. Wavelet transform of NGC 2997 by the à trous algorithm.

tion (middle row), and its wavelet-log representation (bottom). Jets clearly appear in the last representation of the Hale-Bopp Comet image.

2.2.2 Multiscale Transforms Compared to Other Data Transforms

In this section we will discuss in general terms why the wavelet transform has very good noise filtering properties, and how it differs from other data preprocessing transforms in this respect. Among the latter, we can include principal components analysis (PCA) and correspondence analysis, which decompose the input data into a new orthogonal basis, with axes ordered by "variance (or inertia) explained". PCA on images as input observation vectors can be used, for example, for a best synthesis of multiple band images, or for producing eigen-faces in face recognition. Among other data preprocessing transforms, we also include the discrete cosine transform (DCT), which decomposes the data into an orthogonal basis of cosine functions; and the Fourier transform (FT) which uses a basis of sine and cosine functions, each at different frequencies.

PCA, DCT, and FT have the property of *energy packing* (Seales et al., 1996): most of the energy (second order moment) of the input vector is packed into the first few values of the output vector. Thus, one can roughly approxi-

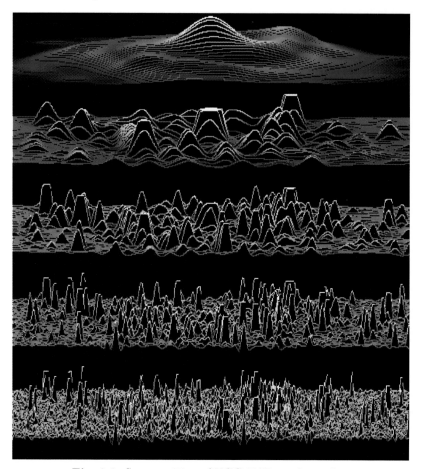

Fig. 2.3. Superposition of NGC 2997 wavelet scales.

mate, or even eliminate, all but the most important values and still preserve most of the input energy.

The wavelet transform (WT), whether orthonormal as in the case of the Haar or Daubechies transforms or non-orthogonal as in the case of the à trous method, is different. It can be viewed as an automatic method for laying bare superimposed scale-related components of the data. Our analysis of the data may be considerably improved by removing noise in *all* scale-related components. This perspective differs from the usual approach of PCA, DCT, and FT: in these methods we remove output scales (or "levels") entirely to filter the data.

We turn attention now to denoising through modification of scale information at all levels. This is the preferred method of denoising using the wavelet transform.

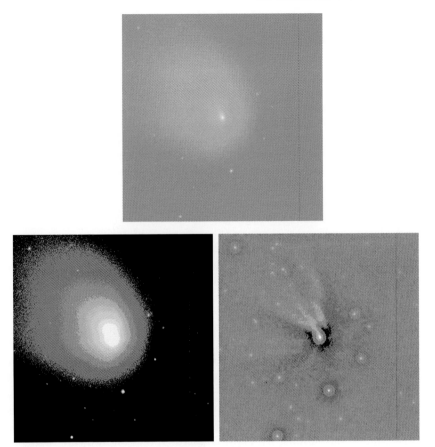

Fig. 2.4. *Top:* Hale-Bopp Comet image. *Bottom left:* histogram equalization results. *Bottom right:* wavelet-log representations.

Donoho and Johnstone (1994) proposed a "universal threshold", $\sqrt{2 \log n}\,\sigma$, used in the additive Gaussian noise case where σ is the known or estimated standard deviation of the data, and n is the size of the input data set. Wavelet coefficients above this threshold are retained, and those below the threshold are set to zero. The authors also propose a soft threshold, referred to as wavelet shrinkage, which reduces wavelet values by a fraction of their initial values.

As an alternative to such hard and soft thresholding, Starck et al. (1994; 1995) assume known or estimated noise properties for the input data, and then derive or make use of wavelet coefficient probability distributions at each level, under a null hypothesis of stochastic input. Other noise modeling work in this direction can be found in Kolaczyk (1997) and Powell et al. (1995), albeit with different wavelet transforms.

In the work described in this chapter we employ thresholding in a data- and noise-driven manner.

2.2.3 Choice of Multiscale Transform

Some important properties of the à trous wavelet transform are as follows.

As already noted, the à trous transform is isotropic. Unlike it, Mallat's widely-used multiresolution algorithm (Mallat, 1989) leads to a wavelet transform with three wavelet functions (at each scale there are three wavelet coefficient subimages) which does not simplify the analysis and the interpretation of the wavelet coefficients. Other anisotropic wavelets include the similarly widely-used Haar and Daubechies wavelet transforms. An isotropic wavelet seems more appropriate for images containing features or objects with no favored orientation.

An important property of the à trous wavelet transform over other wavelet transforms is shift invariance. Lack of independence to pixel shift is a problem in the case of any pyramidal wavelet transform (Haar, Daubechies, Mallat, etc.) due to the down-sampling or decimating. The reason is simply that shift-variance is introduced because Nyquist sampling is violated in each of the (wavelet-decomposed) subbands – wavelets are not ideal filters. By not downsampling the problem is avoided. Various authors have proposed solutions to this problem. The à trous algorithm is in fact a fast implementation of a wavelet transform with no downsampling.

Two inconvenient aspects of many wavelet transforms are negative values and lack of robustness. By definition, the wavelet coefficient mean at each level is zero. Every time we have a positive structure at a scale, we have negative values surrounding it. These negative values often create artifacts during the data reconstruction process, or complicate the analysis. For instance, if we threshold small values (noise, non-significant structures, etc.) in the wavelet transform, and if we reconstruct the image at full resolution, the structure's total intensity will be modified. Furthermore, if an object is associated with high intensity values, the negative values will be significant too and will lead to false structure detections. Point artifacts (e.g. cosmic ray hits in optical astronomy, glitches in the infrared ISO, Infrared Satellite Observatory, detectors) can "pollute" all scales of the wavelet transform. The wavelet transform is non-robust relative to such real or detector faults.

One way around both of these issues – negative wavelet coefficient values, and non-robustness relative to anomalous values – is to keep certain aspects of the multiscale decomposition algorithm provided by the à trous wavelet transform, but to base our algorithm on a function other than the wavelet function. The median smoothing transform provides us with one such possibility. A multiscale pyramidal median transform, for instance, was investigated in Starck et al. (1996), and is discussed in Chapter 5. We conclude that the wavelet transform, à trous or otherwise, is not sacrosanct. Depending on the data, it may well be advisable and necessary to use other multiresolution tools. For instance, if the data presents highly anisotropic features, the ridgelet transform (Candès and Donoho, 1999; Candès and Donoho, 1999) or the curvelet transform (Donoho and Duncan, 2000; Candès and Donoho, 2000b; Starck et al., 2002; Starck et al., 2003a) will outperform the wavelet transform.

2.2.4 The Multiresolution Support

A multiresolution support of a data set describes in a logical or Boolean way if the data s contains information at a given scale j and at a given position l. If $M_{j,k}^{(s)} = 1$ (or $=$ $true$), then s contains information at scale j and at the position k. M depends on several parameters:

- The input data.
- The algorithm used for the multiresolution decomposition.
- The noise.
- All additional constraints we want the support to satisfy.

Such a support results from the data, the treatment (noise estimation, etc.), and from knowledge on our part of the objects contained in the data (size of objects, linearity, etc.). In the most general case, a priori information is not available to us.

The multiresolution support of a data set is computed in several steps:

- Step one is to compute the wavelet transform of the data s: $w = \mathcal{W}s = \{w_1, w_2, ..., w_J, c_J\}$.
- Binarization of each scale leads to the multiresolution support $M = \{M_1, M_2, ..., M_J, M_{J+1}\}$ (the binarization consists of assigning to each pixel a value only equal to 0 or 1). The last scale M_{J+1} relative to the smoothed array is set to 1 ($M_{J+1,k} = 1$ for all k).
- A priori knowledge can be introduced by modifying the support.

This last step depends on the knowledge we have of our data. For instance, if we know there is no interesting object smaller or larger than a given size in our image, we can suppress, in the support, anything which is due to that kind of object. This can often be done conveniently by the use of mathematical morphology. In the most general setting, we naturally have no information to add to the multiresolution support.

The multiresolution support will be obtained by detecting at each scale the significant coefficients. The multiresolution support for $j \leq J$ is defined by:

$$M_{j,k} = \begin{cases} 1 & \text{if } w_{j,k} \text{ is significant} \\ 0 & \text{if } w_{j,k} \text{ is not significant} \end{cases} \quad (2.3)$$

For 2D data set, in order to visualize the support, we can create an image I defined by:

$$I_{k,l} = \sum_{j=1}^{J} 2^j M_{j,k,l} \tag{2.4}$$

The detection of the significant coefficients will be described in the next section. Fig. 2.5 shows such a multiresolution support visualization of an image of galaxy NGC 2997.

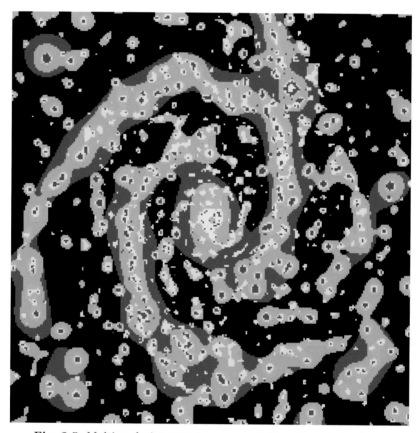

Fig. 2.5. Multiresolution support representation of a spiral galaxy.

2.3 Significant Wavelet Coefficients

2.3.1 Definition

Images and sets of point patterns generally contain noise. Hence the wavelet coefficients are noisy too. In most applications, it is necessary to know if a wavelet coefficient is due to signal (i.e. it is significant) or to noise.

2.3 Significant Wavelet Coefficients

The wavelet transform yields a set of resolution-related views of the input image. A wavelet image scale at level j has coefficients given by $w_{j,k}$. If we obtain the distribution of the coefficient $w_{j,k}$ for each resolution plane, based on the noise, we can introduce a statistical significance test for this coefficient. This procedure is the classical significance-testing one. Let $\mathcal{H}_0$ be the hypothesis that the image is locally constant at scale j. Rejection of hypothesis $\mathcal{H}_0$ depends (for interpretational reasons, restricted to positive coefficient values) on:

$$P = Prob(|\ w_{j,k}\ | < \tau\ |\ \mathcal{H}_0) \tag{2.5}$$

The detection threshold, τ, is defined for each scale. Given an estimation threshold, ϵ, if $P = P(\tau) > \epsilon$ the null hypothesis is not excluded. Although non-null, the value of the coefficient could be due to noise. On the other hand, if $P < \epsilon$, the coefficient value cannot be due to the noise alone, and so the null hypothesis is rejected. In this case, a significant coefficient has been detected. This is illustrated in Fig. 2.6.

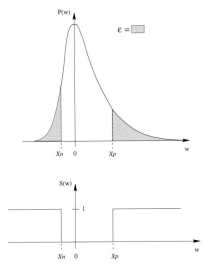

Fig. 2.6. Threshold determination.

2.3.2 Noise Modeling

Gaussian Noise. If the distribution of $w_{j,l}$ is Gaussian, with zero mean and standard deviation σ_j, we have the probability density

$$p(w_{j,l}) = \frac{1}{\sqrt{2\pi}\sigma_j} e^{-w_{j,l}^2/2\sigma_j^2} \tag{2.6}$$

Rejection of hypothesis $\mathcal{H}_0$ depends (for a positive coefficient value) on:

$$P = Prob(w_{j,l} > W) = \frac{1}{\sqrt{2\pi}\sigma_j} \int_{w_{j,l}}^{+\infty} e^{-W^2/2\sigma_j^2} dW \qquad (2.7)$$

and if the coefficient value is negative, it depends on

$$P = Prob(w_{j,l} < W) = \frac{1}{\sqrt{2\pi}\sigma_j} \int_{-\infty}^{w_{j,l}} e^{-W^2/2\sigma_j^2} dW \qquad (2.8)$$

Given stationary Gaussian noise, it suffices to compare $w_{j,l}$ to $k\sigma_j$. Often k is chosen as 3, which corresponds approximately to $\epsilon = 0.002$. If $w_{j,l}$ is small, it is not significant and could be due to noise. If $w_{j,l}$ is large, it is significant:

$$\begin{array}{ll} \text{if } |w_{j,l}| \geq k\sigma_j & \text{then } w_{j,l} \text{ is significant} \\ \text{if } |w_{j,l}| < k\sigma_j & \text{then } w_{j,l} \text{ is not significant} \end{array} \qquad (2.9)$$

So we need to estimate, in the case of Gaussian noise models, the noise standard deviation at each scale. These standard deviations can be determined analytically, but the calculations can become complicated.

The appropriate value of σ_j in the succession of wavelet planes is assessed from the standard deviation of the noise σ_s in the original data s, and from study of the noise in the wavelet space. This study consists of simulating a data set containing Gaussian noise with a standard deviation equal to 1, and taking the wavelet transform of this data set. Then we compute the standard deviation σ_j^e at each scale. We get a curve σ_j^e as a function of j, giving the behavior of the noise in the wavelet space. (Note that if we had used an orthogonal wavelet transform, this curve would be linear.) Due to the properties of the wavelet transform, we have $\sigma_j = \sigma_s \sigma_j^e$. The noise standard deviation at scale j of the data is equal to the noise standard deviation σ_s multiplied by the noise standard deviation at scale j of the simulated data. Table 2.1 gives the σ_j^e values for the 1D, 2D, and 3D à trous wavelet transform using the cubic B_3 spline scaling function.

Table 2.1. σ_j^e table for the first seven resolution levels.

Resolution level j	1	2	3	4	5	6	7
1D	0.700	0.323	0.210	0.141	0.099	0.071	0.054
2D	0.889	0.200	0.086	0.041	0.020	0.010	0.005
3D	0.956	0.120	0.035	0.012	0.004	0.001	0.0005

2.3.3 Automatic Estimation of Gaussian Noise

k-sigma Clipping. The Gaussian noise σ_s can be estimated automatically in a data set s. This estimation is particularly important, because all the noise standard deviations σ_j in the scales j are derived from σ_s. Thus an error associated with σ_s will introduce an error on all σ_j. Noise is therefore more usefully estimated in the high frequencies, where it dominates the signal. The resulting method consists first of filtering the data s with an average filter or the median filter and subtracting from s the filtered signal f: $d = s - f$. In our case, we replace d by the first scale of the wavelet transform ($d = w_1$), which is more convenient from the computation time point of view. The histogram of d shows a Gaussian peak around 0. A k-sigma clipping is then used to reject pixels where the signal is significantly large. We denote $d^{(1)}$ the subset of d which contains only the pixels such that $|d_l| < k\sigma_d$, where σ_d is the standard deviation of d, and k is a constant generally chosen equal to 3. By iterating, we obtain the subset $d^{(n+1)}$ verifying $|d_l^{(n)}| < k\sigma_{d^{(n)}}$, where $\sigma_{d^{(n)}}$ is the noise standard deviation of $d^{(n)}$. Robust estimation of the noise σ_1 in w_1 (as $d = w_1$) is now obtained by calculation of the standard deviation of $d^{(n)}$ ($\sigma_1 = \sigma_{d^{(n)}}$). In practice, three iterations are enough, and accuracy is generally better than 5%. σ_s is finally calculated by:

$$\sigma_s = \frac{\sigma_1}{\sigma_1^e} = \frac{\sigma_{d^{(n)}}}{\sigma_1^e} \qquad (2.10)$$

MAD Estimation. The median absolute deviation, MAD, gives an estimation of the noise standard deviation: $\sigma_m = \text{MED}(|w_1|)/0.6745$, where MED is the median function. Our noise estimate σ_s is obtained by:

$$\sigma_s = \frac{\sigma_m}{\sigma_1^e} \qquad (2.11)$$

Estimation of Gaussian Noise from the Multiresolution Support. The value of σ_s, estimated by the k-sigma clipping or any other method, can be refined by the use of the multiresolution support. Indeed, if we consider the set of pixels $\mathcal{S}$ in the data which are due only to the noise, and if we take the standard deviation of them, we would obtain a good estimate of σ_s. This set is easily obtained from the multiresolution support. We say that a pixel k belongs to the noise if $M_{j,k} = 0$ for all j (i.e. there is no significant coefficient at any scale). The new estimation of σ_s is then computed by the following iterative algorithm:

1. Estimate the standard deviation of the noise in s: we have $\sigma_s^{(0)}$.
2. Compute the wavelet transform (à trous algorithm) of the data s with J scales, providing the additive decomposition.
3. Set n to 0.
4. Compute the multiresolution support M which is derived from the wavelet coefficients and from $\sigma_s^{(n)}$.
5. Select the pixels which belong to the set $\mathcal{S}$: if $M_{j,k} = 0$ for all j in $1 \ldots J$, then the pixel $k \in \mathcal{S}$.
6. For all the selected pixels k, compute the values $s_k - c_{J,k}$ and compute the standard deviation $\sigma_s^{(n+1)}$ of these values (we compute the difference between s and c_J in order not to include the background in the noise estimation).

7. $n = n + 1$
8. If $\frac{|\sigma_s^{(n)} - \sigma_s^{(n-1)}|}{\sigma_s^{(n)}} > \epsilon$ then go to 4.

This method converges in a few iterations, and allows noise estimation to be improved.

The approach is in fact physically meaningful. It consists of detecting the set $\mathcal{N}$ of pixels which does not contain any significant signal (only the background + noise). A pixel k is dominated by the noise if all wavelet coefficients at this position are not significant. The background affects only the last scale of the wavelet transform. We subtract this last scale from the original data, and we compute the standard deviation of the set $\mathcal{N}$ in this background-free data. Wavelet coefficients larger than $3\sigma_j$ are considered as significant, but a small fraction of them will be due to the noise. This introduces a small systematic bias in the final solution, which is easily corrected by dividing the standard deviation by a given constant value, found experimentally as equal to 0.974. Therefore we downgrade the empirical variance in this way. The method is robust and whatever the initial estimation of noise, it converges quickly to a good estimate.

More information on this framework for automated noise estimation can be found in Starck and Murtagh (1998).

Poisson Noise. If the noise in the data s is Poisson, the Anscombe transform $\mathcal{A}$ (Anscombe, 1948)

$$t_l = \mathcal{A}(s_l) = 2\sqrt{s_l + \frac{3}{8}} \tag{2.12}$$

acts as if the data arose from a Gaussian white noise model with $\sigma = 1$, under the assumption that the mean value of s is large.

Gaussian and Poisson Noise. The arrival of photons, and their expression by electron counts, on CCD detectors may be modeled by a Poisson distribution. In addition, there is additive Gaussian read-out noise. The Anscombe transformation (eqn. 2.12) has been extended (Murtagh et al., 1995) to take this combined noise into account. As an approximation, consider the signal's value, s_k, as a sum of a Gaussian variable, γ, of mean g and standard-deviation σ; and a Poisson variable, n, of mean m_0: we set $s_l = \gamma + \alpha n$ where α is the gain.

The generalization of the variance stabilizing Anscombe formula is:

$$t_l = \mathcal{A}_g(s_l) = \frac{2}{\alpha}\sqrt{\alpha s_l + \frac{3}{8}\alpha^2 + \sigma^2 - \alpha g} \tag{2.13}$$

With appropriate values of α, σ and g, this reduces to Anscombe's transformation (eqn. 2.12).

These variance stabilization transformations, it has been shown in Murtagh et al. (1995), are only valid for a sufficiently large number of counts (and of course, for a larger still number of counts, the Poisson distribution becomes Gaussian). The necessary average number of counts is about 20 if bias is to be avoided. Note that errors related to small values carry the risk of removing real objects, but not of amplifying noise. For Poisson parameter values under this threshold acceptable number of counts, the Anscombe transformation loses control over the bias. In this case, an alternative approach to variance stabilization is needed. An approach for very small numbers of counts, including frequent zero cases, has been discussed in (Slezak et al., 1993; Bijaoui et al., 1994; Starck and Pierre, 1998), and will be described below.

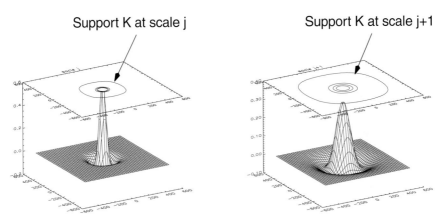

Fig. 2.7. Support K of ψ at two consecutive scales j and $j+1$.

Poisson Noise with Few Photons or Counts. We now consider a data set s_l ($l \in [1 \ldots N]$) of N points in a space of dimension D, and a point at position l is defined by its coordinate $(l_1, \ldots, l_D)$. A wavelet coefficient at a given position l and at a given scale j is

$$w_{j,l} = \langle s, \psi_{j,l} \rangle = \sum_{k \in K} n_k \psi \left(\frac{k_1 - l_1}{2^j}, \ldots, \frac{k_D - l_D}{2^j} \right) \tag{2.14}$$

where K is the support of the wavelet function ψ at scale j (see Fig. 2.7) and n_k is the number of events which contribute to the calculation of $w_{j,l}$ (i.e. the number of events included in the support of the dilated wavelet centered at l).

If all events n_k ($n_k \in K$) are due to noise, $w_{j,l}$ can be considered as a realization of a random variable W_{n_k}, W_{n_k} being defined as the sum of n_k independent random variables. Since independent events are considered, the distribution of the random variable W_n related to n events is given by n autoconvolutions of the distribution function for one event H_1.

$$H_n = H_1 * H_1 * \cdots * H_1 \tag{2.15}$$

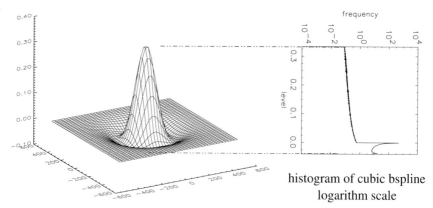

Fig. 2.8. Histogram of ψ_j.

The distribution of one event, H_1, in wavelet space is directly given by the histogram of the wavelet function ψ (see Fig. 2.8). Fig. 2.9 shows the shape of a set of H_n. For a large number of events, H_n converges to a Gaussian.

In order to facilitate the comparisons, the variable W_n of distribution H_n is reduced by

$$c = \frac{W_n - E(W_n)}{\sigma(W_n)} \tag{2.16}$$

and the cumulative distribution functions are

$$F_{W,n}^+(\omega_j) = Prob(W \leq \omega_j) = \int_{-\infty}^{\omega_j} H_n(u) du$$

$$F_{W,n}^-(\omega_j) = Prob(W \geq \omega_j) = \int_{\omega_j}^{\infty} H_n(u) du \tag{2.17}$$

From $F_{W,n}^+$ and $F_{W,n}^-$, we derive two threshold values c_{min} and c_{max} such that

$$p = Prob(W_n \geq c_{min}) = F_{W,n}^-(c_{min}) = 1 - \epsilon$$
$$p = Prob(W_n \leq c_{max}) = F_{W,n}^+(c_{max}) = 1 - \epsilon \tag{2.18}$$

To be compared with the two threshold values, each wavelet coefficient has to be reduced by

$$\begin{aligned} w_{j,l}^r &= \frac{w_{j,l}}{\sqrt{n_k}\sigma_{\psi_j}} \\ &= \frac{w_{j,l}}{\sqrt{n_k}\sigma_\psi} 2^{Dj} \end{aligned} \tag{2.19}$$

where D is the dimension of the input data set, σ_ψ the standard deviation of the wavelet function, σ_{ψ_j} the standard deviation of the dilated wavelet function at scale j ($\sigma_{\psi_j} = \sigma_\psi/2^{Dj}$) and $\sqrt{n_k}$ the normalization factor (n_k events in the support of ψ_j).

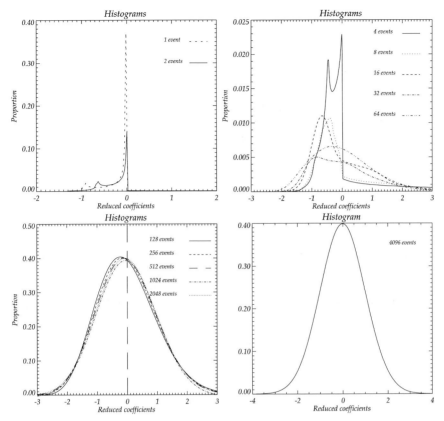

Fig. 2.9. Autoconvolution histograms for the wavelet associated with a B_3 spline scaling function for one and 2 events (*top left*), 4 to 64 events (*top right*), 128 to 2048 (*bottom left*), and 4096 (*bottom right*).

Therefore a reduced wavelet coefficient $w^r_{j,l}$, calculated from $w_{j,l}$, and resulting from n photons or counts is significant if:

$$w^r > c_{max} \tag{2.20}$$

or

$$w^r > c_{min} \tag{2.21}$$

A summary of the few event poisson noise filtering method is:

1. Compute the histogram H_{n_k} for a set of event numbers (for example $N = \{n_k = 2^k\}$).
2. Compute the two threshold levels, c_{min} and c_{max}, for a given ϵ and for all n_k in N.
3. Use the standard filtering method with the new threshold values.

Remarks:

1. If the ϵ value is always the same, threshold values can be computed first and stored.
2. Be aware that the threshold values, used in the standard filtering method, depend on the number of events n_k which contribute to the evaluation of $w_{j,l}$.

Fig. 2.10 shows a simulation. A noisy image containing a faint source was simulated. Fig. 2.10, top left and top right, show respectively the simulated image and the noisy simulated data. Fig. 2.10, middle right, shows the filtered image using the Anscombe transform and hard thresholding in wavelet space. Fig. 2.10, bottom right, shows the filtered image using the thresholding method based on the wavelet function histogram autoconvolution.

Root Mean Square Data Set. If, associated with the data s, we have the root mean square map R (i.e. R_l is the noise standard deviation relative to the value s_l), the noise in s is non-homogeneous. For each wavelet coefficient $w_{j,l}$ of s, the exact standard deviation $\sigma_{j,l}$ needs to be calculated from R. A wavelet coefficient $w_{j,l}$ is obtained by the correlation product between the data s and a function g_j:

$$w_{j,l} = \sum_k s_k g_{j,k-l} \tag{2.22}$$

Then we have

$$\sigma_{j,l}^2 = \sum_k R_k^2 g_{j,k-l}^2. \tag{2.23}$$

In the case of the à trous algorithm, the coefficients $g_{j,l}$ are not known exactly, but they can easily be computed by taking the wavelet transform of a Dirac w^δ: we have $g_{j,l}^2 = (\mathcal{W}w^\delta)_{j,l}^2$ and the set σ_j^2 is calculated by convolving the square of the wavelet scale j of w^δ with R^2 ($\sigma_j^2 = g_j^2 * R^2$).

Other Families of Noise. For any type of noise, an analogous study can be carried out in order to find the detection level at each scale and at each position. The types of noise considered so far in this chapter correspond to the general cases in astronomical imagery. We now describe briefly methods which can be used for non-uniform and multiplicative noise.

– Non-stationary additive noise:
  If the noise is additive, but non-stationary, we cannot estimate a standard deviation for the whole data. However, we can often assume that the noise is locally Gaussian, and we can compute a local standard deviation of the noise for each pixel. In this way, we obtain a standard deviation data set of the noise, $R_\sigma(x)$. Then, the data are treated as for case where the root mean square data set is known (see above).
– Multiplicative noise:
  If the noise is multiplicative, the data can be transformed by taking its logarithm. In the resulting signal, the noise is additive, and a hypothesis

2.3 Significant Wavelet Coefficients

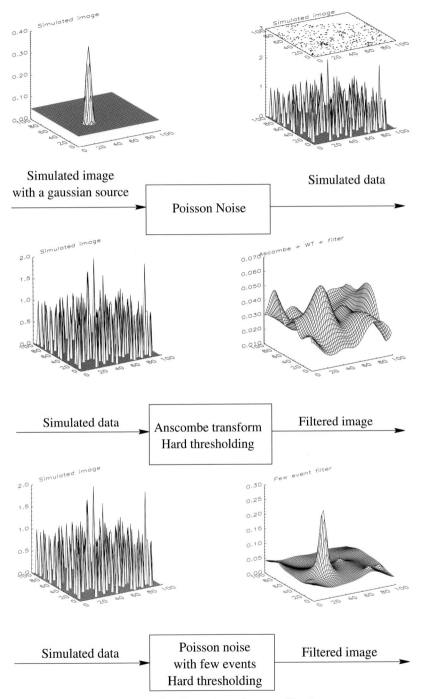

Fig. 2.10. Small number of events filtering.

of Gaussian noise can be used in order to find the detection level at each scale.
- Non-stationary multiplicative noise:
  In this case, we take the logarithm of the data, and the resulting signal is treated as for additive non-stationary noise above.
- Stationary correlated noise:
  If a realization of the noise can be generated, the detection level can be determined by taking the wavelet transform of the noise map, calculating the histogram of each scale, and deriving the thresholds from the normalized histograms. The normalized histograms give us an estimation of the probability density function of a wavelet coefficient due to noise. When the way the data are correlated is unknown, and therefore a noise map cannot be generated, the MAD method or a k-sigma clipping can be used, as for the next case of undefined stationary noise.
- Undefined stationary noise:
  In this case, the data can be treated as for the Gaussian case, but the noise standard deviation σ_j at scale j is calculated independently at each scale. Two methods can be used:
  1. σ_j can be derived from a k-sigma clipping method applied at scale j.
  2. The median absolute deviation, MAD, can be used as an estimator of the noise standard deviation:
  $$\sigma_j = \text{median}(\mid w_j \mid)/0.6745 \tag{2.24}$$
- Unknown noise:
  If the noise does not follow any known distribution, we can consider as significant only wavelet coefficients which are greater than their local standard deviation multiplied by a constant: $w_{j,l}$ is significant if
  $$\mid w_{j,l} \mid > k\sigma(w_{j,x-l...x+l}) \tag{2.25}$$

2.3.4 Detection Level Using the FDR

An alternative approach to the previous detection strategy is the False Discovery Rate method (FDR) (Benjamini and Hochberg, 1995). This technique has recently been introduced for astronomical data analysis (Miller et al., 2001; Hopkins et al., 2002). It allows us to control the average fraction of false detections made over the total number of detections. It also offers an effective way to select an adaptive threshold. The FDR is given by the ratio:

$$FDR = \frac{V_{ia}}{D_a} \tag{2.26}$$

where V_{ia} is the number of pixels that are truly inactive but declared active, and D_a is the number of pixels declared active.

This procedure controlling the FDR specifies a rate α between 0 and 1 and ensures that, *on average*, the FDR is no bigger than α:

$$E(FDR) \leq \frac{T_i}{V}.\alpha \leq \alpha \tag{2.27}$$

2.3 Significant Wavelet Coefficients 49

The unknown factor $\frac{T_i}{V}$ is the proportion of truly inactive pixels where T_i is the number of inactive pixels and V the total number of pixels.

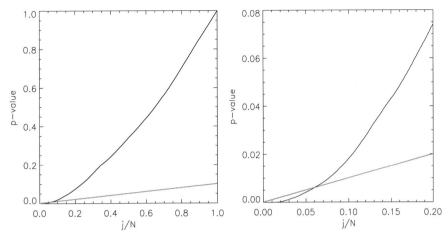

Fig. 2.11. Plot used in FDR algorithm. See text for details. *Right:* zoom.

Let $P_1, \ldots, P_N$ denote the p-values of the N coefficents at a given scale, ordered from smallest to largest. The algorithm to calculate the FDR threshold is the following:

- Plot the curve $F_p(i)$ versus i/N where $F_p(i)$ is the p-value of the ith ordered coefficients (see Fig. 2.11, black curve). We consider as significant (active) all coefficients whose p-values are less than or equal to a given value P_D.
- Plot the line $y = \frac{\alpha j}{N c_N}$, with $c_N = 1$ when the p-values are statistically independent (see Fig. 2.11, red curve).
- The y-axis coordinate of the crossing point between the line and the curve F_p gives the P_D threshold p-value, and the x-axis coordinate gives the related coefficient number i_D.
- The value of the wavelet coefficient i_D is the FDR threshold.

Fig. 2.11, left, illustrates this algorithm. Fig. 2.11, right, is a zoom of the previous one.

A complete description of the FDR method can be found in (Miller et al., 2001). In (Hopkins et al., 2002), it has been shown that FDR outperforms standard methods for source detection. Applying the FDR method at each scale gives us a detection threshold T_j per scale.

2.4 Filtering and Wavelet Coefficient Thresholding

2.4.1 Thresholding

Many filtering methods have been proposed in the last ten years. *Hard thresholding* consists of setting to 0 all wavelet coefficients which have an absolute value lower than a threshold T_j (non-significant wavelet coefficient):

$$\tilde{w}_{j,k} = \begin{cases} w_{j,k} & \text{if } |w_{j,k}| \geq T_j \\ 0 & \text{otherwise} \end{cases}$$

where $w_{j,k}$ is a wavelet coefficient at scale j and at spatial position k.

Soft thresholding consists of replacing each wavelet coefficient by the value $\tilde{w}$ where

$$\tilde{w}_{j,k} = \begin{cases} sgn(w_{j,k})(|w_{j,k}| - T_j) & \text{if } |w_{j,k}| \geq T_j \\ 0 & \text{otherwise} \end{cases}$$

This operation is generally written as:

$$\tilde{w}_{j,k} = \text{soft}(w_{j,k}) = sgn(w_{j,k})(|w_{j,k}| - T_j)_+ \quad (2.28)$$

where $(x)_+ = MAX(0, x)$.

When the discrete orthogonal wavelet transform is used instead of the à trous algorithm, it is interesting to note that the hard and soft thresholded estimators are solutions of the following minimization problems:

$$\tilde{w} = \arg_w \min \frac{1}{2} \| s - \mathcal{W}^{-1} w \|_{l^2}^2 + \lambda \| w \|_{l^0}^2 \quad \textbf{hard threshold}$$

$$\tilde{w} = \arg_w \min \frac{1}{2} \| s - \mathcal{W}^{-1} w \|_{l^2}^2 + \lambda \| w \|_{l^1}^2 \quad \textbf{soft threshold}$$

where s is the input data, $\mathcal{W}$ the wavelet transform operator, and l^0 indicates the limit of l^δ when $\delta \to 0$. This counts in fact the number of non-zero elements in the sequence.

As described before, in the case of Gaussian noise, $T_j = K\sigma_j$, where j is the scale of the wavelet coefficient, σ_j is the noise standard deviation at the scale j, and K is a constant generally chosen equal to 3.

Other threshold methods have been proposed, like the *universal threshold* (Donoho and Johnstone, 1994; Donoho, 1993), or the SURE (Stein Unbiased Risk Estimate) method (Coifman and Donoho, 1995), but they generally do not yield as good results as the hard thresholding method based on the significant coefficients. For astronomical data, soft thresholding should never be used because it leads to a photometry loss associated with all objects, which can easily be verified by looking at the residual map (i.e. data − filtered data). Concerning the threshold level, the universal threshold corresponds to a minimum risk. The larger the number of pixels, the larger is the risk, and it is normal that the threshold T depends on the number of pixels ($T = \sqrt{2 \log n}\sigma_j$, n being the number of pixels). The $K\sigma$ threshold corresponds to a false detection probability, the probability to detect a coefficient as significant when it is due to the noise. The 3σ value corresponds to 0.27 % false detection.

Thresholding methods such the FDR (see previous section) and Adaptive thresholding (Johnstone, 2001) are attractive and can replace the standard k-sigma thresholding.

As described before, a given noise model associated with a data set s produces a multiresolution support M. Hard thresholding can therefore be generalized to any kind of noise when it is derived from the multiresolution support: $\tilde{w}_{j,k} = M_{j,k} w_{j,k}$. The filtered data $\tilde{s}$ are obtained by:

$$\tilde{s} = \mathcal{W}^{-1}(M.\mathcal{W}s) \qquad (2.29)$$

where $\mathcal{W}^{-1}$ is the inverse wavelet transform operator (reconstruction). This notation is not overly helpful in the case of the à trous wavelet transform, which is overcomplete, but it leads to clearer equations.

Hence, wavelet filtering based on hard thresholding consists of taking the wavelet transform of the signal, multiplying the wavelet coefficients by the multiresolution support, and applying the inverse wavelet transform.

2.4.2 Iterative Filtering

When a redundant wavelet transform is used, the result after a simple hard thresholding can still be improved by iterating. We want the wavelet transform of our solution $\tilde{s}$ to reproduce the same significant wavelet coefficients (i.e., coefficients larger than T_j). This can be expressed in the following way:

$$(\mathcal{W}\tilde{s})_{j,k} = w_{j,k} \text{ if } w_{j,k} \text{ is significant} \qquad (2.30)$$

where $w_{j,k}$ are the wavelet coefficients of the input data s. The relation is not necessarily verified in the case of non-orthogonal transforms, and the resulting effect is generally a loss of flux inside the objects. The residual signal (i.e. $s - \tilde{s}$) still contains some information at positions where the objects are.

Denoting M the multiresolution support of s (i.e. $M_{j,k} = 1$ if $w_{j,k}$ is significant, and 0 otherwise), we want:

$$M.\mathcal{W}\tilde{s} = M.\mathcal{W}s$$

The solution can be obtained by the following Van Cittert iteration (Starck et al., 1998a):

$$\begin{aligned}\tilde{s}^{n+1} &= \tilde{s}^n + \mathcal{W}^{-1}(M.\mathcal{W}s - M.\mathcal{W}s^n) \\ &= \tilde{s}^n + \mathcal{W}^{-1}(M.\mathcal{W}R^n)\end{aligned} \qquad (2.31)$$

where $R^n = s - \tilde{s}^n$.

The iterative filtering from the multiresolution support leads therefore to the following algorithm:

1. $n \leftarrow 0$.
2. Initialize the solution, $\tilde{s}^{(0)}$, to zero.
3. Estimate the significance level (e.g. 3-sigma) at each scale.

4. Determine the multiresolution support of the signal.
5. Determine the error, $R^{(n)} = s - \tilde{s}^{(n)}$ (where s is the input signal, to be filtered).
6. Determine the wavelet transform of $R^{(n)}$.
7. Threshold: only retain the coefficients which belong to the support.
8. Reconstruct the thresholded error signal. This yields the signal $\tilde{R}^{(n)}$ containing the significant residuals of the error signal.
9. Add this residual to the solution: $\tilde{s}^{(n)} \leftarrow \tilde{s}^{(n)} + \tilde{R}^{(n)}$.
10. If $|(\sigma_{R^{(n-1)}} - \sigma_{R^{(n)}})/\sigma_{R^{(n)}}| > \epsilon$ then $n \leftarrow n+1$ and go to 4.

Thus the part of the signal which contains significant structures at any level is not modified by the filtering. The residual will contain the value zero over all of these regions. The support can also be enriched by any available a priori knowledge. For example, if artifacts exist around objects, a simple morphological opening of the support can be used to eliminate them. The convergence is fast, generally less than ten iterations.

Dark, Flat and Background Model. In some applications such as gamma-ray image analysis (Integral, GLAST, and others), we may be interested in taking into account a background in order to detect only features not contained in the background. Another standard case is that of dark and flat correction. If we perform this correction before filtering, we lose the noise statistics and the possibility to have robust wavelet coefficient detection. But if we do it after denoising, we may introduce some artifacts into the solution. We describe here how to properly consider these components in the iterative filtering scheme (see eqn. 2.31). Let us denote D the dark image, F the flat image and B the background. The previous case corresponds to $D_k = 0$, $F_k = 1$ and $B_k = 0$. We assume the observed signal s is related to the "sky" component of interest x by the relation:

$$s = F(x + B) + D + n = Fx + D_1 + n \qquad (2.32)$$

where n is the noise and $D_1 = FB + D$. We estimate the detection level from s and the noise model as described in the previous sections, but the significant wavelet coefficients are now detected from the wavelet transform of $s_1 = s - D_1$, and the iteration is:

$$\tilde{x}^{n+1} = \tilde{x}^n + \frac{\mathcal{W}^{-1}(M.\mathcal{W}R^n)}{F} \qquad (2.33)$$

where $R^n = s_1 - F\tilde{x}^n$.

2.4.3 Other Wavelet Denoising Methods

Algorithms that exploit the dependency between wavelet coefficients generally improve the result quality. There are two kind of dependency that can be used: the dependency inside a given band (i.e. the relation between a wavelet

2.4 Filtering and Wavelet Coefficient Thresholding

coefficient and its neighbors) and the dependency between one scale and the next one (i.e. the relation between a wavelet coefficient at the given scale and the wavelet coefficient at the same spatial location but at the next coarsest scale). Taking into account the first dependency can easily be done using a local Wiener filtering in the wavelet domain (Ghael et al., 1997; Choi and Baraniuk, 1998). Therefore, each wavelet coefficient is modified following:

$$\tilde{w}_{j,k} = \frac{s_{j,k}^2 + \sigma_j^2}{s_{j,k}^2} w_{j,k} \qquad (2.34)$$

where σ_j is the noise standard deviation at scale j and $s_{j,k}$ is the standard deviation of the "true" signal at scale j and at position k. This relation is obtained using Bayes' theorem and assuming the wavelet coefficients of the noise-free signal follow a Gaussian distribution. In practice, $s_{j,k}$ is unknown and needs to be estimated. From our model, one gets $d_{j,k}^2 = s_{j,k}^2 + \sigma_j^2$ where $d_{j,k}^2$ is the variance of the noisy data. Since wavelet coefficents have a zero mean, and $d_{j,k}^2$ can be found by:

$$d_{j,k}^2 = \frac{1}{M} \sum_{l \in N(k)} w_{j,l}^2 \qquad (2.35)$$

where $N(k)$ represents the region of neighboring coefficients and M the size of the neighborhood, $s_{j,k}$ is finally derived by computing $s_{j,k}^2 = (d_{j,k}^2 - \sigma_j^2)_+ = MAX(0, d_{j,k}^2 - \sigma_j^2)$. The typical window size is seven by seven or nine by nine for images.

Is has been observed that wavelet coefficients of natural images have highly non-Gaussian statistics (Simoncelli, 1999; Portilla et al., 2003) and the pdf (probability density function) for wavelet coefficients is better modeled using a generalized Gaussian (Portilla et al., 2003):

$$p(w) = K(\alpha, p) \exp\left(-\left|\frac{w}{\alpha}\right|^p\right) \qquad (2.36)$$

where α and p are the model parameters and $K(\alpha, p)$ is a normalization constant. If we consider a Laplacian pdf (i.e. $p=1$), the solution is obtained with a simple soft thresholding (Sendur and Selesnik, 2002):

$$\tilde{w}_{j,k} = \text{soft}\left(w_{j,k}, \frac{\sqrt{2}.\sigma_j^2}{s_{j,k}}\right) \qquad (2.37)$$

The *Bivariate Shrinkage* has been proposed for extending this method in order to take into account the inter-scale relation (Sendur and Selesnik, 2002). Assuming now that the pdf follows:

$$p(w_{j,k}) = \frac{3}{2\pi\sigma^2} \exp\left(-\frac{\sqrt{3}}{\sigma}\sqrt{w_{j,k}^2 + w_{j+1,k}^2}\right) \qquad (2.38)$$

the denoised coefficient is obtained by (Sendur and Selesnik, 2002):

$$\tilde{w}_{j,k} = \frac{\left(\sqrt{d_{j,k}^2 + d_{j+1,k}^2} - \frac{\sqrt{3}\sigma_j^2}{s_{j,k}}\right)_+}{\sqrt{d_{j,k}^2 + d_{j+1,k}^2}} \cdot w_{j,k} \qquad (2.39)$$

Alternative methods, but based on the same concept, can be found in (Crouse et al., 1998; Moulin and Liu, 1999; Sendur and Selesnik, 2002; Portilla et al., 2003; Kazubek, 2003).

2.4.4 Experiments

Simulation 1: 1D Signal Filtering. Fig. 2.12 shows the result after applying the iterative filtering method to a real spectrum. The last plot shows the difference between the original and the filtered spectrum. As we can see, the residual contains only noise.

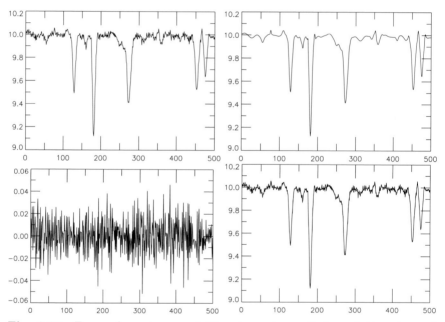

Fig. 2.12. *Top:* real spectrum and filtered spectrum. *Bottom:* both noisy and filtered spectrum overplotted, and difference between the spectrum and the filtered data. As we can see, the residual contains only noise.

Simulation 2: Image with Gaussian Noise. A simulated image containing stars and galaxies is shown in Fig. 2.13 (top left). The simulated noisy image, the filtered image and the residual image are respectively shown in Fig. 2.13 top right, bottom left, and bottom right. We can see that there is

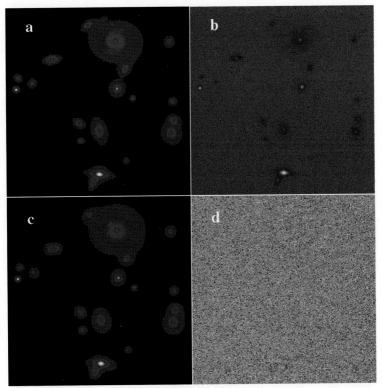

Fig. 2.13. (a) Simulated image, (b) simulated image and Gaussian noise, (c) filtered image, and (d) residual image.

no structure in the residual image. The filtering was carried out using the multiresolution support.

Simulation 3: Image with Poisson Noise. The galaxy cluster A2390 is at a redshift of 0.231. Fig. 2.14, top, shows an image of this cluster, obtained by the ROSAT X-ray spacecraft. The resolution is one arcsecond per pixel, with a total number of photons equal to 13506 for an integration time of 8.5 hours. The background level is about 0.04 photons per pixel.

It is obvious that this image cannot be used directly, and some processing must be performed before any analysis. The standard method consists of convolving the image with a Gaussian. Fig. 2.14, bottom left, shows the result after such processing. (Used was a Gaussian with a full width at half-maximum equal to 5", which is approximately that of the point spread function). The smoothed image shows some structure, but also residual noise, and it is difficult to assign any significance to the structure.

56   2. Filtering

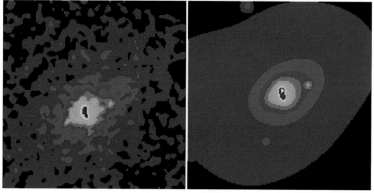

Fig. 2.14. *Top:* ROSAT image of the cluster A2390. *Bottom left:* ROSAT image of the cluster A2390 filtered by the standard method (convolution by a Gaussian). *Bottom right:* ROSAT image of the cluster A2390 filtered by the method based on wavelet coefficients.

Fig. 2.14, bottom right, shows an image filtered by the wavelet transform (Starck and Pierre, 1998; Murtagh et al., 2000). The noise has been eliminated, and the wavelet analysis indicates faint structures in X-ray emission, allowing us to explain gravitational amplification phenomena observed in the visible domain (Pierre and Starck, 1998).

2.4.5 Iterative Filtering with a Smoothness Constraint

A smoothness constraint can be imposed on the solution.

$$\min \|\mathcal{W}\tilde{s}\|_{\ell_1}, \quad \text{subject to} \quad s \in C, \tag{2.40}$$

where C is the set of vectors $\tilde{s}$ which obey the linear constraints

$$\begin{cases} \tilde{s}_k \geq 0, \forall k \\ \mid (\mathcal{W}\tilde{s} - \mathcal{W}s)_{j,k} \mid \leq e_j; \end{cases} \tag{2.41}$$

Here, the second inequality constraint only concerns the set of significant coefficients, i.e. those indices which exceed (in absolute value) a detection threshold t_j. Given a tolerance vector $e = \{e_1, ..., e_j\}$, we seek a solution whose coefficients $(\mathcal{W}\tilde{s})_{j,k}$, at scale and position where significant coefficients were detected, are within e_j of the noisy coefficients $(\mathcal{W}s)_{j,k}$. For example, we can choose $e_j = \sigma_j/2$. In short, our constraint guarantees that the reconstruction be smooth but will take into account any pattern which is detected as significant by the wavelet transform.

Other smoothness penalties can also be used: for instance, an alternative to (2.40) would be (Malgouyres, 2002; Candès and Guo, 2002; Durand and Froment, 2003)

$$\min \|\tilde{s}\|_{TV}, \quad \text{subject to} \quad s \in C.$$

where $\|\cdot\|_{TV}$ is the Total Variation norm, i.e. an edge preservation penalization term defined by:

$$\|\cdot\|_{TV}(\tilde{X}) = \int |\nabla \tilde{X}|^p$$

Minimizing with TV, we force the solution to be closer to a piecewise smooth image. This may however not be good for most astronomical images which contain isotropic features without real edges.

The denoising method described here can be seen as a particuliar case of the Combined Filtering Method described in section 2.5.3.

Results illustrating this method are given in section 8.9, where 3D data sets are filtered using the autoconvolution histograms based method for the detection of the wavelet coefficients (see Fig. 8.19).

2.5 Filtering from the Curvelet Transform

2.5.1 Contrast Enhancement

Because some features are hardly detectable by eye in an image, we often transform it before display. Histogram equalization (cf. section 2.2.1 above) is one the most well-known methods for contrast enhancement. Such an approach is generally useful for images with a poor intensity distribution. Since edges play a fundamental role in image understanding, a way to enhance the contrast is to enhance the edges. For example, we can add to the original image its Laplacian ($I' = I + \gamma \Delta I$, where γ is a parameter). Only features at the finest scale are enhanced (linearly). For a high γ value, only the high frequencies are visible.

Since the curvelet transform is well-adapted to represent images containing edges, it is a good candidate for edge enhancement. Curvelet coefficients can be modified in order to enhance edges in an image. The idea is to not modify curvelet coefficients which are either at the noise level (in order not to amplify the noise) or larger than a given threshold. The largest coefficients correspond to strong edges which do not need to be amplified. Therefore, only curvelet coefficients with an absolute value in $[T_{min}, T_{max}]$ are modified, where T_{min} and T_{max} must be fixed. We define the following function y_c which modifies the values of the curvelet coefficients (Starck et al., 2003b; Starck et al., 2003a):

$$\begin{align}
y_c(x) &= 1 \text{ if } x < T_{min} \\
y_c(x) &= \frac{x - T_{min}}{T_{min}}(\frac{T_{max}}{T_{min}})^p + \frac{2T_{min} - x}{T_{min}} \text{ if } x < 2T_{min} \\
y_c(x) &= (\frac{T_{max}}{x})^p \text{ if } 2T_{min} \leq x < T_{max} \\
y_c(x) &= 1 \text{ if } x \geq T_{max}
\end{align} \quad (2.42)$$

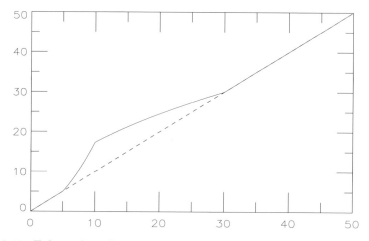

Fig. 2.15. Enhanced coefficients versus original coefficients. Parameters are T_{max} =30, c=5 and p=0.5.

p determines the degree of non-linearity. T_{min} is derived from the noise level, $T_{min} = c\sigma$. A c value larger than 3 guarantees that the noise will not be amplified. The T_{max} parameter can be defined either from the noise standard deviation ($T_{max} = K_m \sigma$) or from the maximum curvelet coefficient M_c of the relative band ($T_{max} = l M_c$, with $l < 1$). The first choice allows the user to define the coefficients to amplify as a function of their signal to noise ratio, while the second one gives an easy and general way to fix the T_{max} parameter independently of the range of the pixel values. Fig. 2.15 shows the curve representing the enhanced coefficients versus the original coefficients. This function is arbitrary and any other function with a similar behavior (i.e. no amplification of the coefficients at the noise level and at very high signal-to-noise ratio, and amplification of the coeffcients slightly above the detection limit) could be used as well.

The curvelet enhancement method consists of the following steps:

1. Estimate the noise standard deviation σ in the input image I.
2. Calculate the curvelet transform of the input image. We get a set of bands w_j, where each band w_j contains N_j coefficients and corresponds to a given resolution level.
3. Calculate the noise standard deviation σ_j for each band j of the curvelet transform (see (Starck et al., 2002) for more details on this step).
4. For each band j do
   – Calculate the maximum M_j of the band.
   – Multiply each curvelet coefficient $w_{j,k}$ by $y_c(|w_{j,k}|)$.
5. Reconstruct the enhanced image from the modified curvelet coefficients.

Example: Saturn Image

Figs. 2.16 show respectively a part of the Saturn image, the histogram equalized image, the Laplacian enhanced image and the curvelet multiscale edge enhanced image (parameters were $p = 0.5$, $c = 3$, and $l = 0.5$). The curvelet multiscale edge enhanced image shows clearly better the rings and edges of Saturn.

2.5.2 Curvelet Denoising

Curvelet transform denoising is completely similar to wavelet denoising. It consists of the following:

– Apply the curvelet transform.
– Correct the curvelet coefficients from the noise.
– Apply the inverse curvelet transform.

Most of the methods proposed for wavelets can be used as well with curvelet coefficients. In our experiments, we used a hard thresholding with a scale-dependent value for the thresholding parameter k; we have $k = 4$ for the first scale ($j = 1$) while $k = 3$ for the others ($j > 1$). We used the same thresholding strategy with the wavelet transform.

Gaussian white noise with a standard deviation fixed to 20 was added to the Saturn image. We employed several methods to filter the noisy image:

1. Thresholding of the curvelet transform.
2. Bi-orthogonal undecimated wavelet de-noising methods using the Dauchechies-Antonini 7/9 filters (FWT-7/9) and hard thresholding.
3. À trous wavelet transform algorithm and hard thresholding.

Fig. 2.16. *Top:* Saturn image and its histogram equalization. *Bottom:* Saturn image enhancement by the Laplacian method and by the curvelet transform.

Our experiments are reported on in Fig. 2.17. The curvelet reconstruction does not contain the same quantity of disturbing artifacts along edges that one sees in wavelet reconstructions. An examination of the details of the restored images is instructive. One notices that the decimated wavelet transform exhibits distortions of the boundaries and suffers substantial loss of important detail. The à trous wavelet transform gives better boundaries, but completely omits to reconstruct certain ridges. In addition, it exhibits numerous small-scale embedded blemishes; setting higher thresholds to avoid these blemishes would cause even more of the intrinsic structure to be missed.

2.5 Filtering from the Curvelet Transform

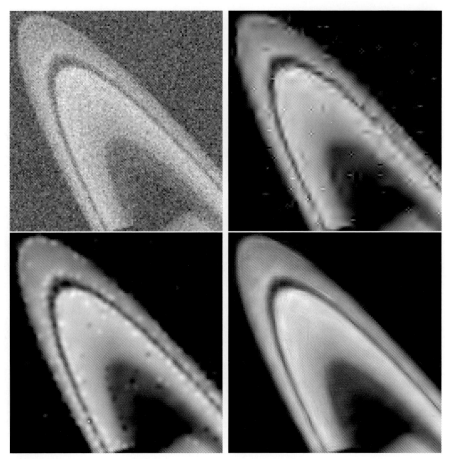

Fig. 2.17. *Top left:* part of Saturn image with Gaussian noise. *Top right:* filtered image using the undecimated bi-orthogonal wavelet transform. *Bottom left and right:* filtered image by the à trous wavelet transform algorithm and the curvelet transform.

2.5.3 The Combined Filtering Method

Wavelets do not restore long features with high fidelity while curvelets are seriously challenged by isotropic or small features. Each transform has its own area of expertise and this complementarity is of potential benefit. The Combined Filtering Method (CFM) (Starck et al., 2001) allows us to benefit from the advantages of both transforms. This iterative method detects the significant coefficients in both the wavelet domain and the curvelet domain and guarantees that the reconstructed map will take into account any pattern which is detected as significant by either of the transforms.

In general, suppose that we are given K linear transforms $T_1, \ldots, T_K$ and let α_k be the coefficient sequence of an object x after applying the transform

T_k, i.e. $\alpha_k = T_k x$. We will assume that for each transform T_k we have available a reconstruction rule that we will denote by T_k^{-1} although this is clearly an abuse of notation. Finally, T will denote the block diagonal matrix with the T_ks as building blocks and α the amalgamation of the α_ks.

A hard thresholding rule associated with the transform T_k synthesizes an estimate $\tilde{s}_k$ via the formula

$$\tilde{s}_k = T_k^{-1} \delta(\alpha_k) \tag{2.43}$$

where δ is a rule that sets to zero all the coordinates of α_k whose absolute value falls below a given sequence of thresholds (such coordinates are said to be non-significant).

Given data y of the form $y = s + \sigma z$, where s is the image we wish to recover and z is standard white noise, we propose solving the following optimization problem (Starck et al., 2001):

$$\min \|T\tilde{s}\|_{\ell_1}, \quad \text{subject to} \quad s \in C, \tag{2.44}$$

where C is the set of vectors $\tilde{s}$ which obey the linear constraints

$$\begin{cases} \tilde{s} \geq 0, \\ |T\tilde{s} - Ty| \leq e \end{cases} \tag{2.45}$$

Here, the second inequality constraint only concerns the set of significant coefficients, i.e. those indices μ such that $\alpha_\mu = (Ty)_\mu$ exceeds (in absolute value) a threshold t_μ. Given a vector of tolerance (e_μ), we seek a solution whose coefficients $(T\tilde{s})_\mu$ are within e_μ of the noisy empirical α_μs. Think of α_μ as being given by

$$y = \langle y, \varphi_\mu \rangle,$$

so that α_μ is normally distributed with mean $\langle f, \varphi_\mu \rangle$ and variance $\sigma_\mu^2 = \sigma^2 \|\varphi_\mu\|_2^2$. In practice, the threshold values range typically between three and four times the noise level σ_μ and in our experiments we will put $e_\mu = \sigma_\mu/2$. In summary, our constraints guarantee that the reconstruction will take into account any pattern which is detected as significant by any of the K transforms.

The Minimization Method. We propose solving (2.44) using the method of hybrid steepest descent (HSD) (Yamada, 2001). HSD consists of building the sequence

$$s^{n+1} = P(s^n) - \lambda_{n+1} \nabla_J (P(s^n)); \tag{2.46}$$

Here, P is the ℓ_2 projection operator onto the feasible set C, ∇_J is the gradient of equation 2.40, and $(\lambda_n)_{n \geq 1}$ is a sequence obeying $(\lambda_n)_{n \geq 1} \in [0, 1]$ and $\lim_{n \to +\infty} \lambda_n = 0$.

The combined filtering algorithm is:

1. Initialize $L_{\max} = 1$, the number of iterations N_i, and $\delta_\lambda = \frac{L_{\max}}{N_i}$.
2. Estimate the noise standard deviation σ, and set $e_k = \frac{\sigma}{2}$.
3. For k = 1, .., K calculate the transform: $\alpha_k^{(s)} = T_k s$.
4. Set $\lambda = L_{\max}$, $n = 0$, and $\tilde{s}^n$ to 0.
5. While $\lambda >= 0$ do
   - $u = \tilde{s}^n$.
   - For k = 1, .., K do
     - Calculate the transform $\alpha_k = T_k u$.
     - For all coefficients $\alpha_{k,l}$ do
       - Calculate the residual $r_{k,l} = \alpha_{k,l}^{(s)} - \alpha_{k,l}$
       - if $\alpha_{k,l}^{(s)}$ is significant and $\mid r_{k,l} \mid > e_{k,l}$ then $\alpha_{k,l} = \alpha_{k,l}^{(s)}$
       - $\alpha_{k,l} = sgn(\alpha_{k,l})(\mid \alpha_{k,l} \mid - \lambda)_+$.
     - $u = T_k^{-1} \alpha_k$
   - Threshold negative values in u and $\tilde{s}^{n+1} = u$.
   - $n = n + 1$, $\lambda = \lambda - \delta_\lambda$, and goto 5.

2.6 Haar Wavelet Transform and Poisson Noise

Several authors (Kolaczyk, 1997; Kolaczyk and Dixon, 2000; Timmermann and Nowak, 1999; Nowak and Baraniuk, 1999; Bijaoui and Jammal, 2001; Fryźlewicz and Nason, 2004) have recently suggested independently that the Haar wavelet transform is very well-suited for treating data with Poisson noise. Since a Haar wavelet coefficient is just the difference between two random variables following a Poisson distribution, it is easier to derive mathematical tools to remove the noise than with any other wavelet method. An isotropic wavelet transform seems more adapted to astronomical data. However, there is a trade-off to be made between an algorithm which optimally represents the information, and another which furnishes a robust way to treat the noise. The approach used for noise filtering differs depending on the authors. In (Nowak and Baraniuk, 1999), a type of Wiener filter was implemented. Timmermann and Nowak (1999) used a Bayesian approach with an a priori model on the original signal, and Kolaczyk and Dixon (2000) and Jammal and Bijaoui (1999; 2001) derived different thresholds resulting from the probability density function of the wavelet coefficients. The Fisz transform (Fryźlewicz and Nason, 2004) is a variance stabilization method. It is used to "gaussianize" the noise. Then the standard wavelet thresholding can be applied to the transformed signal. After the thresholding, the inverse Fisz transform has to be applied.

2.6.1 Haar Wavelet Transform

Kolaczyk (1997) proposed the Haar transform for gamma-ray burst detection in one-dimensional signals, and extended his method to images (Kolaczyk and

Dixon, 2000). The reason why the Haar transform is used is essentially its simplicity, providing a resilient mathematical tool for noise removal.

As far back as 1910, Haar described the following function as providing an orthonormal basis. The analyzing wavelet of a continuous variable is a step function.

$$\begin{array}{ll} \psi(x) = 1 & \text{if } 0 \leq x < \frac{1}{2} \\ \psi(x) = -1 & \text{if } \frac{1}{2} \leq x < 1 \\ \psi(x) = 0 & \text{otherwise} \end{array}$$

The Haar wavelet constitutes an orthonormal basis. Two Haar wavelets of the same scale (i.e. value of m) never overlap, so we have scalar product $< \psi_{m,n}, \psi_{m,n'} > = \delta_{n,n'}$. Overlapping supports are possible if the two wavelets have different scales, e.g. $\psi_{1,1}$ and $\psi_{3,0}$: see (Daubechies, 1992), pp. 10–11. However, if $m < m'$, then the support of $\psi_{m,n}$ lies wholly in the region where $\psi_{m',n'}$ is constant. It follows that $< \psi_{m,n}, \psi_{m',n'} >$ is proportional to the integral of $\psi_{m,n}$, i.e. zero.

2.6.2 Poisson Noise and Haar Wavelet Coefficients

For 1D data, Kolaczyk (1997) proposed to use a detection level for the scale j equal to:

$$t_j = 2^{-(j+2)/2} \left(2 \log(n_j) + \sqrt{(4 \log n_j)^2 + 8 \log n_j \lambda_j} \right) \qquad (2.47)$$

where $n_j = N/2^j$, N is the number of samples, and λ_j is the background rate per bin in the n_j bin. We now describe the 2D case.

Thresholding Assuming a Uniform Background. Assuming a constant background with a background rate λ, Kolaczyk and Dixon (2000) proposed to use the normalized Haar transform (L2-normalization, i.e. the h is defined by $h = \left(\frac{1}{\sqrt{2}}, \frac{1}{\sqrt{2}} \right)$) with the following threshold, corresponding to a false detection rate of α:

$$t_j = 2^{-(j+1)} \left[z_{\alpha/2}^2 + \sqrt{z_{\alpha/2}^4 + 4\lambda_j z_{\alpha/2}^2} \right] \qquad (2.48)$$

where j is the scale level ($j = 1 \ldots J$, J the number of scales), $\lambda_j = 2^{2j}\lambda$ is the background rate in $n_j = N/2^{2j}$ bins (N is the number of pixels in the input image), and $z_{\alpha/2}$ is the point under the Gaussian density function for which there falls $\alpha/2$ mass in the tails beyond each $z_{\alpha/2}$. An upper bound limit for the threshold limit, valid also with other filters, is (Kolaczyk and Dixon, 2000):

$$t_j = 2^{-j} \left[\log(n_j) + \sqrt{\log^2(n_j) + 2\lambda_j \log(n_j)} \right] \qquad (2.49)$$

This formula results from substituting $z = \sqrt{2\log(n_j)}$ into the previous equation.

Bijaoui and Jammal (2001) calculated the probability density function of an unnormalized Haar wavelet coefficient as

$$p(w_j = \nu) = e^{-2^{2j}\lambda} I_\nu(2^{2j}\lambda) \qquad (2.50)$$

where $I_\nu(x)$ is the modified Bessel function of integer order ν. For a given false detection rate α the threshold t_j can be derived from this probability density function.

Thresholding with Non-uniform Background. In many cases, the background cannot be considered to be constant, and an estimation of $\lambda_{j,k,l}$ (the background rate at scale j and position k, l) is needed. Several approaches to background variation can be used:

– Image model: If a model image M can be provided by the user, then $\lambda_{j,k,l}$ is easily obtained by integrating M over the correct surface area.
– Lower resolution: The filtering must be started from the coarser scale S_N. The solution is refined scale by scale in order to obtain $\tilde{s}_J, \tilde{s}_{J-1}, \ldots, \tilde{s}_1$. The $\lambda_{j,k,l}$ values are obtained from $\lambda_{j,k,l}$ ($\lambda_{j,k,l} = \frac{\lambda_{j+1,k/2,l/2}}{4}$).
– Iterative procedure: Filtering can first be performed assuming a constant background (with rate equal to the mean of the image), and the filtered image can be used as a model image. This process is repeated several times until convergence.

Reconstruction. The Haar transform is known to produce block artifacts. Two approaches have been proposed to resolve this problem.

– Cycle-spinning method (Coifman and Donoho, 1995):
  The cycle-spinning method processes the restoration algorithm (Haar transform + thresholding + reconstruction) using every version of the original image data obtainable by combinations of left-right and upwards-downwards translations. The final image is simply the average of all images resulting from the translation sequence.
– Iterative constraint reconstruction (Bobichon and Bijaoui, 1997):
  The iterative constraint reconstruction consists of considering the reconstruction as an inverse problem, where the solution must respect some constraints. Constraints on the solution are:
  – the positivity,
  – the range of variation of its Haar coefficients, and
  – the smoothness at all resolution levels.
  If $w_{j,k,l}$ and $\tilde{w}_{j,k,l}$ are respectively a Haar coefficient at scale j and at position k, l of the data and the solution, then $\tilde{w}_{j,k,l}$ must verify:

$$\begin{cases} \tilde{w}_{j,k,l} \in [-t_j, 0] & \text{if } w_{j,k,l} \in [-t_j, 0] \\ \tilde{w}_{j,k,l} \in [0, t_j,] & \text{if } w_{j,k,l} \in [0, t_j] \\ \tilde{w}_{j,k,l} \in [w_{j,k,l} - t_j/2, w_{j,k,l} + t_j/2] & \text{if } |w_{j,k,l}| > t_j \end{cases} \qquad (2.51)$$

The smoothness constraint consists of minimizing the gradient of the solution $\tilde{s}_j$ at the scale j in both vertical and horizontal directions:

$$C(\tilde{s}_j) = \| D_x \tilde{s}_{j,x,y} \|^2 + \| D_y \tilde{s}_{j,x,y} \|^2 \qquad (2.52)$$

where D_x and D_y are the gradient operators in both directions. A full description of the algorithm can be found in (Bobichon and Bijaoui, 1997).

The results can also be improved by replacing the Haar wavelet transform with the bi-orthogonal Haar wavelet transform. In this case, the analysis filters h and g (i.e. filters used for the transformation) are still the Haar filters, but the synthesis filters $\tilde{h}$ and $\tilde{g}$ (i.e. filters used for the reconstruction) are different. Their support is larger, and they are more regular. Table 2.2 gives two examples of synthesis filters $\tilde{h}_2$ and $\tilde{h}_4$ allowing an exact reconstruction, while keeping the same Haar analysis filter.

Table 2.2. Examples of bi-orthogonal Haar filters.

	h	$\tilde{h}_2$	$\tilde{h}_4$
-4	0	0	0.016572815184
-3	0	0	-0.016572815184
-2	0	-0.088388347648	-0.121533978016
-1	0	0.088388347648	0.121533978016
0	0.707106781187	0.707106781187	0.707106781187
1	0.707106781187	0.707106781187	0.707106781187
2	0	0.088388347648	0.121533978016
3	0	-0.088388347648	-0.121533978016
4	0	0	-0.016572815184
5	0	0	0.016572815184

MMI Model. The Multiscale Multiplicative Innovations (MMI) model was proposed in Timmermann and Nowak (1999), and introduces a prior model f_Λ, which is a beta-mixture density function of the form:

$$f(\delta) = \sum_{i=1}^{M} p_i \frac{(1-\delta^2)^{s_i-1}}{B(s_i, s_i) 2^{2s_i-1}} \qquad (2.53)$$

for $-1 \leq \delta \leq 1$, where B is the Euler beta function, $-1 \leq p_i \leq 1$ is the weight of the ith beta density $\frac{(1-\delta^2)^{s_i-1}}{B(s_i,s_i)2^{2s_i-1}}$ with parameter $s_i \geq 1$, and $\sum_{i=1}^{M} p_i = 1$. The final algorithm consists of multiplying each wavelet coefficient $w_{j,k,l}$ by a term which is derived from the model.

PRESS-optimal Filter. The PRESS-optimal filter (Nowak and Baraniuk, 1999) shrinks the noisy wavelet coefficient toward zero according to the estimated signal to signal-plus-noise ratio. The PRESS-optimal filter is given by:

$$h_{j,k,l} = \frac{\tilde{w}_{j,k,l}^2}{\tilde{w}_{j,k,l}^2 + \sigma_{j,k,l}^2} \qquad (2.54)$$

where $\sigma_{j,k,l}^2$ and $\tilde{w}_{j,k,l}$ are the noise and signal power. The noise power is proportional to an estimate of the local intensity of the image falling under the support of the Haar coefficient $w_{j,k,l}$. The signal power can be estimated from a model, or directly from the data by:

$$\tilde{w}_{j,k,l}^2 = w_{j,k,l}^2 - \sigma_{j,k,l}^2 \qquad (2.55)$$

2.6.3 Experiments

Two test images were used in our experiments. The first one is a simulated XMM image containing a large number of sources of different sizes and of different intensities (see Fig. 2.18). Fig. 2.18, top right, bottom left and right, show respectively the filtered images by Haar-Kolaczyk, Haar-Jammal-Bijaoui, and the à trous algorithm. A 4-sigma level ($\epsilon = 6.34e^{-05}$) was used with the three methods. The undecimated Haar transform was used with the bi-orthogonal Haar filters (filter $\tilde{h}_2$).

The second experiment is a simulated galaxy cluster. Two point sources are superimposed (on the left of the cluster), a cooling flow is at the center, a sub-structure on its left, and a group of galaxies at the top (see Fig. 2.19 (a)). From this image, a "noisy" image was created (see Fig. 2.19 (b)). The mean background level is equal to 0.1 events per pixel. This corresponds typically to X-ray cluster observations. In the noisy galaxy image, the maximum value is equal to 25 events. The background is not very relevant. The problem in this kind of image is the small number of photons per object. It is very difficult to extract any information from them. Fig. 2.19 (c) and (d) show the filtered images using the Haar-Kolaczyk thresholding approach and, respectively, an undecimated Haar transform and the undecimated bi-orthogonal Haar transform. Blocking artifacts due to the shape of the Haar function are visible when using the undecimated Haar transform, but not with the undecimated bi-orthogonal Haar transform. Fig. 2.19 (e) and (f) show the filtered image using the Haar-Jammal-Bijaoui method and the undecimated bi-orthogonal Haar transform. Fig. 2.19 (f) shows the filtered image using the iterative filtering method based on the à trous algorithm described in section 2.4.2 and the significant wavelet coefficients were detected using the autoconvolution histograms based method (see section 2.3.3). This is the only method to correctly detect the two faint objects.

From our experiments, we conclude the following.

- The à trous algorithm is significantly better than any of the Haar-based methods, especially for the detection of faint objects.
- The PRESS-optimal filter produces very poor results.
- The lower resolution furnishes a good estimation of the background. Iterating does not improve significantly the results.

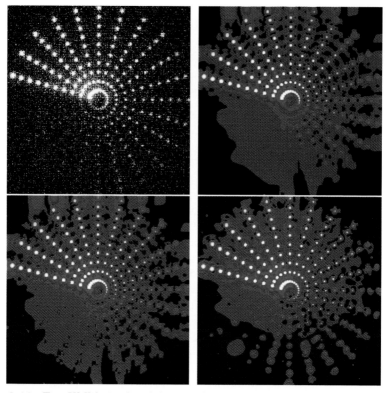

Fig. 2.18. *Top:* XMM simulated data, and Haar-Kolaczyk filtered image. *Bottom:* Haar-Jammal-Bijaoui and à trous filtered images.

– The Jammal-Bijaoui threshold is a little better than the Kolaczyk one for compact source detection, and is equivalent for more extended sources. But the Kolaczyk threshold requires less computation time and is easier to implement.
– The Fisz transform has the drawback of being time-consuming in its translation invariant version. Indeed, in the Fisz transform, as opposed to the other methods, we cannot use an undecimated Haar transform instead of the Haar transform. The only way to be invariant here through translation is to use the cycle-spanning procedure (Coifman and Donoho, 1995) which multiplies the number of operations by the number of pixels contained in the data.

This study shows clearly that the Haar transform is less effective for restoring X-ray astronomical images than the à trous algorithm. But its simplicity, and the computation time required, may be attractive alternatives. In the framework of the XMM-LSS project (Pierre et al., 2004), after a set of simulations (Valtchanov et al., 2001), the denoising method chosen to be

2.6 Haar Wavelet Transform and Poisson Noise

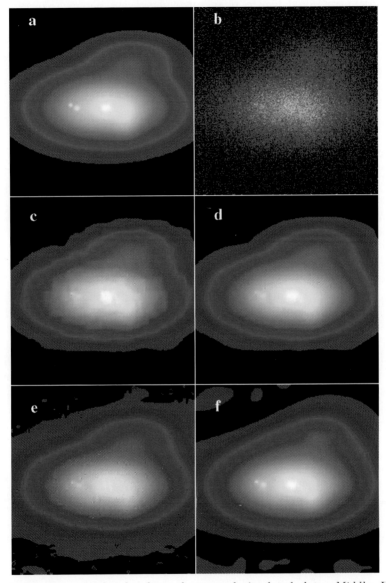

Fig. 2.19. *Top:* simulated galaxy cluster and simulated data. *Middle:* Haar-Kolaczyk filtered images using the undecimated Haar transform and the undecimated bi-orthogonal Haar transform. *Bottom:* Haar-Jammal-Bijaoui filtered image and the à trous algorithm.

included in the pipeline was the iterative method (see section 2.4.2) using noise modeling based on histogram autoconvolutions (see section 2.3.3).

Finally we would like to point out some recent papers (Kolaczyk and Nowak, 2004; Kolaczyk and Nowak, 2005; Willet and Nowak, 2005) which propose a dyadic partitioning of the image. This dyadic partitioning concept may however not be very well suited to astrophysical data. This should be investigated in the future.

2.7 Chapter Summary

Image and signal filtering can be quite a basic operation. But then a visualization or some alternative appraisal of results is needed. By taking account of the noise properties in the data, we go a long way towards having the algorithmic procedure tell us, the user, what filtering is required.

Wavelet/curvelet transform spaces, by virtue of their energy compacting properties, provide an ideal framework for noise filtering. A redundant transform may in addition provide valuable retention of information on the objects or features at different resolution scales which would otherwise be lost with a non-redundant transform.

A wide range of statistical noise modeling and filtering approaches were examined in this chapter, which are appropriate for a range of different types of data.

The wavelet function used in the Haar wavelet transform is both simple and interesting from the point of view of analytically defining Poisson noise in transform space. Does the Haar wavelet transform score highly for Poisson noise in analogy with the redundant à trous wavelet transform scoring highly for Gaussian noise? In fact we showed that the latter can still do better, even if the former has a number of useful properties.

In this chapter, we have considered noisy data obtained from an instrument. But noise is also a problem of major concern for N-body simulations of structure formation in the early Universe and it has been shown that using wavelets for removing noise from N-body simulations is equivalent to simulations with two orders of magnitude more particles (Romeo et al., 2003; Romeo et al., 2004).

3. Deconvolution

3.1 Introduction

Deconvolution is a key area in signal and image processing. It can include deblurring of an observed signal to remove atmospheric effects. More generally, it means correcting for instrumental effects or observing conditions.

Research in image deconvolution has recently seen considerable work, partly triggered by the HST optical aberration problem at the beginning of its mission that motivated astronomers to improve current algorithms or develop new and more efficient ones. Since then, deconvolution of astronomical images has proven in some cases to be crucial for extracting scientific content. For instance, IRAS images can be efficiently reconstructed thanks to a new pyramidal maximum entropy algorithm (Bontekoe et al., 1994). Io volcanism can be studied with a lower resolution of 0.15 arcsec, or 570 km on Io (Marchis et al., 2000). Deconvolved mid-infrared images at 20 μm revealed the inner structure of the AGN in NGC1068, hidden at lower wavelength because of the high extinction (Alloin et al., 2000): see Fig. 3.1. Research on gravitational lenses is easier and more efficient when applying deconvolution methods (Courbin et al., 1998). A final example is the high resolution (after deconvolution) of mid-infrared images revealing the intimate structure of young stellar objects (Zavagno et al., 1999): see also Fig. 3.2. Deconvolution will be even more crucial in the future in order to fully take advantage of increasing numbers of high-quality ground-based telescopes, for which images are strongly limited in resolution by the seeing.

HST provided a leading example of the need for deconvolution, in the period before the detector system was refurbished. Two proceedings (White and Allen, 1991; Hanisch and White, 1994) provide useful overviews of this work, and a later reference is (Adorf et al., 1995). While an atmospheric seeing point spread function (PSF) may be relatively tightly distributed around the mode, this was not the case for the spherically aberrated HST PSF. Whenever the PSF "wings" are extended and irregular, deconvolution offers a straightforward way to mitigate the effects of this and to upgrade the core region of a point source. One usage of deconvolution of continuing importance is in information fusion from different detectors. For example, Faure et al. (2002) deconvolve HST images when correlating with ground-based observations. In Radomski et al. (2002), Keck data are deconvolved, for study with HST data.

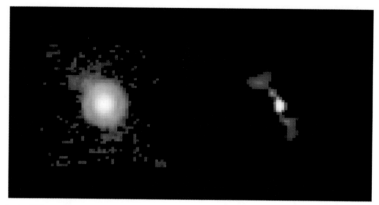

Fig. 3.1. The active galaxy nucleus of NGC1068 observed at 20 μm. *Left:* the raw image is highly blurred by telescope diffraction. *Right:* the restored image using the multiscale entropy method reveals the inner structure in the vicinity of the nucleus.

VLT (Very Large Telescope) data are deconvolved in (Burud et al., 2002), with other ESO and HST data used as well. In planetary work, Coustenis et al. (2001) discuss CFHT data as well as HST and other observations.

What emerges very clearly from this small sample – which is in no way atypical – is that a major use of deconvolution is to help in cross-correlating image and signal information.

An observed signal is never in pristine condition, and improving it involves inverting the spoiling conditions, i.e. finding a solution to an inverse equation. Constraints related to the type of signal we are dealing with play an important role in the development of effective and efficient algorithms. The use of constraints to provide for a stable and unique solution is termed regularization.

Our review opens in section 2 with a formalization of the problem. Section 3 considers the issue of regularization. In section 4, the CLEAN method which is central to radio astronomy is described. Bayesian modeling and inference in deconvolution is reviewed in section 5.

Section 6 further considers regularization, surveying more complex and powerful regularization methods. Section 7 introduces wavelet-based methods as used in deconvolution. These methods are based on multiple resolution or scale. In sections 8 and 9, important issues related to resolution of the output result image are discussed. Section 8 is based on the fact that it is normally not worthwhile to target an output result with better resolution than some limit, for instance a pixel size. Section 9 investigates when, where and how missing information can be inferred to provide a super-resolution output image.

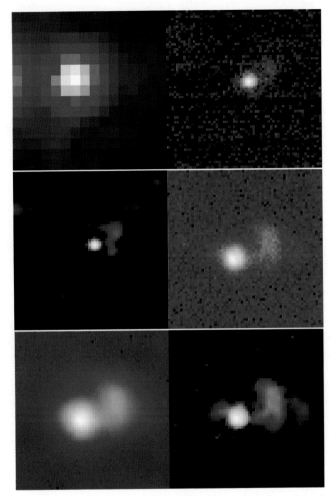

Fig. 3.2. The young stellar object AFGL4029 observed with various instruments of different resolutions, at the same approximate wavelength (11 µm). *Upper left:* the ISOCAM image highly blurred by the diffraction, *upper right:* raw observation made at the NOT (Nordic Optical Telescope, La Palma, Canarias, Spain), a 2m class telescope using the CAMIRAS mid-infrared camera; the diffraction still blurs the resolution quite heavily. *Middle left:* image obtained at NOT, deconvolved using the multiscale entropy method. *Middle right:* the raw image obtained at the CFHT. During the observations, the seeing conditions at CFHT were better than at the NOT and the weaker diffraction explains the better quality of the resolution, as seen in the filtered image (*lower left panel*) using the multiscale entropy filtering method (Pantin and Starck, 1996). *Lower right:* CFHT, the image at the best resolution achieved, showing the sharpest detail of the cavity carved by the bipolar outflow. This example from a series of images at different resolutions shows that results obtained from deconvolution methods are confirmed by raw images at higher intrinsic resolution, and we see also how increasing the resolution through the use of deconvolution methods helps in extracting the scientific content from the data.

3.2 The Deconvolution Problem

Consider an image characterized by its intensity distribution (the "data") I, corresponding to the observation of a "real image" O through an optical system. If the imaging system is linear and shift-invariant, the relation between the data and the image in the same coordinate frame is a convolution:

$$\begin{aligned} I(x,y) &= \int_{x_1=-\infty}^{+\infty} \int_{y_1=-\infty}^{+\infty} P(x-x_1, y-y_1) O(x_1, y_1) dx_1 dy_1 \\ &\quad + N(x,y) \\ &= (P*O)(x,y) + N(x,y) \end{aligned} \quad (3.1)$$

P is the point spread function, PSF, of the imaging system, and N is additive noise.

In Fourier space we have:

$$\hat{I}(u,v) = \hat{O}(u,v)\hat{P}(u,v) + \hat{N}(u,v) \quad (3.2)$$

We want to determine $O(x,y)$ knowing I and P. This inverse problem has led to a large amount of work, the main difficulties being the existence of: (i) a cut-off frequency of the point spread function, and (ii) the additive noise. See for example (Cornwell, 1989; Katsaggelos, 1993; Bertero and Boccacci, 1998; Molina et al., 2001).

A solution can be obtained by computing the Fourier transform of the deconvolved object $\hat{\tilde{O}}$ by a simple division between the image $\hat{I}$ and the PSF $\hat{P}$

$$\hat{\tilde{O}}(u,v) = \frac{\hat{I}(u,v)}{\hat{P}(u,v)} = \hat{O}(u,v) + \frac{\hat{N}(u,v)}{\hat{P}(u,v)} \quad (3.3)$$

This method, sometimes called the *Fourier-quotient method* is very fast. We only need to do a Fourier transform and an inverse Fourier transform. For frequencies close to the frequency cut-off, the noise term becomes important, and the noise is amplified. Therefore in the presence of noise, this method cannot be used.

Eqn. 3.1 is usually in practice an ill-posed problem. This means that there is no unique and stable solution.

Other topics related to deconvolution are:

- Super-resolution: object spatial frequency information outside the spatial bandwidth of the image formation system is recovered.
- Blind deconvolution: the PSF P is unknown.
- Myopic deconvolution: the PSF P is partially known.
- Image reconstruction: an image is formed from a series of projections (computed tomography, positron emission tomography or PET, and so on).

We will discuss only the deconvolution and super-resolution problems in this chapter.

In the deconvolution problem, the PSF is assumed to be known. In practice, we have to construct a PSF from the data, or from an optical model of the imaging telescope. In astronomy, the data may contain stars, or one can point towards a reference star in order to reconstruct a PSF. The drawback is the "degradation" of this PSF because of unavoidable noise or spurious instrument signatures in the data. So, when reconstructing a PSF from experimental data, one has to reduce very carefully the images used (background removal for instance) or otherwise any spurious feature in the PSF would be repeated around each object in the deconvolved image. Another problem arises when the PSF is highly variable with time, as is the case for adaptive optics images. This means usually that the PSF estimated when observing a reference star, after or before the observation of the scientific target, has small differences from the perfect one. In this particular case, one has to turn towards myopic deconvolution methods (Christou et al., 1999) in which the PSF is also estimated in the iterative algorithm using a first guess deduced from observations of reference stars.

Another approach consists of constructing a synthetic PSF. Several studies (Buonanno et al., 1983; Moffat, 1969; Djorgovski, 1983; Molina et al., 1992) have suggested a radially symmetric approximation to the PSF:

$$P(r) \propto (1 + \frac{r^2}{R^2})^{-\beta} \tag{3.4}$$

The parameters β and R are obtained by fitting the model with stars contained in the data.

3.3 Linear Regularized Methods

3.3.1 Least Squares Solution

It is easy to verify that the minimization of $\| I(x,y) - P(x,y) * O(x,y) \|^2$ leads to the solution:

$$\hat{\tilde{O}}(u,v) = \frac{\hat{P}^*(u,v) \hat{I}(u,v)}{| \hat{P}(u,v) |^2} \tag{3.5}$$

which is defined only if $\hat{P}(u,v)$ is different from zero. The problem is generally ill-posed and we need to introduce *regularization* in order to find a unique and stable solution.

3.3.2 Tikhonov Regularization

Tikhonov regularization (Tikhonov et al., 1987) consists of minimizing the term:

$$J_T(O) = \| I(x,y) - (P*O)(x,y) \| + \lambda \| H*O \| \tag{3.6}$$

where H corresponds to a high-pass filter. This criterion contains two terms. The first, $\| I(x,y) - P(x,y)*O(x,y) \|^2$, expresses fidelity to the data $I(x,y)$, and the second, $\lambda \| H*O \|^2$, expresses smoothness of the restored image. λ is the regularization parameter and represents the trade-off between fidelity to the data and the smoothness of the restored image.

The solution is obtained directly in Fourier space:

$$\hat{\tilde{O}}(u,v) = \frac{\hat{P}^*(u,v)\hat{I}(u,v)}{\mid \hat{P}(u,v) \mid^2 + \lambda \mid \hat{H}(u,v) \mid^2} \tag{3.7}$$

Finding the optimal value λ necessitates use of numerical techniques such as cross-validation (Golub et al., 1979; Galatsanos and Katsaggelos, 1992). This method works well, but computationally it is relatively lengthy and produces smoothed images. This second point can be a real problem when we seek compact structures such as is the case in astronomical imaging.

3.3.3 Generalization

This regularization method can be generalized, and we write:

$$\hat{\tilde{O}}(u,v) = \hat{W}(u,v) \frac{\hat{I}(u,v)}{\hat{P}(u,v)} \tag{3.8}$$

where W must satisfy the following conditions (Bertero and Boccacci, 1998). We give here the window definition in 1D.

1. $\mid \hat{W}(\nu) \mid \leq 1$, for any $\nu > 0$
2. $\lim_{\nu \to 0} \hat{W}(\nu) = 1$ for any ν such that $\hat{P}(\nu) \neq 0$
3. $\hat{W}(\nu)/\hat{P}(\nu)$ bounded for any $\nu > 0$

Any function satisfying these three conditions defines a regularized linear solution. Commonly used windows are:

- Truncated window function: $\hat{W}(\nu) = \begin{cases} 1 & \text{if } \mid \hat{P}(\nu) \mid \geq \sqrt{\epsilon} \\ 0 & \text{otherwise} \end{cases}$

    where ϵ is the regularization parameter.
- Rectangular window: $\hat{W}(\nu) = \begin{cases} 1 & \text{if } \mid \nu \mid \leq \Omega \\ 0 & \text{otherwise} \end{cases}$

    where Ω defines the bandwidth.
- Triangular window: $\hat{W}(\nu) = \begin{cases} 1 - \frac{\nu}{\Omega} & \text{if } \mid \nu \mid \leq \Omega \\ 0 & \text{otherwise} \end{cases}$
- Hamming window: $\hat{W}(\nu) = \begin{cases} 0.54 + 0.46\cos(\frac{2\pi\nu}{\Omega}) & \text{if } \mid \nu \mid \leq \Omega \\ 0 & \text{otherwise} \end{cases}$
- Hanning window: $\hat{W}(\nu) = \begin{cases} \cos(\frac{\pi\nu}{\Omega}) & \text{if } \mid \nu \mid \leq \Omega \\ 0 & \text{otherwise} \end{cases}$
- Gaussian window: $\hat{W}(\nu) = \begin{cases} \exp(-4.5\frac{\nu^2}{\Omega^2}) & \text{if } \mid \nu \mid \leq \Omega \\ 0 & \text{otherwise} \end{cases}$

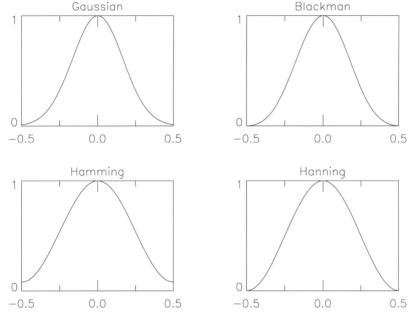

Fig. 3.3. Example of windows with support in the interval $[-1/2, 1/2]$.

- Blackman window:
$$\hat{W}(\nu) = \begin{cases} 0.42 + 0.5\cos(\frac{\pi\nu}{\Omega}) + 0.08\cos(\frac{2\pi\nu}{\Omega}) & \text{if } |\nu| \le \Omega \\ 0 & \text{otherwise} \end{cases}$$

Fig. 3.3 shows examples of windows. The function can also be derived directly from the PSF (Pijpers, 1999). Linear regularized methods have the advantage of being very attractive from a computation point of view. Furthermore, the noise in the solution can easily be derived from the noise in the data and the window function. For example, if the noise in the data is Gaussian with a standard deviation σ_d, the noise in the solution is $\sigma_s^2 = \sigma_d^2 \sum W_k^2$. But this noise estimation does not take into account errors relative to inaccurate knowledge of the PSF, which limits its interest in practice.

Linear regularized methods present also a number of severe drawbacks:

- Creation of Gibbs oscillations in the neighborhood of the discontinuities contained in the data. The visual quality is therefore degraded.
- No a priori information can be used. For example, negative values can exist in the solution, while in most cases we know that the solution must be positive.
- Since the window function is a low-pass filter, the resolution is degraded. There is trade-off between the resolution we want to achieve and the noise level in the solution. Other methods such as wavelet-based methods do not have such a constraint.

3.4 CLEAN

The CLEAN method is a mainstream one in radio-astronomy. This approach assumes the object is composed of point sources. It tries to decompose the image (called the dirty map) into a set of δ-functions. This is done iteratively by finding the point with the largest absolute brightness and subtracting the PSF (dirty beam) scaled with the product of the loop gain and the intensity at that point. The resulting residual map is then used to repeat the process. The process is stopped when some prespecified limit is reached. The convolution of the δ-functions with an ideal PSF (clean beam) plus the residual equals the restored image (clean map). This solution is only possible if the image does not contain large-scale structures. The algorithm is:

1. Compute the dirty map $I^{(0)}(x, y)$ and the dirty beam $A(x, y)$
2. Find the maximum value, and the coordinate (x_{max}, y_{max}) of the corresponding pixel in $I^{(i)}(x, y)$.
3. Compute $I^{(i+1)}(x, y) = I^{(i)}(x, y) - \gamma I_{max} A_m(x, y)$
   with $A_m(x, y) = A(x - x_{max}, y - y_{max})$
   and the loop gain γ inside $[0,1]$.
4. If the residual map is at the noise level, then go to step 5.
   Else $i \longleftarrow i + 1$ and go to step 2.
5. The clean map is the convolution of the list of maxima with the clean beam (which is generally a Gaussian).
6. Addition of the clean map and the residual map produces the deconvolved image.

In the work of Champagnat et al. (1996) and Kaaresen (1997), the restoration of an object composed of peaks, called *sparse spike trains*, has been treated in a rigorous way.

3.5 Bayesian Methodology

3.5.1 Definition

The Bayesian approach consists of constructing the conditional probability density relationship:

$$p(O \mid I) = \frac{p(I \mid O) p(O)}{p(I)} \qquad (3.9)$$

The Bayes solution is found by maximizing the right part of the equation. The maximum likelihood solution (ML) maximizes only the density $p(I \mid O)$ over O:

$$\text{ML}(O) = \max_{O} p(I \mid O) \qquad (3.10)$$

The maximum-a-posteriori solution (MAP) maximizes over O the product $p(I \mid O)p(O)$ of the ML and a prior:

$$\text{MAP}(O) = \max_O p(I \mid O)p(O) \tag{3.11}$$

$p(I)$ is considered as a constant value which has no effect in the maximization process, and is ignored. The ML solution is equivalent to the MAP solution assuming a uniform probability density for $p(O)$.

3.5.2 Maximum Likelihood with Gaussian Noise

The probability $p(I \mid O)$ is

$$p(I \mid O) = \frac{1}{\sqrt{2\pi}\sigma_N} \exp -\frac{(I - P * O)^2}{2\sigma_N^2} \tag{3.12}$$

and, assuming that $p(O)$ is a constant, maximizing $p(O \mid I)$ is equivalent to minimizing

$$J(O) = \frac{\parallel I - P * O \parallel^2}{2\sigma_n^2} \tag{3.13}$$

Using the steepest descent minimization method, a typical iteration is

$$O^{n+1} = O^n + \gamma P^* * (I - P * O^n) \tag{3.14}$$

where $P^*(x, y) = P(-x, -y)$. P^* is the transpose of the PSF, and $O^{(n)}$ is the current estimate of the desired "real image". This method is usually called the Landweber method (1951), but sometimes also the *successive approximations* or Jacobi method (Bertero and Boccacci, 1998).

The solution can also be found directly using the FFT by

$$\hat{O}(u,v) = \frac{\hat{P}^*(u,v)\hat{I}(u,v)}{\hat{P}^*(u,v)\hat{P}(u,v)} \tag{3.15}$$

3.5.3 Gaussian Bayes Model

If the object and the noise are assumed to follow Gaussian distributions with zero mean and variance respectively equal to σ_O and σ_N, then a Bayes solution leads to the Wiener filter:

$$\hat{O}(u,v) = \frac{\hat{P}^*(u,v)\hat{I}(u,v)}{\mid \hat{P}(u,v) \mid^2 + \frac{\sigma_N^2(u,v)}{\sigma_O^2(u,v)}} \tag{3.16}$$

Wiener filtering has serious drawbacks (artifact creation such as ringing effects), and needs spectral noise estimation. Its advantage is that it is very fast.

3.5.4 Maximum Likelihood with Poisson Noise

The probability $p(I \mid O)$ is

$$p(I \mid O) = \prod_{x,y} \frac{((P * O)(x,y))^{I(x,y)} \exp\{-(P * O)(x,y)\}}{I(x,y)!} \qquad (3.17)$$

The maximum can be computed by taking the derivative of the logarithm:

$$\frac{\partial \ln p(I \mid O)(x,y)}{\partial O(x,y)} = 0 \qquad (3.18)$$

which leads to the result (assuming the PSF is normalized to unity)

$$\frac{I(x,y)}{P * O(x,y)} * P^* = 1 \qquad (3.19)$$

Multiplying both sides by $O(x,y)$

$$O(x,y) = \left[\frac{I(x,y)}{(P * O)(x,y)} * P^*(x,y) \right] O(x,y) \qquad (3.20)$$

and using Picard iteration (Issacson and Keller, 1966) (see Appendix B for more details) leads to

$$O^{n+1}(x,y) = \left[\frac{I(x,y)}{(P * O^n)(x,y)} * P^*(x,y) \right] O^n(x,y) \qquad (3.21)$$

which is the Richardson-Lucy algorithm (Richardson, 1972; Lucy, 1974; Shepp and Vardi, 1982), also sometimes called the *expectation maximization* or EM method (Dempster et al., 1977). This method is commonly used in astronomy. Flux is preserved and the solution is always positive. The positivity of the solution can be obtained too with Van Cittert's and the one-step gradient methods by thresholding negative values in O^n at each iteration.

Example: Application to Deep Impact Data. Deep Impact flew the largest telescope in history to deep space (subsequently exceeded by the HIRISE instrument on the Mars Reconnaisance Observer (MRO)), an f/35 Cassegrain design with 10m focal length. A flaw in the pre-launch calibration of this High Resolution Instrument (HRI) resulted in the inability to accurately focus the instrument during flight. Fortunately, because of the nature of the PSF, much of the higher spatial frequency information is retained and use of image deconvolution can recover much of the originally expected spatial resolution. Fig. 3.4, left, shows an image taken approximately 8 minutes after impact with the CLEAR6 filter and Fig. 3.4, right, shows the result of thirty iterations of the Richardson-Lucy algorithm.

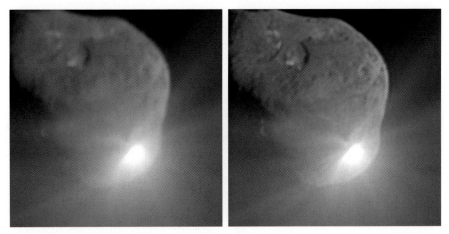

Fig. 3.4. Deep Impact image (*left*) and deconvolved image (*right*) using the Richardson-Lucy method.

3.5.5 Poisson Bayes Model

We formulate the object PDF (probability density function) as

$$p(O) = \prod_{x,y} \frac{M(x,y)^{O(x,y)} \exp\{-M(x,y)\}}{O(x,y)!} \qquad (3.22)$$

The MAP solution is

$$O(x,y) = M(x,y) \exp\left\{\left[\frac{I(x,y)}{(P*O)(x,y)} - 1\right] * P^*(x,y)\right\} \qquad (3.23)$$

and choosing $M = O^n$ and using Picard iteration leads to

$$O^{n+1}(x,y) = O^n(x,y) \exp\left\{\left[\frac{I(x,y)}{(P*O^n)(x,y)} - 1\right] * P^*(x,y)\right\} \qquad (3.24)$$

3.5.6 Maximum Entropy Method

In the absence of any information on the solution O except its positivity, a possible course of action is to derive the probability of O from its entropy, which is defined from information theory. Then if we know the entropy H of the solution, we derive its probability as

$$p(O) = \exp(-\alpha H(O)) \qquad (3.25)$$

The most commonly used entropy functions are:

- Burg (1978): $H_b(O) = -\sum_x \sum_y \ln(O(x,y))$
- Frieden (1978a): $H_f(O) = -\sum_x \sum_y O(x,y) \ln(O(x,y))$

– Gull and Skilling (1991):

$$H_g(O) = \sum_x \sum_y O(x,y) - M(x,y) - O(x,y) \ln(O(x,y)|M(x,y))$$

The last definition of the entropy has the advantage of having a zero maximum when O equals the model M, usually taken as a flat image.

3.5.7 Other Regularization Models

Molina et al. (2001) present an excellent review of taking the spatial context of image restoration into account. Some appropriate prior is used for this. One such regularization constraint is:

$$\|CI\|^2 = \sum_x \sum_y I(x,y) - \frac{1}{4}(I(x,y+1) + I(x,y-1) + I(x+1,y) + I(x-1,y)) \tag{3.26}$$

Similar to the discussion above in section 5.2, this is equivalent to defining the prior

$$p(O) \propto \exp\left\{-\frac{\alpha}{2}\|CI\|^2\right\} \tag{3.27}$$

Given the form of equation (3.26), such regularization can be viewed as setting a constraint on the Laplacian of the restoration. In statistics this model is a simultaneous autoregressive model, SAR (Ripley, 1981).

Alternative prior models can be defined, related to the SAR model of equation (3.26). In

$$p(O) \propto \exp\left\{-\alpha \sum_x \sum_y (I(x,y) - I(x,y+1))^2 + (I(x,y) - I(x+1,y))^2\right\} \tag{3.28}$$

constraints are set on first derivatives.

Blanc-Feraud and Barlaud (1996), and Charbonnier et al. (1997b) consider the following prior:

$$p(O) \propto \exp\left\{-\alpha \sum_x \sum_y \phi(\|\nabla I\|(x,y))\right\}$$

$$\propto \exp\left\{-\alpha \sum_x \sum_y \left(\phi(I(x,y) - I(x,y+1))^2 + \phi(I(x,y) - I(x+1,y))^2\right)^{\frac{1}{2}}\right\}$$

$$\tag{3.29}$$

The function ϕ, called *potential function*, is an edge preserving function. The term $\alpha \sum_x \sum_y \phi(\|\nabla I\|(x,y))$ can also be interpreted as the Gibbs energy of a Markov Random Field.

Generally, functions ϕ are chosen with a quadratic part which ensures good smoothing of small gradients (Green, 1990), and a linear behavior which cancels the penalization of large gradients (Bouman and Sauer, 1993):

1. $\lim_{t \to 0} \frac{\phi'(t)}{2t} = 1$, smooth faint gradiants.
2. $\lim_{t \to \infty} \frac{\phi'(t)}{2t} = 0$, preserve strong gradiants.
3. $\frac{\phi'(t)}{2t}$ is strictly decreasing.

Such functions are often called L_2-L_1 functions. Examples of ϕ functions:

1. $\phi_q(x) = x^2$: quadratic function.
2. $\phi_{TV}(x) = |x|$: Total Variation.
3. $\phi_2(x) = 2\sqrt{1+x^2} - 2$: Hyper-Suface (Charbonnier et al., 1997a).
4. $\phi_3(x) = x^2/(1+x^2)$ (Geman and McClure, 1985).
5. $\phi_4(x) = 1 - e^{-x^2}$ (Perona and Malik, 1990).
6. $\phi_5(x) = \log(1+x^2)$ (Hebert and Leahy, 1989).

Fig. 3.5 shows different ϕ functions.

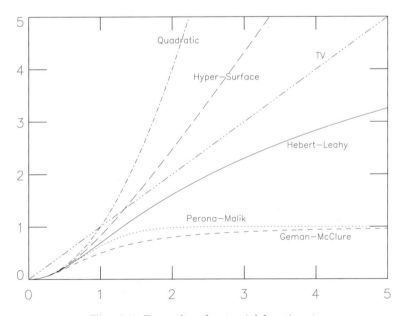

Fig. 3.5. Examples of potential function ϕ.

The ARTUR method (1997b), which has been used for helioseismic inversion (Corbard et al., 1999), uses the function $\phi(t) = \log(1 + t^2)$. Anisotropic diffusion (Perona and Malik, 1990; Alvarez et al., 1992) uses similar functions, but in this case the solution is computed using *partial differential equations*.

The function $\phi(t) = t$ leads to the *total variation* method (Rudin et al., 1992; Acar and Vogel, 1994), the constraints are on first derivatives, and the model is a special case of a conditional autoregressive or CAR model. Molina et al. (2001) discuss the applicability of CAR models to image restoration involving galaxies. They argue that such models are particularly appropriate for the modeling of luminosity exponential and $r^{1/4}$ laws.

The priors reviewed above can be extended to more complex models. In Molina et al. (1996; 2000), a compound Gauss Markov random field (CGMRF) model is used, one of the main properties of which is to target the preservation and improvement of line processes.

Another prior again was used in Molina and Cortijo (1992) for the case of planetary images.

3.6 Iterative Regularized Methods

3.6.1 Constraints

We assume now that there exists a general operator, $\mathcal{P}_\mathcal{C}(.)$, which enforces a set of constraints on a given object O, such that if O satisfies all the constraints, we have: $O = \mathcal{P}_\mathcal{C}(O)$. Commonly used constraints are:

- Positivity: the object must be positive.

$$\mathcal{P}_{C_p}(O(x,y)) = \begin{cases} O(x,y) & \text{if } O(x,y) \geq 0 \\ 0 & \text{otherwise} \end{cases} \quad (3.30)$$

This constraint can be generalized when upper and lower bounds U and L are known:

$$\mathcal{P}_{C_p}(O(x,y)) = \begin{cases} U & \text{if } O(x,y) > U \\ O(x,y) & \text{if } L \leq O(x,y) \leq U \\ L & \text{otherwise} \end{cases} \quad (3.31)$$

For example, if the background image B is known, we can fix $L = B$ and $U = +\infty$.

- Support constraint: the object belongs to a given spatial domain $\mathcal{D}$.

$$\mathcal{P}_{C_s}(O(x,y)) = \begin{cases} O(x,y) & \text{if } k \in \mathcal{D} \\ 0 & \text{otherwise} \end{cases} \quad (3.32)$$

- Band-limited: the Fourier transform of the object belongs to a given frequency domain. For instance, if F_c is the cut-off frequency of the instrument, we want to impose on the object that it be band-limited:

$$\mathcal{P}_{C_f}(\hat{O}(u,v)) = \begin{cases} \hat{O}(u,v) & \text{if } \nu < F_c \\ 0 & \text{otherwise} \end{cases} \quad (3.33)$$

These constraints can be incorporated easily in the basic iterative scheme.

3.6.2 Jansson-Van Cittert Method

Van Cittert (1931) restoration is relatively easy to write. We start with $n = 0$ and $O^{(0)} = I$ and we iterate:

$$O^{n+1} = O^n + \alpha(I - P * O^n) \quad (3.34)$$

where α is a convergence parameter generally taken as 1. When n tends to infinity, we have $O = O + I - P * O$, so $I = P * O$. In Fourier space, the convolution product becomes a product

$$\hat{O}^{n+1} = \hat{O}^n + \alpha(\hat{I} - \hat{P}\hat{O}^n) \quad (3.35)$$

In this equation, the object distribution is modified by adding a term proportional to the residual. The algorithm converges quickly, often after only 5 or 6 iterations. But the algorithm generally diverges in the presence of noise. Jansson (1970) modified this technique in order to give it more robustness by considering constraints on the solution. If we wish that $A \leq O_k \leq B$, the iteration becomes

$$O^{n+1}(x,y) = O^n(x,y) + r(x,y)\left[I(x,y) - (P * O^n)(x,y)\right] \quad (3.36)$$

with:

$$r(x,y) = C\left[1 - 2(B-A)^{-1} \mid O^n(x,y) - 2^{-1}(A+B) \mid\right]$$

and with C constant.

More generally the constrained Van-Cittert method is written as:

$$O^{n+1} = \mathcal{P}_C\left[O^n + \alpha(I - P * O^n)\right] \quad (3.37)$$

3.6.3 Other Iterative Methods

Other iterative methods can be constrained in the same way:

– Landweber:
$$O^{n+1} = \mathcal{P}_C\left[O^n + \gamma P^*(I - P * O^n)\right] \quad (3.38)$$

– Richardson-Lucy method:
$$O^{n+1}(x,y) = \mathcal{P}_C\left[O^n(x,y)\left[\frac{I(x,y)}{(P * O^n)(x,y)} * P^*(x,y)\right]\right] \quad (3.39)$$

– Tikhonov: the Tikhonov solution can be obtained iteratively by computing the gradient of equation (3.6):

$$\nabla(J_T(O)) = [P^* * P + \mu H^* * H] * O - P^* * I \qquad (3.40)$$

and applying the following iteration:

$$O^{n+1}(x,y) = O^n(x,y) - \gamma \nabla(J_T(O))(x,y) \qquad (3.41)$$

The constrained Tikhonov solution is therefore obtained by:

$$O^{n+1}(x,y) = \mathcal{P}_C\left[O^n(x,y) - \gamma \nabla(J_T(O))(x,y)\right] \qquad (3.42)$$

The number of iterations plays an important role in these iterative methods. Indeed, the number of iterations can be considered as another regularization parameter. When the number of iterations increases, the iterates first approach the unknown object, and then potentially go away from it (Bertero and Boccacci, 1998).

3.7 Wavelet-Based Deconvolution

3.7.1 Introduction

Deconvolution and Fourier Domain. The Fourier domain diagonalizes the convolution operator, and we can identify and reduce the noise which is amplified during the inversion. When the signal can be modeled as stationary and Gaussian, the Wiener filter is optimal. But when the signal presents spatially localized features such as singularities or edges, these features cannot be well-represented with Fourier basis functions, which extend over the entire spatial domain. Other basis functions, such as wavelets, are better-suited to represent a large class of signals.

Towards Multiresolution. The concept of multiresolution was first introduced for deconvolution by Wakker and Schwarz (1988) when they proposed the Multiresolution CLEAN algorithm for interferometric image deconvolution. During the last ten years, many developments have taken place in order to improve the existing methods (CLEAN, Landweber, Lucy, MEM, and so on), and these results have led to the use of different levels of resolution.

The Lucy algorithm was modified (Lucy, 1994) in order to take into account a priori information about stars in the field where both position and brightness are known. This is done by using a two-channel restoration algorithm, one channel representing the contribution relative to the stars, and the second to the background. A smoothness constraint is added on the background channel. This method, called *PLUCY*, was then refined firstly (and called *CPLUCY*) for considering subpixel positions (Hook, 1999), and a second time (Pirzkal et al., 2000) (and called *GIRA*) for modifying the smoothness constraint.

A similar approach has been followed by Magain (1998), but more in the spirit of the CLEAN algorithm. Again, the data are modeled as a set of point sources on top of spatially varying background, leading to a two-channel algorithm.

The MEM method has also been modified by several authors (Weir, 1992; Bontekoe et al., 1994; Pantin and Starck, 1996; Núñez and Llacer, 1998; Starck et al., 2001). First, Weir proposed the *Multi-channel MEM* method, in which an object is modeled as the sum of objects at different levels of resolution. The method was then improved by Bontekoe et al. (1994) with the *Pyramid MEM*. In particular, many regularization parameters were fixed by the introduction of the dyadic pyramid. The link between *Pyramid MEM* and wavelets was underlined in (Pantin and Starck, 1996; Starck et al., 2001), and it was shown that all the regularization parameters can be derived from the noise modeling. Wavelets were also used in (Núñez and Llacer, 1998) in order to create a segmentation of the image, each region being then restored with a different smoothness constraint, depending on the resolution level where the region was found. This last method has however the drawback of requiring user interaction for deriving the segmentation threshold in the wavelet space.

The *Pixon* method (Dixon et al., 1996; Puetter and Yahil, 1999) is relatively different to the previously described methods. This time, an object is modeled as the sum of pseudo-images smoothed locally by a function with position-dependent scale, called the pixon shape function. The set of pseudo-images defines a dictionary, and the image is supposed to contain only features included in this dictionary. But the main problem lies in the fact that features which cannot be detected directly in the data, nor in the data after a few Lucy iterations, will not be modeled with the pixon functions, and will be strongly regularized as background. The result is that the faintest objects are over-regularized while strong objects are well restored. This is striking in the example shown in Fig. 3.8.

The *total variation* method has a close relation with the Haar transform (Cohen et al., 1999; Steidl et al., 2003), and more generally, it has been shown that potential functions, used in Markov Random Field and PDE methods, can be applied on the wavelet coefficients as well. This leads to multicale regularization, and the original method becomes a specific case where only one decomposition level is used in the wavelet transform.

Wavelets offer a mathematical framework for the multiresolution processing. Furthermore, they furnish an ideal way to include noise modeling in the deconvolution methods. Since the noise is the main problem in deconvolution, wavelets are very well adapted to the regularization task.

3.7.2 Wavelet-Vaguelette Decomposition

The Wavelet-Vaguelette decomposition, proposed by Donoho (1995; 2004), consists of first applying an inverse filtering:

$$F = P^{-1} * I + P^{-1} * N = O + Z \tag{3.43}$$

where P^{-1} is the inverse filter ($\hat{P}^{-1}(u,v) = \frac{1}{\hat{P}(u,v)}$). The noise $Z = P^{-1} * N$ is not white but remains Gaussian. It is amplified when the deconvolution problem is unstable. Then, a wavelet transform is applied to F, the wavelet coefficients are soft- or hard-thresholded (Donoho, 1993), and the inverse wavelet transform furnishes the solution.

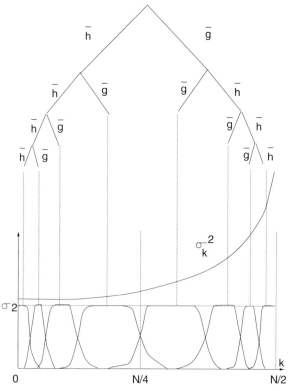

Fig. 3.6. Wavelet packet decomposition by a mirror basis. The variance of the noise has a hyperbolic growth.

The method has been refined by adapting the wavelet basis to the frequency response of the inverse of P (Kalifa, 1999; Kalifa et al., 2003). This *Mirror Wavelet Basis* has a time-frequency tiling structure different from conventional wavelets, and isolates the frequency where $\hat{P}$ is close to zero, because a singularity in $\hat{P}^{-1}(u_s, v_s)$ influences the noise variance in the wavelet scale corresponding to the frequency band which includes (u_s, v_s). Fig. 3.6 and Fig. 3.7 show the decomposition of the Fourier space respectively in 1D and 2D.

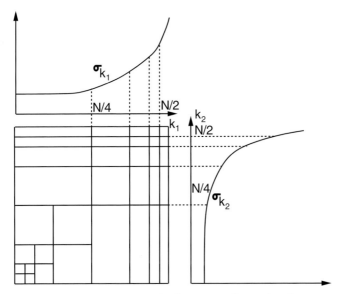

Fig. 3.7. The mirror wavelet basis in 2-dimensional space.

Because it may be not possible to isolate all singularities, Neelamani (1999; 2004) advocated a hybrid approach, proposing to still use the Fourier domain so as to restrict excessive noise amplification. Regularization in the Fourier domain is carried out with the window function W_λ:

$$\hat{W}_\lambda(u,v) = \frac{\mid \hat{P}(u,v) \mid^2}{\mid \hat{P}(u,v) \mid^2 + \lambda \mathcal{T}(u,v)} \tag{3.44}$$

where $\mathcal{T}(u,v) = \frac{\sigma^2}{\hat{S}(u,v)}$, S being the power spectral density of the observed signal.

$$F = W_\lambda * P^{-1} * I + W_\lambda * P^{-1} * N \tag{3.45}$$

The regularization parameter λ controls the amount of Fourier-domain shrinkage, and should be relatively small (< 1) (Neelamani et al., 2004). The estimate F still contains some noise, and a wavelet transform is performed to remove the remaining noise. The optimal λ is determined using a given cost function. See (Neelamani et al., 2004) for more details.

This approach is fast and competitive compared to linear methods, and the wavelet thresholding removes the Gibbs oscillations. It presents however several drawbacks:

1. The regularization parameter is not so easy to find in practice (Neelamani et al., 2004), and requires some computation time, which limits the usefulness of the method.
2. The positivity a priori is not used.

3. The power spectrum of the observed signal is generally not known.
4. It is not trivial to consider non-Gaussian noise.
5. Choosing the basis to conform to the operator (PSF) may not be optimal because the wavelet packets, matched to the frequency behavior of the PSF, may not match image structures as well as conventional wavelet basis (Figueiredo and Nowak, 2003).

The second point is important for astronomical images. It is well-known that the positivity constraint has a strong influence on the solution quality (Kempen and van Vliet, 2000). We will see in the following that it is straightforward to modify the standard iterative methods in a way that they benefit from the capacity of wavelets to separate the signal from the noise.

3.7.3 Regularization from the Multiresolution Support

Noise Suppression Based on the Wavelet Transform. We have noted how, in using an iterative deconvolution algorithm such as Van Cittert or Richardson-Lucy, we define $R^{(n)}(x,y)$, the residual at iteration n:

$$R^n(x,y) = I(x,y) - (P * O^n)(x,y) \qquad (3.46)$$

By using the à trous wavelet transform algorithm, R^n can be defined as the sum of its J wavelet scales and the last smooth array:

$$R^n(x,y) = c_J(x,y) + \sum_{j=1}^{J} w_{j,x,y} \qquad (3.47)$$

where the first term on the right is the last smoothed array, and w denotes a wavelet scale.

The wavelet coefficients provide a mechanism to extract only the significant structures from the residuals at each iteration. Normally, a large part of these residuals are statistically non-significant. The significant residual (Murtagh and Starck, 1994; Starck and Murtagh, 1994) is then:

$$\bar{R}^n(x,y) = c_{J,x,y} + \sum_{j=1}^{J} M(j,x,y) w_{j,x,y} \qquad (3.48)$$

where $M(j,x,y)$ is the multiresolution support, and is defined by:

$$M(j,x,y) = \begin{cases} 1 & \text{if } w_{j,x,y} \text{ is significant} \\ 0 & \text{if } w_{j,x,y} \text{ is non-significant} \end{cases} \qquad (3.49)$$

This describes in a logical or Boolean way if the data contains information at a given scale j and at a given position (x,y) (see section 2.2.4).

An alternative approach was outlined in (Murtagh et al., 1995) and (Starck et al., 1995): the support was initialized to zero, and built up at each iteration of the restoration algorithm. Thus in equation (3.48) above,

$M(j, x, y)$ was additionally indexed by n, the iteration number. In this case, the support was specified in terms of significant pixels at each scale, j; and in addition pixels could become significant as the iterations proceeded, but could not be made non-significant.

Regularization of Van Cittert's Algorithm. Van Cittert's iteration (1931) is:

$$O^{n+1}(x, y) = O^n(x, y) + \alpha R^n(x, y) \tag{3.50}$$

with $R^n(x, y) = I^n(x, y) - (P * O^n)(x, y)$. Regularization using significant structures leads to:

$$O^{n+1}(x, y) = O^n(x, y) + \alpha \bar{R}^n(x, y) \tag{3.51}$$

The basic idea of this regularization method consists of detecting, at each scale, structures of a given size in the residual $R^n(x, y)$ and putting them in the restored image $O^n(x, y)$. The process finishes when no more structures are detected. Then, we have separated the image $I(x, y)$ into two images $\tilde{O}(x, y)$ and $R(x, y)$. $\tilde{O}$ is the restored image, which ought not to contain any noise, and $R(x, y)$ is the final residual which ought not to contain any structure. R is our estimate of the noise $N(x, y)$.

Regularization of the One-step Gradient Method. The one-step gradient iteration is:

$$O^{n+1}(x, y) = O^n(x, y) + P^*(x, y) * R^n(x, y) \tag{3.52}$$

with $R^n(x, y) = I(x, y) - (P * O^n)(x, y)$. Regularization by significant structures leads to:

$$O^{n+1}(x, y) = O^n(x, y) + P^*(x, y) * \bar{R}^n(x, y) \tag{3.53}$$

Regularization of the Richardson-Lucy Algorithm. From equation (3.1), we have $I^n(x, y) = (P * O^n)(x, y)$. Then $R^n(x, y) = I(x, y) - I^n(x, y)$, and hence $I(x, y) = I^n(x, y) + R^n(x, y)$.
The Richardson-Lucy equation is:

$$O^{n+1}(x, y) = O^n(x, y) \left[\frac{I^n(x, y) + R^n(x, y)}{I^n(x, y)} * P^*(x, y) \right]$$

and regularization leads to:

$$O^{n+1}(x, y) = O^n(x, y) \left[\frac{I^n(x, y) + \bar{R}^n(x, y)}{I^n(x, y)} * P^*(x, y) \right]$$

Convergence. The standard deviation of the residual decreases until no more significant structures are found. Convergence can be estimated from the residual. The algorithm stops when a user-specified threshold is reached:

$$(\sigma_{R^{n-1}} - \sigma_{R^n})/(\sigma_{R^n}) < \epsilon \tag{3.54}$$

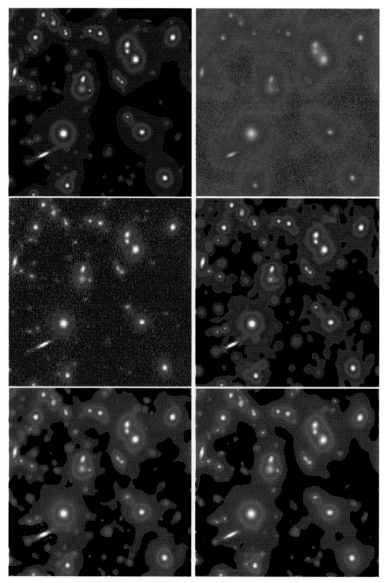

Fig. 3.8. Simulated Hubble Space Telescope Wide Field Camera image of a distant cluster of galaxies. Six quadrants. *Upper left:* original, unaberrated and noise-free. *Upper right:* input, aberrated, noise added. *Middle left:* restoration, Richardson-Lucy. *Middle right:* restoration, Pixon method. *Lower left:* restoration wavelet-vaguelette. *Lower right:* restoration wavelet-Lucy.

3.7 Wavelet-Based Deconvolution

Examples. A simulated Hubble Space Telescope Wide Field Camera image of a distant cluster of galaxies is shown in Fig. 3.8, upper left. The image used was one of a number described in (Caulet and Freudling, 1993; Freudling and Caulet, 1993). The simulated data are shown in Fig. 3.8, upper right. Four deconvolution methods were tested: Richardson-Lucy, Pixon, wavelet-vaguelette, Wavelet-Lucy. Deconvolved images are presented respectively in Fig. 3.8 middle left, middle right, bottom left and right. The Richardson-Lucy method amplifies the noise, which implies that the faintest objects disappear in the deconvolved image. The Pixon method introduces regularization, and the noise is under control, while objects where "pixons" have been detected are relatively well-protected from the regularization effect. Since the "pixon" features are detected from noisy partially deconvolved data, the faintest objects are not in the pixon map and are strongly regularized. The wavelet-vaguelette method is very fast and produces relatively high quality results when compared to Pixon or Richardson-Lucy, but the Wavelet-Lucy method seems clearly the best of the four methods. There are fewer spurious objects than in the wavelet-vaguelette method, it is stable for any kind of PSF, and any kind of noise modeling can be considered.

3.7.4 Wavelet CLEAN

Multiresolution CLEAN. The CLEAN solution is only available if the image does not contain large-scale structures. Wakker and Schwarz (1988) introduced the concept of Multiresolution Clean (MRC) in order to alleviate the difficulties occurring in CLEAN for extended sources. The MRC approach consists of building two intermediate images, the first one (called the smooth map) by smoothing the data to a lower resolution with a Gaussian function, and the second one (called the difference map) by subtracting the smoothed image from the original data. Both these images are then processed separately. By using a standard CLEAN algorithm on them, the smoothed clean map and difference clean map are obtained. The recombination of these two maps gives the clean map at the full resolution.

In order to describe how the clean map at the full resolution is obtained from the smoothed and difference clean map, a number of symbols must be defined:

- G = the normalized ($\int G(x)dx = 1$) smoothing function; the width of the function is chosen such that the full width at half-maximum of the smoothed dirty beam is f times larger than the full width at half-maximum of the original dirty beam.
- A = dirty beam
- D = dirty map
- δ = δ-functions
- R = residual after using CLEAN on the map
- B = clean beam with peak value 1

- C = clean map
- s = the scale factor of the dirty beam needed to rescale the smooth dirty beam back to a peak value 1
- r = the scale factor of the dirty beam needed to rescale the smooth clean beam back to a peak value 1
- A_s = normalized smooth dirty beam = $sA * G$
- A_d = normalized difference dirty beam = $1/(1 - \frac{1}{s})(A - \frac{A_s}{s})$
- B_s = normalized smooth clean beam = $rB * G$
- B_d = normalized difference clean beam = $1/(1 - \frac{1}{r})(B - \frac{B_s}{r})$

From the delta-functions found by the CLEAN algorithm, one can restore the dirty map by convolving with the dirty beam and adding the residuals:

$$D = D_s + D_d = \delta_s * A_s + R_s + \delta_d * A_d + R_d \tag{3.55}$$

which can be written also as:

$$D = [s\delta_s * G + \frac{s}{s-1}\delta_d * (1 - G)] * A + R_s + R_d \tag{3.56}$$

If we replace the dirty beam by the clean beam, we obtain the clean map:

$$C = \frac{s}{r}\delta_s * B_s + \frac{s(r-1)}{r(s-1)}\delta_d * B_d + R_s + R_d \tag{3.57}$$

The MRC algorithm needs three parameters. The first fixes the smoothing function G, and the other two are the loop gain and the extra loop gain which are used by CLEAN respectively on the smooth dirty map and difference dirty map.

This algorithm may be viewed as an artificial recipe, but it has been shown (Starck and Bijaoui, 1991) that it is linked to multiresolution analysis. Wavelet analysis leads to a generalization of MRC from a set of scales.

Wavelet and CLEAN. We have seen that there are many wavelet transforms. For interferometric deconvolution, we choose the wavelet transform based on the FFT (Starck et al., 1994; Starck and Bijaoui, 1994; Starck et al., 1998a) for the following reasons:

- The convolution product is kept at each scale.
- The data are already in Fourier space, so this decomposition is natural.
- There is a pyramidal implementation available which does not take much memory.

Hence until the end of this section on Wavelet CLEAN, we will consider the use of the pyramidal transform based on the FFT (see Appendix C for more details).

3.7 Wavelet-Based Deconvolution

Deconvolution by CLEAN in Wavelet Space. If $w_j^{(I)}$ are the wavelet coefficients of the image I at scale j, we get:

$$\hat{w}_j^{(I)}(u,v) = \hat{w}_j^{(P)} \hat{O}(u,v) \tag{3.58}$$

where $w_j^{(P)}$ are the wavelet coefficients of the point spread function at the scale j. The wavelet coefficients of the image I are the convolution product of the object O by the wavelet coefficients of the point spread function.

At each scale j, the wavelet plane $w_j^{(I)}$ can be decomposed by CLEAN ($w_j^{(I)}$ represents the dirty map and $w_j^{(P)}$ the dirty beam) into a set, denoted δ_j, of weighted δ-functions.

$$\delta_j = \{A_{j,1}\delta(x-x_{j,1}, y-y_{j,1}), A_{j,2}\delta(x-x_{j,2}, y-y_{j,2}),\ldots, \tag{3.59}$$
$$A_{j,n_j}\delta(x-x_{j,n_j}, y-y_{j,n_j})\}$$

where n_j is the number of δ-functions at the scale j and $A_{j,k}$ represents the height of the peak k at the scale j.

By repeating this operation at each scale, we get a set $\mathcal{W}_\delta$ composed of weighted δ-functions found by CLEAN ($\mathcal{W}_\delta = \{\delta_1, \delta_2, \ldots\}$). If B is the ideal point spread function (clean beam), the estimation of the wavelet coefficients of the object at the scale j is given by:

$$w_{j,x,y}^{(E)} = \delta_j * w_{j,x,y}^{(B)} + w_{j,x,y}^{(R)} = \tag{3.60}$$
$$\sum_k A_{j,k} w_{j,x-x_{j,k}, y-y_{j,k}}^{(B)} + w_{j,x,y}^{(R)}$$

where $w_j^{(R)}$ is the residual map. The clean map at the full resolution is obtained by the reconstruction algorithm. If we take a Gaussian function as the scaling function, and the difference between two resolutions as the wavelet ($\frac{1}{2}\psi(\frac{x}{2},\frac{y}{2}) = \phi(x,y) - \frac{1}{2}\phi(\frac{x}{2},\frac{y}{2})$), we find the algorithm proposed by Wakker and Schwarz (1988). The MRC algorithm in wavelet space is as follows:

1. We compute the wavelet transforms of the dirty map, the dirty beam and the clean beam.
2. For each scale j, we decompose by CLEAN the wavelet coefficients of the dirty map into a list of weighted δ-functions δ_j.
3. For each scale j, we convolve δ_j by the wavelet coefficients of the clean beam and we add the residual map $w_j^{(R)}$ to the result in order to obtain the wavelet coefficients of the clean map.
4. We compute the clean map at the full resolution by using the reconstruction algorithm.

Improvements to Multiresolution CLEAN. We apply CLEAN to each plane of the wavelet transform. This allows us to detect at each scale the significant structure. The reconstructed image gives the estimation $\tilde{O}$ found by MRC for the object. But MRC does not assume that this estimation is compatible with the measured visibilities. We want:

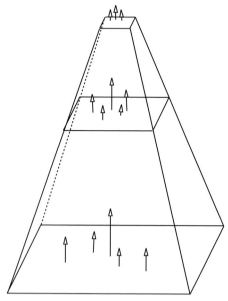

Fig. 3.9. Example of detection of peaks by CLEAN at each scale.

$$| \hat{O}(u,v) - V_m(u,v) | < \Delta_m(u,v) \qquad (3.61)$$

where $\Delta_m(u,v)$ is the error associated with the measure V_m.

To achieve this, we use the position of the peaks determined by the MRC algorithm. We have seen that after the use of CLEAN, we get a list of positions δ_j in each plane j, with approximate heights A_j. In fact, we get a nice description of the significant structures in the wavelet space (see Fig. 3.9). The height values are not sufficiently accurate, but CLEAN enhances these structures. So we have to determine heights which reduce the error. We do so using Van Cittert's algorithm (1931) which converges, even in the presence of noise, because our system is well-regularized. Then, heights of the peaks contained in $\mathcal{W}_\delta$ will be modified by the following iterative algorithm:

1. Set $n = 0$ and $\mathcal{W}_\delta^{(0)} = \mathcal{W}_\delta$.
2. Compute $A_{j,l}^{(n+1)} = A_{j,l}^{(n)} + \mathcal{Q}_{j,l}.\mathcal{W}_\delta^{(n)}$ so that we then have:

$$\delta_j^{(n+1)} = \{A_{j,1}^{(n+1)} \delta(x - x_{j,1}, y - y_{j,1}),$$

and:

$$\mathcal{W}_\delta^{(n+1)} = \{\delta_1^{(n+1)}, \delta_2^{(n+1)}, \ldots\}$$

3. $n = n + 1$ and go to step 1.

$\mathcal{Q}$ is the operator which:

- computes the wavelet coefficients of the clean map $w^{(C)}$ by convolving at each scale $\delta_j^{(n)}$ by the clean beam wavelet $w_j^{(B)}$

$$w_j^{(C)} = \delta_j^{(n)} * w_j^{(B)}$$

- reconstructs the estimated object $O^{(n)}$ at full resolution from $w^{(C)}$
- thresholds the negative values of $O^{(n)}$
- computes the residual $r^{(n)}$ by:

$$\hat{r}^{(n)} = p(V - \hat{O}^{(n)})$$

where p is a weight function which depends on the quality of the measurement V (error bars). A possible choice for p is:
- $p(u,v) = 0$ if we do not have any information at this frequency (i.e. a frequency hole).
- $p(u,v) = 1 - 2\frac{\Delta_m(u,v)}{V_m(0,0)}$ if $\Delta_m(u,v)$ is the error associated with the measurement $V_m(u,v)$.
- computes the wavelet transform $w^{(r^{(n)})}$ of $r^{(n)}$
- extracts the wavelet coefficient of $w^{(r^{(n)})}$ which is at the position of the peak $A_{j,l}\delta(x - x_l, y - y_l)$.

The final deconvolution algorithm is:

1. Convolution of the dirty map and the dirty beam by the scaling function.
2. Computation of the wavelet transform of the dirty map which yields $w^{(I)}$.
3. Computation of the wavelet transform of the dirty beam which yields $w^{(D)}$.
4. Estimation of the standard deviation of the noise N_0 of the first plane from the histogram of w_0. Since we process oversampled images, the values of the wavelet image corresponding to the first scale $(w_0^{(I)})$ are nearly always due to the noise. The histogram shows a Gaussian peak around 0. We compute the standard deviation of this Gaussian function, with a 3σ clipping, rejecting pixels where the signal could be significant.
5. Computation of the wavelet transform of the clean beam. We get $w^{(B)}$. If the clean beam is a Dirac, then $\hat{w}_j^{(B)}(u,v) = \frac{\psi(2^j u, 2^j v)}{\phi(u,v)}$.
6. Set j to 0.
7. Estimation of the standard deviation of the noise N_j from N_0. This is done from the study of the variation of the noise between two scales, with the hypothesis of a white Gaussian noise.
8. Detection of significant structures by CLEAN: we get δ_j from $w_j^{(I)}$ and $w_j^{(D)}$. The CLEAN algorithm is very sensitive to the noise. Step 1 of this algorithm offers more robustness. CLEAN can be modified in order to optimize the detection.
9. $j = j + 1$ and go to step 7.

10. Reconstruction of the clean map from $\mathcal{W}_\delta = \{\delta_1, \delta_2, \cdots\}$ by the iterative algorithm using Van Cittert's method.

The limited support constraint is implicit because we put information only at the position of the peaks, and the positivity constraint is introduced in the iterative algorithm. We have made the hypothesis that MRC, by providing the coordinates of the peaks, gives the exact position of the information in the wavelet space and we limited the deconvolution problem by looking for the height of the peaks which give the best results. It is a very strong limited support constraint which allows our problem to be regularized. CLEAN is not used as a deconvolution algorithm, but only as a tool to detect the position of structures.

Examples. Hen 1379 is a post-Asymptotic Giant Branch star in a phase of intense mass loss. The circumstellar dust distribution for post-AGB generally departs from spherical geometry. This is the case for Hen 1379, the high polarization measured at visible and near-infrared wavelengths indicating that the envelope is strongly non-spherical.

Fig. 3.10 shows the uv plane coverage of Hen 1379 (left), and the inverse Fourier transform of the data (right). The high angular resolution observations of this source were performed using the ESO one-dimensional (1D) slit-scanning near-infrared specklegraph attached to the ESO 3.6m telescope Cassegrain focus.

Fig. 3.11 shows the reconstruction by the wavelet transform of the evolved star Hen 1379. The ratio of point-source to the maximum amplitude of the envelope is 290. Contour levels are 12, 24, 36, ..., 96% of the maximum amplitude of the envelope.

3.7.5 The Wavelet Constraint

We have seen previously that many regularized deconvolution methods (MEM, Tikhonov, total variation, etc.) can be expressed by two terms (i.e. $\| I - P * O \|^2 + \lambda \mathcal{C}(O)$), the first representing the fidelity to the data and the second (i.e. $\mathcal{C}(O)$) the smoothness constraint on the solution. The parameter λ fixes the trade-off between the fit to the data and the smoothness. Using a wavelet based penalizing term $\mathcal{C}_w$, we want to minimize

$$J(O) = \| I - P * O \|^2 + \lambda \mathcal{C}_w(O) \tag{3.62}$$

If ϕ is a potential function which was applied on the gradients (see section 3.5.7), it can also be applied to the wavelet coefficients and the constraint on the solution is now expressed in the wavelet domain by (Jalobeanu, 2001):

$$J(O) = \| I - P * O \|^2 + \lambda \sum_{j,k,l} \phi(\| (\mathcal{W}O)_{j,k,l} \|_p) \tag{3.63}$$

When $\phi(x) = x$ and $p = 1$, it corresponds to the l_1 norm of the wavelet coefficients. In this framework, the multiscale entropy deconvolution method

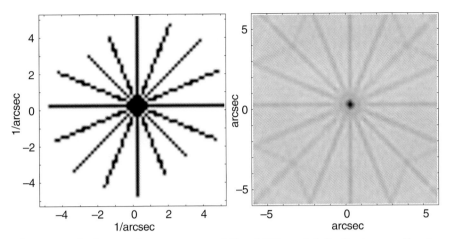

Fig. 3.10. *Right:* uv plane coverage of Hen 1379, and *left:* the inverse Fourier transform of the data.

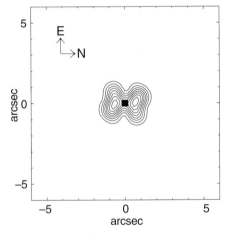

Fig. 3.11. Reconstructed image.

(see below and Chapter 7) is only one special case of the wavelet constraint deconvolution method.

Multiscale Entropy. In (Starck et al., 1998b; Starck and Murtagh, 1999; Starck et al., 2001), the benchmark properties for a good "physical" definition of entropy were discussed. The multiscale entropy, which fulfills these properties, consists of considering that the entropy of a signal is the sum of the information at each scale of its wavelet transform (Starck et al., 1998b), and the information of a wavelet coefficient is related to the probability of it being due to noise.

For Gaussian noise, the multiscale entropy penalization function is:

100   3. Deconvolution

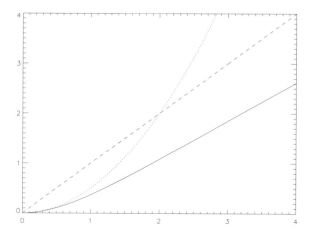

Fig. 3.12. Penalization functions: dashed, l_1 norm (i.e. $\phi(w) = |w|$); dotted l_2 norm $\phi(w) = \frac{w^2}{2}$; continuous, multiscale entropy function.

$$h_n(w_{j,k}) = \frac{1}{\sigma_j^2} \int_0^{|w_{j,k}|} u \operatorname{erfc}\left(\frac{|w_{j,k}| - u}{\sqrt{2}\sigma_j}\right) du \qquad (3.64)$$

A complete description of this method is given in Chapter 7. Fig. 3.12 shows the multiscale entropy penalization function. The dashed line corresponds to a l_1 penalization (i.e. $\phi(w) = |w|$), the dotted line to a l_2 penalization $\phi(w) = \frac{w^2}{2}$, and the continuous line to the multiscale entropy function. We can immediately see that the multiscale entropy function presents quadratic behavior for small values, and is closer to the l_1 penalization function for large values. Penalization functions with a l_2-l_1 behavior are generally a good choice for image restoration.

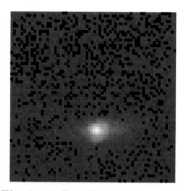

Fig. 3.13. Beta Pictoris raw data.

The Beta Pictoris image (Pantin and Starck, 1996) was obtained by integrating 5 hours on-source using a mid-infrared camera, TIMMI, placed on the 3.6 ESO telescope (La Silla, Chile). The raw image has a peak signal-to-noise ratio of 80. It is strongly blurred by a combination of seeing, diffraction (0.7 arcsec on a 3m class telescope) and additive Gaussian noise. The initial disk shape in the original image has been lost after the convolution with the PSF (see Fig. 3.13). Thus we need to deconvolve such an image to get the best information on this object i.e. the exact profile and thickness of the disk, and subsequently to compare the results to models of thermal dust emission.

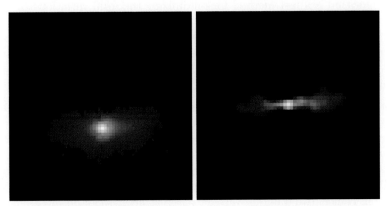

Fig. 3.14. Filtered image (*left*), and deconvolved one (*right*).

After filtering (see Fig. 3.14, left), the disk appears clearly. For detection of faint structures (the disk here), one can calculate that the application of such a filtering method to this image provides a gain of observing time of a factor of around 60. The deconvolved image (Fig. 3.14, right) shows that the disk is extended at 10 μm and asymmetrical. The multiscale entropy method is more effective for regularizing than other standard methods, and leads to good reconstruction of the faintest structures of the dust disk.

TV and Undecimated Haar Transform. A link between the TV and the undecimated Haar wavelet soft thresholding has been studied in (Cohen et al., 1999; Steidl et al., 2003), arguing that in the 1D case the TV and the undecimated single resolution Haar are equivalent. When going to 2D, this relation does not hold anymore, but the two approaches share some similarities. Whereas the TV introduces translation- and rotation-invariance, the undecimated 2D Haar presents translation- and scale-invariance (being multi-scale).

Minimization Algorithm. Recent works (Figueiredo and Nowak, 2003; Daubechies et al., 2004; Combettes and Vajs, 2005) show that the solution of eqn. 3.63 can be obtained in a very efficient way, by applying a wavelet denoising on the solution at each step of the Landweber iteration.

$$O^{(n+1)} = \mathbf{WDen}_\lambda \left(O^{(n)} + P^* * \left(I - P * O^{(n)} \right) \right) \tag{3.65}$$

where **WDen** is the operator which performs wavelet denoising, i.e. applies the wavelet transform, corrects the wavelet coefficients from the noise, and applies the inverse wavelet transform.

If $\phi(x) = x$ and $p = 1$ (i.e. l_1 norm), the solution is obtained by the following iteration:

$$O^{(n+1)} = \mathbf{soft}_\lambda (O^{(n)} + P^* * (I - P * O^{(n)})) \tag{3.66}$$

where **soft** is the soft thresholding. If the Haar wavelet transform is chosen, this algorithm is a fast method for minimizing the total variation.

The penalty function needs to be continuous in order to guarantee the convergence. Therefore, a hard threshold cannot be used but a soft threshold as well as many other shrinkage techniques can be used. If the penalty function is stricly convex (as in soft thresholding), then it converges to a global minimum (Figueiredo and Nowak, 2003).

Constraints in the Object or Image Domains. Let us define the *object domain* $\mathcal{O}$ as the space to which the solution belongs, and the *image domain* $\mathcal{I}$ as the space to which the data belongs (i.e. if $X \in \mathcal{O}$ then $P * X \in \mathcal{I}$). In section 3.7.3, it was shown that the multiresolution support constraint leads to a powerful regularization method. In this case, the constraint was applied in the image domain. Here, we have considered constraints on the solution. Hence, two different wavelet based strategies can be chosen in order to regularize the deconvolution problem.

The constraint in the image domain through the multiresolution support leads a very robust way to control the noise. Indeed, whatever the nature of the noise, we can always derive robust detection levels in the wavelet space and determine scales and positions of the important coefficients. A drawback of the image constraints is that there is no guarantee that the solution is free of artifacts such as ringing around point sources. A second drawback is that image constraints can be used only if the point spread function is relatively compact, i.e. does not smear the information over the whole image. In it does do so, the concept of localization of information does not make sense any more.

The property of introducing a robust noise modeling is lost when applying the constraint in the object domain. For example, in the case of Poisson noise, there is no way (except using time consuming Monte Carlo techniques) to estimate the level of the noise in the solution and to adjust properly the thresholds. The second problem with this approach is that we try to solve two problems (noise amplification and artifact control in the solution) with one parameter (i.e. λ). The choice of this parameter is crucial, while such a parameter does not exist when using the multiresolution support.

Constraints can be added in both the object and image domains in order to better control the noise by using the multiresolution support. This gives us a warranty that the solution is free of artifacts when using the wavelet constraint on the solution (Pantin and Starck, 1996; Starck et al., 2001; Starck et al., 2003c). This leads to the following equation to be minimized:

$$J(O) = \| M.\mathcal{W}_1(I - P * O) \|^2 + \lambda \sum_{j,k,l} \phi(\| (\mathcal{W}_2 O)_{j,k,l} \|_p) \qquad (3.67)$$

where M is the multiresolution support derived from I and $\mathcal{W}_1$. $\mathcal{W}_1$ and $\mathcal{W}_2$ are the wavelet transforms used in the object and image domains. We may want to use two different wavelet decompositions: $\mathcal{W}_1$ for detecting the significant coefficients and $\mathcal{W}_2$ for removing the artifacts in the solution. Since the noise is controlled by the multiscale transforms, the regularization parameter λ does not have the same importance as in standard deconvolution methods. A much lower value is enough to remove the artifacts relative to the use of the wavelets. The positivity constraint can be applied at each iteration. The iterative scheme is now:

$$O^{(n+1)} = \mathbf{WDen}_\lambda\left(\left(O^{(n)} + P^* * \bar{R}^n\right)\right) \qquad (3.68)$$

where $\bar{R}^n$ is the significant residual, i.e. $\bar{R}^n = \mathcal{W}_1^{-1} M[\mathcal{W}_1(I - P * O^{(n)})]$ (see eqn. 3.48)).

The Combined Deconvolution Method. As for the filtering (see section 2.5.3), we may want to benefit from the advantages of both the wavelet and the curvelet transforms for detecting the significant features contained in the data. More generally, we assume we use K transforms $T_1, \ldots, T_K$, and we derive K multiresolution supports $M_1, \ldots, M_K$ from the input image I using noise modeling. Following determination of a set of multiresolution supports, we can solve the following optimization problem (Starck et al., 2003c):

$$\min_{\tilde{O}} \mathcal{C}(\tilde{O}), \quad \text{subject to} \quad M_k T_k [P * \tilde{O}] = M_k T_k I \quad \text{for all } k, \qquad (3.69)$$

where $\mathcal{C}$ is the smoothness constraint.

The constraint imposes fidelity on the data, or more exactly, on the significant coefficients of the data, obtained by the different transforms. Non-significant (i.e. noisy) coefficients are not taken into account, preventing any noise amplification in the final algorithm.

A solution for this problem could be obtained by relaxing the constraint to become an approximate one:

$$\min_{\tilde{O}} \sum_k \| M_k T_k I - M_k T_k [P * \tilde{O}] \|_2 + \lambda \mathcal{C}(\tilde{O}) \qquad (3.70)$$

The solution is computed by using the projected Landweber method (Bertero and Boccacci, 1998):

$$\tilde{O}^{n+1} = \tilde{O}^n + \alpha \left(P^* * \bar{R}^n - \lambda \frac{\partial \mathcal{C}}{\partial O}(\tilde{O}^n) \right) \qquad (3.71)$$

where $\bar{R}^n$ is the significant residual which is obtained using the following algorithm:

- Set $I_0^n = I^n = P * \tilde{O}^n$.
- For $k = 1, \ldots, K$ do $I_k^n = I_{k-1}^n + \mathcal{R}_k \left[M_k (\mathcal{T}_k I - \mathcal{T}_k I_{k-1}^n) \right]$
- The significant residual $\bar{R}^n$ is obtained by: $\bar{R}^n = I_K^n - I^n$.

This can be interpreted as a generalization of the multiresolution support constraint to the case where several transforms are used. The order in which the transforms are applied has no effect on the solution. We extract in the residual the information at scales and pixel indices where significant coefficients have been detected.

α is a convergence parameter, chosen either by a line-search minimizing the overall penalty function or as a fixed step-size of moderate value that guarantees convergence.

If the $\mathcal{C}$ is a wavelet based penalization function, then the minimization can again be done using the previous wavelet denoising approach:

$$\tilde{O}^{n+1} = \mathbf{WDen}\left(\tilde{O}^n + (P^* * \bar{R}^n)\right) \tag{3.72}$$

The positivity is introduced in the following way:

$$O^{n+1} = \mathcal{P}_c \left[\mathbf{WDen}\left(\tilde{O}^n + (P^* * \bar{R}^n)\right) \right] \tag{3.73}$$

3.8 Deconvolution and Resolution

In many cases, there is no sense in trying to deconvolve an image at the resolution of the pixel (especially when the PSF is very large). The idea to limit the resolution is relatively old, because it is already this concept which is used in the CLEAN algorithm (Högbom, 1974). Indeed the clean beam fixes the resolution in the final solution. This principle was also developed by Lannes (1987) in a different form. This concept was re-invented, first by Gull and Skilling (1991) who called the clean beam the *Intrinsic Correlation Function* (ICF), and more recently by Magain (1998) and Pijpers (1999).

The ICF is usually a Gaussian, but in some cases it may be useful to take another function. For example, if we want to compare two images I_1 and I_2 which are obtained with two wavelengths or with two different instruments, their PSFs P_1 and P_2 will certainly be different. The classic approach would be to deconvolve I_1 with P_2 and I_2 with P_1, so we are sure that both are at the same resolution. But unfortunately we lose some resolution in doing this. Deconvolving both images is generally not possible because we can never be sure that both solutions O_1 and O_2 will have the same resolution.

A solution would be to deconvolve only the image which has the worse resolution (say I_1), and to limit the deconvolution to the second image resolution (I_2). Then, we just have to take P_2 for the ICF. The deconvolution problem is to find $\tilde{O}$ (hidden solution) such that:

$$I_1 = P_1 * P_2 * \tilde{O} \tag{3.74}$$

and our real solution O_1 at the same resolution as I_2 is obtained by convolving $\tilde{O}$ with P_2. O_1 and I_2 can then be compared.

Introducing an ICF G in the deconvolution equation leads to just considering a new PSF P' which is the convolution of P and G. The deconvolution is carried out using P', and the solution must be reconvolved with G at the end. In this way, the solution has a constrained resolution, but aliasing may occur during the iterative process, and it is not sure that the artifacts will disappear after the re-convolution with G. Magain (1998) proposed an innovative alternative to this problem, by assuming that the PSF can be considered as the convolution product of two terms, the ICF G and an unknown S, $P = G * S$. Using S instead of P in the deconvolution process, and a sufficiently large FWHM value for G, implies that the Shannon sampling theorem (Shannon, 1948) is never violated. But the problem is now to calculate S, knowing P and G, which is again a deconvolution problem. Unfortunately, this delicate point was not discussed in the original paper. Propagation of the error on the S estimation in the final solution has also until now not been investigated, even if this issue seems to be quite important.

3.9 Super-Resolution

3.9.1 Definition

Super-resolution consists of recovering object spatial frequency information outside the spatial bandwidth of the image formation system. In other terms, frequency components where $\hat{P}(\nu) = 0$ have to be recovered. It has been demonstrated (Donoho et al., 1992) that this is possible under certain conditions. The observed object must be *nearly black*, i.e. nearly zero in all but a small fraction of samples. Denoting n the number of samples, m the number of non-zero values in the Fourier transform of the PSF, and $\epsilon = \frac{m}{n}$ the incompleteness ratio, it has been shown that an image (Donoho et al., 1992):

- Must admit super-resolution if the object is $\frac{1}{2}\epsilon$-black.
- Might admit super-resolution if the object is ϵ-black. In this case, it depends on the noise level and the spacing of non-zero elements in the object. Well-spaced elements favor the possibility of super-resolution.
- Cannot admit super-resolution if the object is not ϵ-black.

Near blackness is both necessary and sufficient for super-resolution. Astronomical images often present such data sets, where the real information (stars and galaxies) is contained in very few pixels. If the $\frac{1}{2}\epsilon$-blackness of the object is not verified, a solution is to limit the Fourier domain Ω of the restored object. Several methods have been proposed in different contexts for achieving super-resolution.

3.9.2 Gerchberg-Saxon Papoulis Method

The Gerchberg-Saxon-Papoulis (Gerchberg, 1974) method is iterative, and uses the a priori information on the object, which is its positivity and its support in the spatial domain. It was developed for interferometric image reconstruction, where we want to recover the object O from some of its *visibilities*, i.e. some of its frequency components. Hence, the object is supposed to be known in a given Fourier domain Ω and we need to recover the object outside this domain. The problem can also be seen as a deconvolution problem, $I = P * O$, where

$$P(u,v) = \begin{cases} 1 & \text{if } (u,v) \in \Omega \\ 0 & \text{otherwise} \end{cases} \quad (3.75)$$

We denote $\mathcal{P}_{C_s}$ and $\mathcal{P}_{C_f}$ the projection operators in the spatial and the Fourier domain:

$$\mathcal{P}_{C_s}(X(x,y)) = \begin{cases} X(x,y) & \text{if } (x,y) \in \mathcal{D} \\ 0 & \text{otherwise} \end{cases}$$

$$\mathcal{P}_{C_f}(\hat{X}(u,v)) = \begin{cases} \hat{I}(u,v) = \hat{O}(u,v) & \text{if } (u,v) \in \Omega \\ 0 & \text{otherwise} \end{cases} \quad (3.76)$$

The projection operator $\mathcal{P}_{C_s}$ replaces by zero all pixel values which are not in the spatial support defined by $\mathcal{D}$, and $\mathcal{P}_{C_f}$ replaces all frequencies in the Fourier domain Ω by the frequencies of the object O. The Gerchberg algorithm is:

1. Compute $\tilde{O}^0 = $ inverse Fourier transform of $\hat{I}$, and set $i = 0$.
2. Compute $X_1 = \mathcal{P}_{C_s}(\tilde{O}^i)$.
3. Compute $\hat{X}_1 = $ Fourier transform of X_1.
4. Compute $\hat{X}_2 = \mathcal{P}_{C_f}(\hat{X}_1)$.
5. Compute $X_2 = $ inverse Fourier transform of $\hat{X}_2$.
6. Compute $\tilde{O}^{i+1} = \mathcal{P}_{C_s}(\hat{X}_2)$.
7. Set $X_1 = \tilde{O}^{i+1}$, $i = i+1$ and go to 2.

The algorithm consists just of forcing iteratively the solution to be zero outside the spatial domain $\mathcal{D}$, and equal to the observed visibilities inside the Fourier domain Ω. It has been shown that this algorithm can be derived from the Landweber method (Bertero and Boccacci, 1998), and therefore its convergence and regularization properties are the same as for the Landweber method. It is straightforward to introduce the positivity constraint by replacing $\mathcal{P}_{C_s}$ by $\mathcal{P}_{C_s}^+$

$$\mathcal{P}_{C_s}^+(X(x,y)) = \begin{cases} \max(X(x,y),0) & \text{if } (x,y) \in \mathcal{D} \\ 0 & \text{otherwise} \end{cases}$$

The Gerchberg method can be generalized (Bertero and Boccacci, 1998) using the Landweber iteration:

$$O^{n+1} = \mathcal{P}_{C_s}^+ [O^n + \alpha(P^* * L - P^* * P * O^n)] \tag{3.77}$$

where $L = \mathcal{P}_{C_s}^+(I)$.

3.9.3 Deconvolution with Interpolation

The MAP Poisson algorithm, combined with an interpolation, can be used to achieve super-resolution (Hunt, 1994):

$$O^{n+1} = O^n \exp\left\{\left(\frac{I}{(P*O^n)_\downarrow} - 1\right)_\uparrow * P^*\right\} \tag{3.78}$$

where uparrow and downarrow notation describes respectively the oversampling and downsampling operators. The PSF P must be sampled on the same grid as the object.

3.9.4 Undersampled Point Spread Function

Some observations are made with an undersampled PSF. When the observation is repeated several times with a small shift between two measurements, we can reconstruct a deconvolved image on a smaller grid. We denote $D(i,j,k)$ the kth observation (k = 1 ... n), $\Delta_{i,k}$, $\Delta_{j,k}$ the shift in both directions relative to the first frame, $\mathcal{L}_\uparrow$ the operator which coadds all the frame on a smaller grid, and $\mathcal{L}_\downarrow^{-1}$ the operator which estimates D from $\mathcal{L}_\uparrow D$ using shifting and averaging operations. The $\Delta_{i,k}$, $\Delta_{j,k}$ shifts are generally derived from the observations using correlation methods, or a PSF fitting (if a star is in the field), but can also be the jitter information if the data is obtained from space. Note also that $\mathcal{L}_\downarrow^{-1}\mathcal{L}_\uparrow D \neq D$. The point spread function P can generally be derived on a finer grid using a set of observations of a star, or using an optical modeling of the instrument. The deconvolution iteration becomes:

$$O^{n+1} = O^n + \alpha P^* \left[\mathcal{L}_\uparrow(D - \mathcal{L}_\downarrow^{-1}(P*O^n))\right] \tag{3.79}$$

and the positivity and spatial constraints can also be used:

$$O^{n+1} = \mathcal{P}_{C_s}^+ \left[O^n + \alpha P^* \left[\mathcal{L}_\uparrow(D - \mathcal{L}_\downarrow^{-1}(P*O^n))\right]\right] \tag{3.80}$$

The coaddition operator $\mathcal{L}_\uparrow$ can be implemented in different ways. All frames can first be interpolated to the finer grid size, shifted using an interpolation function, and then coadded.

"Dithering" or "jitter" have been terms applied to purposeful use of offsets in imaging (Hook and Fruchter, 2000). An ad-hoc method called "drizzling" is developed by Hook and Fruchter (2000) and implemented in IRAF, based on mapping pixels to a finer grid and assuming knowledge of geometric distortion.

Dithering was first described in (Bennet, 1948). Any discrete image necessarily implies quantization, and therefore some distortion (the presence of frequencies other than found in the original data) and loss of signal detail (related to the quantization step). Quantization error in many cases is not white, but instead is highly correlated with the signal. Adding a small amount of noise to the signal before quantization can mitigate this. Such "signal shaping" has been used at least since the 1960s. Gammaitoni et al. (1998) relate this to the phenomenon of "stochastic resonance", i.e. the fact that weak signals can be amplified and optimized by the assistance of small quantities of noise. They summarize guidelines for addition of noise as follows: the addition of dither can statistically reduce the quantization error; uniformly distributed dither is best; and there exists an optimal value of random dither amplitude which coincides with the amplitude of the quantization step. Why then does one use multiple images with random image offsets, when it would seem that adding a small amount of noise to the signal would suffice? If quantization were the only objective this would be the case. However the use of interpolation in combining images (justified by a resultant higher signal-to-noise ratio) would re-introduce signal dependencies, i.e. further sources of distortion.

Lauer (1999) ignores geometric distortion and instead addresses the problem of aliasing resulting from combining undersampled images. A linear combination of Fourier transforms of the offset images is used, which mitigates aliasing artifacts in direct space.

Multiresolution Support Constraint

The constraint operator $\mathcal{P}_{C_s}^+$ may not always be easy to determine, especially when the observed object is extended. Furthermore, if the object is very extended, the support will be very large and the support constraint may have very small influence on the final result. For this reason, it may be convenient to replace the support constraint by the multiresolution support constraint. The advantages are the following:

- It can be automatically calculated from the noise modeling in the wavelet space.
- Extended objects are generally also smooth. This means that the support on the small scales will be small, and therefore this introduces a smoothness constraint on extended objects, and no constraint on point-like objects.

Wavelet Constraint

As in section 3.7.5, a wavelet smoothness constraint can be added to the solution, and the same kind of minimization method can be used (Willett et al., 2004). The solution is obtained using the following two step algorithms (Willett et al., 2004):

$$Z^{n+1} = \mathcal{P}^+_{C_s}\left[O^n + \alpha P^*\left[\mathcal{L}_\uparrow(D - \mathcal{L}_\downarrow^{-1}(P * O^n))\right]\right]$$
$$O^{n+1} = \text{soft}(Z^{n+1}) \qquad (3.81)$$

where soft is the wavelet soft thresholding operation. Other wavelet denoising procedures could be used. Using the soft thresholding means that we a priori assume that the wavelet coefficients of the solution follow a Laplacian distribution.

3.10 Conclusions and Chapter Summary

As in many fields, simple methods can be availed of – for example the solution provided by equation (3.16) – but at the expense of quality in the solution and a full understanding of one's data. Often a simple solution can be fully justified. However, if our data or our problems are important enough, then appropriate problem solving approaches have to be adopted. The panoply of methods presented in this review provide options for high quality image and signal restoration.

We have noted how the wavelet transform offers a powerful mathematical and algorithmic framework for multiple resolution analysis. Furthermore noise modeling is very advantageously carried out in wavelet space. Finally, and of crucial importance in this chapter, noise is the main problem in deconvolution.

Progress has been significant in a wide range of areas related to deconvolution. One thinks of Bayesian methods, the use of entropy, and issues relating to super-resolution, for example.

We will conclude with a short look at how multiscale methods used in deconvolution are evolving and maturing.

We have seen that the recent improvement in deconvolution methods has led to use of a multiscale approach. This could be summarized in the following way:

– Linear inverse filtering → wavelet-vaguelette decomposition

– CLEAN → wavelet-CLEAN

– Fixed step gradient, Lucy, Van Cittert → Regularized by the multiresolution support

– MEM, MRF, PDE → Penalization based on the wavelet coefficients

Finally, wavelet based constraints can be added in both domains (Starck et al., 2001). This allows us to separate the deconvolution problem into two separate problems: noise control from one side, and solution smoothness control on the other side. The advantage is that noise control is better carried out in the image domain, while smoothness control can only be carried out in the object domain.

The reason for the success of wavelets is due to the fact that wavelet bases represent well a large class of signals, especially astronomical data where most of the objects are more or less isotropic. When the data contains anisotropic features (solar, planerary images, etc.), other multiscale methods, such as the ridgelet or the curvelet transform (Candès and Donoho, 1999; Candès and Donoho, 2000b; Donoho and Duncan, 2000; Starck et al., 2002), are good candidates for replacing the wavelet transform. The ultimate step is the combination of the different multiscale decomposition.

4. Detection

4.1 Introduction

Information extraction from images is a fundamental step for astronomers. For example, to build catalogs, stars and galaxies must be identified and their position and photometry must be estimated with good accuracy. Various methods have been proposed in the past to achieve such results. One of the most widely used software packages is SExtractor (Bertin and Arnouts, 1996). Its ability to deal with very large images (up to 60000×60000 pixels) and its robustness make for its success. A standard source detection approach, such as in SExtractor, consists of the following steps:

1. Background estimation.
2. Convolution with a mask.
3. Detection.
4. Deblending/merging.
5. Photometry.
6. Classification.

These different steps are described in the next section. Astronomical images contain typically a large set of point-like sources (the stars), some quasi point-like objects (faint galaxies, double stars), and some complex and diffuse structures (galaxies, nebulae, planetary stars, clusters, etc.). These objects are often hierarchically organized: a star in a small nebula, itself embedded in a galaxy arm, itself included in a galaxy, and so on.

Faint extended objects may be lost by the standard approach. Fig. 4.1 shows a typical example where a faint extended object is under the detection limit. In order to detect faint objects, whatever their sizes, Bijaoui (1993) proposed the Multiscale Vision Model, MVM. A *vision model* is defined as the sequence of operations required for automated image analysis. Taking into account the scientific purposes, the characteristics of the objects and the existence of hierarchical structures, astronomical images need specific vision models. This is also the case in many other fields, such as remote sensing, hydrodynamic flows, or biological studies. Specific vision models have been implemented for these kinds of images.

We introduce the Multiscale Vision Model as defined in Bijaoui and Rué (1995) in section 3. Then we show how deconvolution can be combined with

112   4. Detection

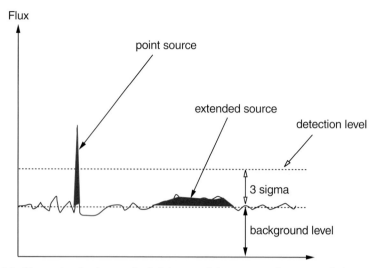

Fig. 4.1. Example of astronomical data: a point source and an extended source are shown, with noise and background. The extended object, which can be detected by eye, is undetected by a standard detection approach.

object reconstruction, how this helps with object identification, and how it can be very useful for deconvolution with a space-variant point spread function (PSF).

Section 4.5 presents the specific case of source and non-Gaussianity detection in the Cosmological Background.

4.2 From Images to Catalogs

Background Estimation

In most cases, objects of interest are superimposed on a relatively flat signal, called *background signal*. The background must be accurately estimated, or otherwise it will introduce bias in flux estimation. In (Bijaoui, 1980; Irwin, 1985), the image is partitioned into blocks, and the local sky level in each block is estimated from its histogram. The pixel intensity histogram $p(I)$ is modeled using three parameters, the true sky level S, the RMS (root mean square) noise σ, and a parameter describing the asymmetry in $p(I)$ due to the presence of objects, and is defined by (Bijaoui, 1980):

$$p(I) = \frac{1}{a} \exp(\sigma^2/2a^2) \exp\left[-(I-s)/a\right] \operatorname{erfc}\left(\frac{\sigma}{a} - \frac{(I-s)}{\sigma}\right) \qquad (4.1)$$

Median filtering can be applied to the 2D array of background measurements in order to correct for spurious background values. Finally the background map is obtained by a bi-linear or a cubic interpolation of the 2D array. The blocksize is a crucial parameter. If it is too small, the background estimation map will be affected by the presence of objects, and if too large it will not take into account real background variations.

In (Costa, 1992; Bertin and Arnouts, 1996), the local sky level is calculated differently. A 3-sigma clipping around the median is performed in each block. If the standard deviation is changed by less than 20% in the clipping iterations, the block is uncrowded, and the background level is considered to be equal to the mean of the clipped histogram. Otherwise, it is calculated by $c_1 \times$ median $- c_2 \times$ mean, where $c_1 = 3, c_2 = 2$ in (Costa, 1992), and $c_1 = 2.5, c_2 = 1.5$ in (Bertin and Arnouts, 1996). This approach has been preferred to histogram fitting for two reasons: it is more efficient from the computation point of view, and more robust with small sample size.

Convolution

In order to optimize the detection, the image must be convolved with a filter. The shape of this filter optimizes the detection of objects with the same shape. Therefore, for star detection, the optimal filter is the PSF. For extended objects, a larger filter size is recommended. In order to have optimal detection for any object size, the detection must be repeated several times with different filter sizes, leading to a kind of multiscale approach.

Detection

Once the image is convolved, all pixels I_l at location l with a value larger than T_l are considered as significant, i.e. belonging to an object. T_l is generally chosen as $B_l + k\sigma$, where B_l is the background estimation at the same position, σ is the noise standard deviation, and k is a given constant (typically chosen between 3 and 5). The thresholded image is then segmented, i.e. a label is assigned to each group of connected pixels. The next step is to separate the blended objects which are connected and have the same label.

An alternative to the thresholding/segmentation procedure is to find peaks. This is only well-suited to star detection and not to extended objects. In this case, the next step is to merge the pixels belonging to the same object.

Deblending/Merging

This is the most delicate step. Extended objects must be considered as single objects, while multiple objects must be well separated. In SExtractor, each group of connected pixels is analyzed at different intensity levels (30), starting from the highest down to the lowest level. The pixel group can be seen as a surface, with mountains and valleys. At the beginning (highest level), only the highest peak is visible. When the level decreases several other peaks may become visible, defining therefore several structures. At a given level, two structures may become connected, and the decision whether they form only one (i.e. merging) or several objects (i.e. deblending) must be taken. This is done by comparing the integrated intensities inside the peaks. If the ratio between them is too low, then the two structures must be merged.

Photometry and Classification

Photometry. Several methods can be used to derive the photometry of a detected object (Bijaoui, 1980; Kron, 1980). Adaptive aperture photometry uses the first image moment to determine the elliptical aperture from which the object flux is integrated. Kron (1980) proposed an aperture size of twice the radius of the first image moment radius r_1, which leads to recovery of most of the flux (> 90 %). In (Bertin and Arnouts, 1996), the value of $2.5r_1$ is discussed, leading to loss of less than 6% of the total flux. Assuming that the intensity profiles of the faint objects are Gaussian, flux estimates can be refined (Maddox et al., 1990; Bertin and Arnouts, 1996). When the image contains only stars, specific methods can be developed which take the PSF into account (Debray et al., 1994; Naylor, 1998).

Star-galaxy Separation. In the case of star–galaxy classification, following the scanning of digitized images, Kurtz (1983) lists the following parameters which have been used:

1. mean surface brightness;
2. maximum intensity, area;
3. maximum intensity, intensity gradient;
4. normalized density gradient;
5. areal profile;
6. radial profile;
7. maximum intensity, 2nd and 4th order moments, ellipticity;
8. the fit of galaxy and star models;
9. contrast versus smoothness ratio;
10. the fit of a Gaussian model;
11. moment invariants;
12. standard deviation of brightness;
13. 2nd order moment;
14. inverse effective squared radius;
15. maximum intensity, intensity weighted radius;
16. 2nd and 3rd order moments, number of local maxima, maximum intensity.

References for all of these may be found in the cited work. Clearly there is room for differing views on parameters to be chosen for what is essentially the same problem. It is of course the case also that aspects such as the following will help to orientate us towards a particular set of parameters in a particular case: the quality of the data; the computational ease of measuring certain parameters; the relevance and importance of the parameters measured relative to the data analysis output (e.g. the classification, or the planar graphics); and, similarly, the importance of the parameters relative to theoretical models under investigation.

Galaxy Morphology Classification. The inherent difficulty of characterizing spiral galaxies especially when not face-on has meant that most work focuses on ellipticity in the galaxies under study. This points to an inherent bias in the potential multivariate statistical procedures. In the following, it will not be attempted to address problems of galaxy photometry per se (Davoust and Pence, 1982; Pence and Davoust, 1985), but rather to draw some conclusions on what types of parameters or features have been used in practice.

From the point of view of multivariate statistical algorithms, a reasonably homogeneous set of parameters is required. Given this fact, and the available literature on quantitative galaxy morphological classification, two approaches to parameter selection appear to be strongly represented:

1. The luminosity profile along the major axis of the object is determined at discrete intervals. This may be done by the fitting of elliptical contours, followed by the integrating of light in elliptical annuli (Lefèvre et al., 1986). A similar approach was used in the ESO-Upsalla survey. Noisiness and faintness require attention to robustness in measurement: the radial profile may be determined taking into account the assumption of a face–on optically–thin axisymmetric galaxy, and may be further adjusted to yield values for circles of given radius (Watanabe et al., 1982). Alternatively, isophotal contours may determine the discrete radial values for which the profile is determined (Thonnat, 1985).
2. Specific morphology-related parameters may be derived instead of the profile. The integrated magnitude within the limiting surface brightness of 25 or 26 mag. arcsec^{-2} in the visual is popular (Takase et al., 1984; Lefèvre et al., 1986). The logarithmic diameter (D_{26}) is also supported by Okamura (1985). It may be interesting to fit to galaxies under consideration model bulges and disks using, respectively, $r^{\frac{1}{4}}$ or exponential laws (Thonnat, 1985), in order to define further parameters. Some catering for the asymmetry of spirals may be carried out by decomposing the object into octants; furthermore the taking of a Fourier transform of the intensity may indicate aspects of the spiral structure (Takase et al., 1984).

The following remarks can be made relating to image data and reduced data.

– The range of parameters to be used should be linked to the subsequent use to which they might be put, such as to underlying physical aspects.

- Parameters can be derived from a carefully-constructed luminosity profile, rather than it being possible to derive a profile from any given set of parameters.
- The presence of both partially reduced data such as luminosity profiles, and more fully reduced features such as integrated flux in a range of octants, is of course not a hindrance to analysis. However it is more useful if the analysis is carried out on both types of data separately.

Parameter data can be analyzed by clustering algorithms, by principal components analysis or by methods for discriminant analysis. Profile data can be sampled at suitable intervals and thus analyzed also by the foregoing procedures. It may be more convenient in practice to create dissimilarities between profiles, and analyze these dissimilarities: this can be done using clustering algorithms with dissimilarity input.

4.3 Multiscale Vision Model

4.3.1 Introduction

The multiscale transform of an image by the à trous algorithm produces at each scale j a set $\{w_j\}$. This has the same number of pixels as the image. The original image I can be expressed as the sum of all the wavelet scales and the smoothed array c_J by the expression

$$I(k,l) = c_{J,k,l} + \sum_{j=1}^{J} w_{j,k,l}. \tag{4.2}$$

Hence, we have a *multiscale pixel representation*, i.e. each pixel of the input image is associated with a set of pixels of the multiscale transform. A further step is to consider a *multiscale object representation*, which would associate with an object contained in the data a volume in the multiscale transform. Such a representation obviously depends on the kind of image we need to analyze, and we present here a model which has been developed for astronomical data. It may however be used for other kinds of data, to the extent that such data are similar to astronomical data. We assume that an image I can be decomposed into a set of components:

$$I(k,l) = \sum_{i=1}^{N_o} O_i(k,l) + B(k,l) + N(k,l) \tag{4.3}$$

where N_o is the number of objects, O_i are the objects contained in the data (stars galaxies, etc.), B is the background image, and N is the noise.

To perform such a decomposition, we have to detect, to extract, to measure and to recognize the significant structures. This is done by first computing the multiresolution support of the image, and by applying a segmentation scale by scale. The wavelet space of a 2D direct space is a 3D one. An object has to be defined in this space. A general idea for object definition lies in the connectivity property. An object occupies a physical region, and in this region we can join any pixel to other pixels based on significant adjacency. Connectivity in direct space has to be transported into wavelet transform space, WTS. In order to define the objects we have to identify the WTS pixels we can attribute to the objects. We describe in this section the different steps of this method.

4.3.2 Multiscale Vision Model Definition

The multiscale vision model, MVM (Bijaoui and Rué, 1995), described an object as a hierarchical set of structures. It uses the following definitions:
- *Significant wavelet coefficient*: a wavelet coefficient is significant when its absolute value is above a given detection limit. The detection limit depends on the noise model (Gaussian noise, Poisson noise, and so on).
- *Structure*: a structure $S_{j,k}$ is a set of significant connected wavelet coefficients at the same scale j:

$$S_{j,k} = \{w_{j,x_1,y_1}, w_{j,x_2,y_2}, \cdots, w_{j,x_p,y_p}\} \tag{4.4}$$

where p is the number of significant coefficients included in the structure $S_{j,k}$, and w_{j,x_i,y_i} is a wavelet coefficient at scale i and at position (x_i, y_i).
- *Object*: an object is a set of structures:

$$O_l = \{S_{j_1,k_1}, \cdots, S_{j_n,k_n}\} \tag{4.5}$$

We define also the operator $\mathcal{L}$ which indicates to which object a given structure belongs: $\mathcal{L}(S_{j,k}) = l$ is $S_{j,k} \in O_l$, and $\mathcal{L}(S_{j,k}) = 0$ otherwise.
- *Object scale*: the scale of an object is given by the scale of the maximum of its wavelet coefficients.
- *Interscale relation*: the criterion allowing us to connect two structures into a single object is called the "interscale relation".
- *Sub-object*: a sub-object is a part of an object. It appears when an object has a local wavelet maximum. Hence, an object can be composed of several sub-objects. Each sub-object can also be analyzed.

4.3.3 From Wavelet Coefficients to Object Identification

Multiresolution Support Segmentation. Once the multiresolution support has been calculated, we have at each scale a boolean image (i.e. pixel intensity equals 1 when a significant coefficient has been detected, and 0 otherwise). The segmentation consists of labeling the boolean scales. Each group of connected pixels having a "1" value gets a label value between 1 and $L_{\max}$, $L_{\max}$ being the number of groups. This process is repeated at each scale of the multiresolution support. We define a "structure" $S_{j,i}$ as the group of connected significant pixels which has the label i at a given scale j.

118   4. Detection

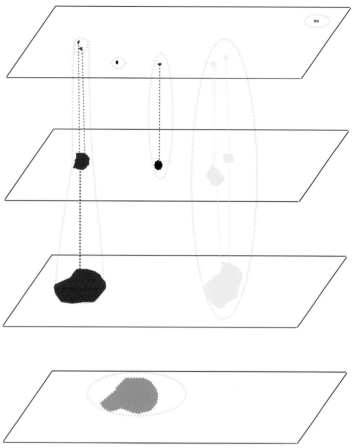

Fig. 4.2. Example of connectivity in wavelet space: contiguous significant wavelet coefficients form a structure, and following an interscale relation, a set of structures forms an object. Two structures S_j, S_{j+1} at two successive scales belong to the same object if the position pixel of the maximum wavelet coefficient value of S_j is included in S_{j+1}.

Interscale Connectivity Graph. An object is described as a hierarchical set of structures. The rule which allows us to connect two structures into a single object is called "interscale relation". Fig. 4.2 shows how several structures at different scales are linked together, and form objects. We have now to define the interscale relation: let us consider two structures at two successive scales, $S_{j,k}$ and $S_{j+1,l}$. Each structure is located in one of the individual images of the decomposition and corresponds to a region in this image where the signal is significant. Denoting (x_m, y_m) the pixel position of the maximum wavelet coefficient value of $S_{j,k}$, $S_{j,k}$ is said to be connected to $S_{j+1,l}$ if $S_{j+1,l}$ contains the pixel position (x_m, y_m) (i.e. the pixel position of the maximum wavelet coefficient of the structure $S_{j,k}$ must also be contained in the structure $S_{j+1,l}$). Several structures appearing in successive wavelet coefficient images can be connected in such a way, which we call an object in the interscale connectivity graph. Hence, we identify n_o objects in the wavelet space, each object O_i being defined by a set of structures, and we can assign to each structure a label i, with $i \in [1, n_o]$: $\mathcal{L}(S_{j,k}) = i$ if the structure $S_{j,k}$ belongs to the ith object.

Filtering. Statistically, some significant structures can be due to the noise. They contain very few pixels and are generally isolated, i.e. connected to no field at upper and lower scales. So, to avoid false detection, the isolated fields can be removed from the initial interscale connection graph. Structures at the border of the images may also have been detected because of the border problem, and can be removed.

Merging/Deblending. As in the standard approach, true objects which are too close may generate a set of connected structures, initially associated with the same object, and a decision must be taken whether to consider such a case as one or two objects. Several cases may be distinguished:

- Two (or more) close objects, approximately of the same size, generate a set of structures. At a given scale j, two separate structures $S_{j,1}$ and $S_{j,2}$ are detected while at the scale $j+1$, only one structure is detected $S_{j+1,1}$, which is connected to the $S_{j,1}$ and $S_{j,2}$.
- Two (or more) close objects of different sizes generate a set of structures, from scale j to scale k ($k > j$).

In the wavelet space, the merging/deblending decision will be based on the local maxima values of the different structures belonging to this object. A new object (i.e. deblending) is derived from the structure $S_{j,k}$ if there exists at least one other structure at the same scale belonging to the same object (i.e. there exists one structure $S_{j+1,a}$ and at least one structure $S_{j,b}$ such that $\mathcal{L}(S_{j+1,a}) = \mathcal{L}(S_{j,b}) = \mathcal{L}(S_{j,k})$), and if the following relationship is verified: $w_j^m > w_{j-1}^m$ and $w_j^m > w_{j+1}^m$, where:

- w_j^m is the maximum wavelet coefficient of the structure $S_{j,k}$: $w_j^m = \text{MAX}(S_{j,k})$.
- $w_{j-1}^m = 0$ if $S_{j,k}$ is not connected to any structure at scale $j-1$.
- w_{j-1}^m is the maximum wavelet coefficient of the structure $S_{j-1,l}$, where $S_{j-1,l}$ is such that $\mathcal{L}(S_{j-1,l}) = \mathcal{L}(S_{j,k})$ and the position of its highest wavelet coefficient is the closest to the position of the maximum of $S_{j,k}$.
- $w_{j+1}^m = \text{MAX}\{w_{j+1,x_1,y_1}, \cdots, w_{j+1,x_n,y_n}\}$, where all wavelet coefficients $w_{j+1,x,y}$ are at a position which belongs also to $S_{j,k}$ (i.e. $w_{j,x,y} \in S_{j,k}$).

When these conditions are verified, $S_{j,k}$ and all structures at smaller scales which are directly or indirectly connected to $S_{j,k}$ will define a new object.

Object Identification. We can now summarize this method allowing us to identify all the objects in a given image I:

1. We compute the wavelet transform with the à trous algorithm, which leads to a set $W = \mathcal{W}I = \{w_1, \ldots, w_J, c_J\}$. Each scale w_j has the same size as the input image.
2. We determine the noise standard deviation in w_1.
3. We deduce the thresholds at each scale from the noise modeling.
4. We threshold scale-by-scale and we do an image labeling.
5. We determine the interscale relations.
6. We identify all the wavelet coefficient maxima of the WTS.
7. We extract all the connected trees resulting from each WTS maximum.

4.3.4 Partial Reconstruction

Partial Reconstruction as an Inverse Problem. A set of structures $\mathcal{S}_i$ ($\mathcal{S}_i = \{S_{j,k}, \cdots, S_{j',k'}\}$) defines an object O_i which can be reconstructed separately from other objects. The coaddition of all reconstructed objects is a filtered version of the input data. We will denote W_i the set of wavelet coefficients belonging to the object O_i. Therefore, W_i is a subset of the wavelet transform of O_i, $\tilde{W}_i = \mathcal{W}O_i$. Indeed, the last scale of $\tilde{W}_i$ is unknown, as well as many wavelet coefficients which have not been detected. Then the reconstruction problem consists of searching for an image O_i such that its wavelet transform reproduces the coefficients W_i (i.e they are the same as those of $\mathcal{S}_i$, the detected structures). If $\mathcal{W}$ describes the wavelet transform operator, and P_w the projection operator in the subspace of the detected coefficients (i.e. having set to zero all coefficients at scales and positions where nothing was detected), the solution is found by minimization of (Bijaoui and Rué, 1995)

$$J(O_i) = \| W_i - A(O_i) \| \qquad (4.6)$$

where the operator A is defined by: $A = P_w \circ \mathcal{W}$.

We have to solve the inverse problem which consists of determining O_i knowing A and W_i. The solution of this problem depends on the regularity of A. The size of the restored image O_i is arbitrary and it can be easily set greater than the number of known coefficients. It is certain that there exists at least one image O_i which gives exactly W_i, i.e. the original one. But generally we have an infinity of solutions, and we have to choose among them the one which is considered as correct. An image is always a positive function, which leads us to constrain the solution, but this is not sufficient to get a unique solution.

Reconstruction Algorithm. The least squares method can be used to solve the relation $W_i = A(O_i)$ which leads to seeking the image O_i which minimizes the distance $\|W_i - A(O_i)\|$. $\|W_i - A(O_i)\|$ is minimum if and only if O_i is a solution of the following equation:

$$\tilde{A}(W_i) = (\tilde{A} \circ A)(O_i) \tag{4.7}$$

and $\tilde{A}$ is defined by (Bijaoui and Rué, 1995):

$$\tilde{A}(W_i) = \sum_{j=1}^{J}(h_1 * \cdots * h_{j-1})W_i(j) \tag{4.8}$$

where h is the low-pass filter used in the wavelet decomposition, and $W_i(j)$ is a subset of W_i, i.e. its wavelet coefficients of scale j.

The reconstruction algorithm is:

1. Initialization step: the estimated image O_i^n, the residual wavelet w_r^n and residual image R^n are initialized.

$$\begin{cases} O_i^0 = \mathcal{W}^{-1}W_i \\ w_r^0 = W_i - A(O_i^0) \\ R^0 = \tilde{A}(w_r^0) \end{cases} \tag{4.9}$$

$\mathcal{W}^{-1}$ is the wavelet reconstruction operator. From a wavelet structure W_i, an image O_i is restored corresponding to the sum of the wavelet scales and the last smoothed image. W_i is not necessarily the wavelet transform of an image, so $\mathcal{W}^{-1}\mathcal{W}O_i$ may not be equal to O_i.

2. Computation of the convergence parameter α^n:

$$\alpha^n = \frac{\|\tilde{A}(w_r^n)\|^2}{\|A(R^n)\|^2} \tag{4.10}$$

3. An iterative correction is applied to O_i^n to get the intermediate image O_i^{n+1}:

$$O_i^{n+1} = O_i^n + \alpha^n R^n \tag{4.11}$$

4. Positivity constraint: negative values in O_i^{n+1} are set to zero.
5. Wavelet residual computation:

$$w_r^{n+1} = W_i - A(O_i^{n+1}) \tag{4.12}$$

6. Test on the wavelet residual: if $\|w_r^{n+1}\|$ is less than a given threshold, the desired precision has been reached and the procedure is stopped.
7. Computation of the convergence parameter β^{n+1}:

$$\beta^{n+1} = \frac{\|\tilde{A}(w_r^{n+1})\|^2}{\|\tilde{A}(w_r^n)\|^2} \tag{4.13}$$

8. Residual image computation

$$R^{n+1} = \tilde{A}(w_r^{n+1}) + \beta^{n+1}R^n \tag{4.14}$$

9. Return to step 1.

4.3.5 Examples

Example 1: Band Extraction. We simulated a spectrum which contains an emission band at 3.50 μm and non-stationary noise superimposed on a smooth continuum. The band is a Gaussian of width FWHM = 0.01 μm (FWHM = full width at half-maximum), and normalized such that its maximum value equals ten times the local noise standard deviation.

Fig. 4.3 (top) contains the simulated spectrum. The wavelet analysis results in the detection of an emission band at 3.50 μm above 3σ. Fig. 4.3 (middle) shows the reconstruction of the detected band in the simulated spectrum. The real feature is over-plotted as a dashed line. Fig. 4.3 (bottom) contains the original simulation with the reconstructed band subtracted. It can be seen that there are no strong residuals near the location of the band, which indicates that the band is well-reconstructed. The center position of the band, its FWHM, and its maximum, can then be estimated via a Gaussian fit. More details about the use of MVM for spectral analysis can be found in (Starck et al., 1997b).

Example 2: Star Extraction in NGC2997. We applied MVM to the galaxy NGC2997 (Fig. 4.4, top left). Two images were created by coadding objects detected from scales 1 and 2, and from scales 3 to 6. They are displayed respectively in Fig. 4.4, top right, and bottom left. Fig. 4.4, bottom right, shows the difference between the input data and the image which contained the objects from scales 1 and 2. As we can see, all small objects have been removed, and the galaxy can be better analyzed.

Example 3: Galaxy Nucleus Extraction. Fig. 4.5 shows the extracted nucleus of NGC2997 using the MVM method, and the difference between the galaxy image and the nucleus image.

4.3.6 Application to ISOCAM Data Calibration

The ISOCAM infrared camera is one of the four instruments on board the ISO (Infrared Space Observatory) spacecraft which ended its life in May 1998. It operated in the 2.5-17 μm range, and was developed by the ISOCAM consortium led by the Service d'Astrophysique of CEA Saclay, France (Cesarsky et al., 1996).

The main difficulty in dealing with ISOCAM faint source detection is the combination of the cosmic ray impacts (glitches) and the transient behavior of the detectors (Siebenmorgen et al., 1996; Starck et al., 1999a). For glitches producing a single fast increase and decrease of the signal, simple median filtering allows fairly good deglitching, while for other glitches, memory effects can produce false detections. Consequently, the major source of error here is not the detection limit of the instrument, which is quite low, but the large number of glitches which create false detection.

4.3 Multiscale Vision Model

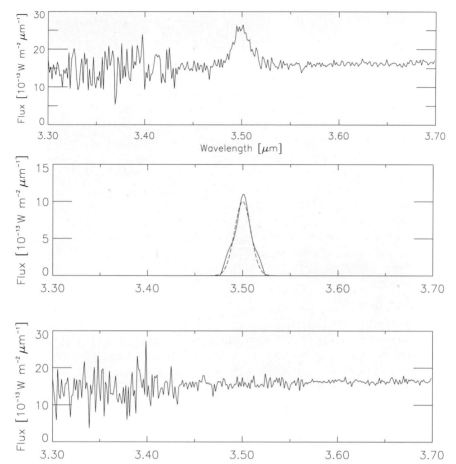

Fig. 4.3. *Top:* simulated spectrum. *Middle:* reconstructed simulated band (*full line*) and original band (*dashed line*). *Bottom:* simulated spectrum minus the reconstructed band.

Three types of glitches can be distinguished (Claret et al., 2000; Starck, 2000):

− a positive strong and short feature (lasting one readout),
− a positive tail (called *fader*, lasting a few readouts),
− a negative tail (called *dipper*, lasting several tens of readouts).

Fig. 4.6 is a plot of the camera units (ADU: analog to digital units) measured by a single pixel as a function of the number of readouts, i.e. time, which shows the three types of glitches. At the top (a), three sharp type "1" glitches are clearly visible. In the middle plot (b), another pixel history shows a "fader" (at about 80 readouts and lasting about 20 readouts). In the bottom plot (c), a "dipper" is present at readout 230, which lasts about 150 readouts.

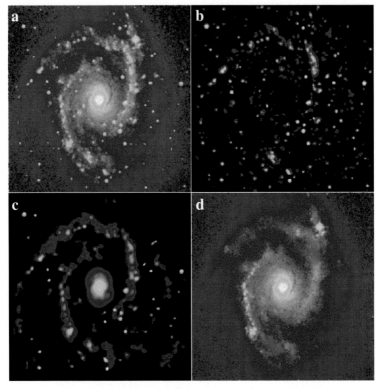

Fig. 4.4. (a) Galaxy NGC2997, (b) objects detected from scales 1 and 2, (c) objects detected from scales 3 to 6, and (d) difference between (a) and (b).

The signal measured by a single pixel as a function of time is the combination of memory effects, cosmic ray impacts and real sources. Memory effects begin with the first readouts, since the detector faces a flux variation from an offset position to the target position (stabilization). Subsequently memory effects appear with long-lasting glitches and following real sources. One needs to clearly separate all these constituents of the signal in each pixel before building a final raster map. One must also keep information on the associated noise before applying a source detection algorithm. Indeed, since the glitches do not follow Gaussian statistics, it is clear that an analysis of the final raster map would lead to poor results, since the standard detection criteria (detection above N times the standard deviation of the noise) would no longer be valid.

The calibration from pattern recognition (Starck et al., 1997a; Starck et al., 1999b) consists of searching only for objects which verify given conditions. For example, finding glitches of the first type is equivalent to finding objects which are positive, strong, and with a temporal size lower than that of the sources.

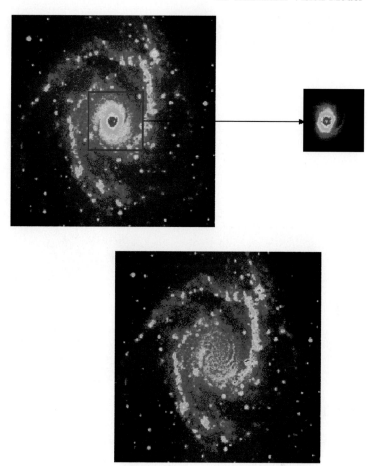

Fig. 4.5. *Upper left:* galaxy NGC2997; *upper right:* extracted nucleus; *bottom:* difference between the two previous images.

Fig. 4.7 (bottom) presents the result after carrying out such processing. Original data are shown in Fig. 4.7 (top). Fig. 4.8 shows the decomposition of the original signal (see Fig. 4.7 top) into its main components: (a), (b), and (d) are features (short glitch, glitch negative tail, and baseline) which present no direct interest for faint source detection, and (c) and (e) (source and noise) must be considered further. The noise must also be kept because faint sources could be undetectable in a single temporal signal, but detectable after co-addition of the data. The simple sum of the five components is exactly equal to the original data (see Fig. 4.7 top). The calibrated background free data (see Fig. 4.7 bottom) are then obtained by addition of (c) and (e).

126   4. Detection

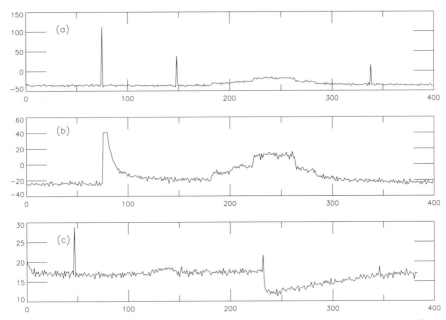

Fig. 4.6. Examples of glitches. (**a**) This signal contains three glitches of the first type. (**b**) A glitch with a long tail appears (named fader) around the position 80. The glitch has been truncated, and its real amplitude is 2700 ADUs. (**c**) A glitch with a negative tail (named dipper) appears around position 240.

4.4 Detection and Deconvolution

The PSF is not needed with MVM. This is an advantage when the PSF is unknown, or difficult to estimate, which happens relatively often when it is space-variant. However, when the PSF is well-determined, it becomes a drawback because known information is not used for the object reconstruction. This can lead to systematic errors in the photometry, which depends on the PSF and on the source signal-to-noise ratio. In order to preempt such a bias, a kind of calibration must be performed using simulations (Starck et al., 1999b). This section shows how the PSF can be used in the MVM, leading to a deconvolution.

Object Reconstruction using the PSF

A reconstructed and deconvolved object can be obtained by searching for a signal O such that the wavelet coefficients of $P * O$ are the same as those of the detected structures. If $\mathcal{W}$ describes the wavelet transform operator, and P_w the projection operator in the subspace of the detected coefficients, the solution is found by minimization of

$$J(O) = \parallel W - (P_w \circ \mathcal{W}) P * O \parallel \tag{4.15}$$

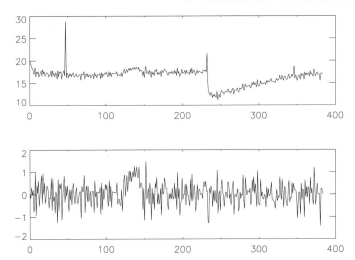

Fig. 4.7. *Top:* original data, and *bottom:* calibrated data (background free). The flux in ADUs (analog to digital units) is plotted against time given by the number of exposures. Note the gain variation of about 5 ADUs which appears after the second glitch.

where W represents the detected wavelet coefficients of the data, and P is the PSF. In this approach, each object is deconvolved separately. The flux related to the extent of the PSF will be taken into account. For point sources, the solution will be close to that obtained by PSF fitting. This problem is also different from global deconvolution in the sense that it is well-constrained. Except for the positivity of the solution which is always true and must be used, no other constraint is needed. This is due to the fact that the reconstruction is performed from a small set of wavelet coefficients (those above a detection limit). The number of objects is the same as those obtained by the MVM, but the photometry and the morphology are different. The astrometry may also be affected.

The Algorithm

Any minimizing method can be used to obtain the solution O. Since we did not find any problem of convergence, noise amplification, or ringing effect, we chose the Van Cittert method on the grounds of its simplicity. For each detected object, we apply the following algorithm:

$$O^{n+1} = O^n + \mathcal{W}^{-1}(W - (P_w \circ \mathcal{W}) P * O^n) \qquad (4.16)$$

where $\mathcal{W}^{-1}$ is the inverse wavelet transform.

1. Set n to 0.
2. Find the initial estimation O^n by applying an inverse wavelet transform to the set W corresponding to the detected wavelet coefficients in the data.

128   4. Detection

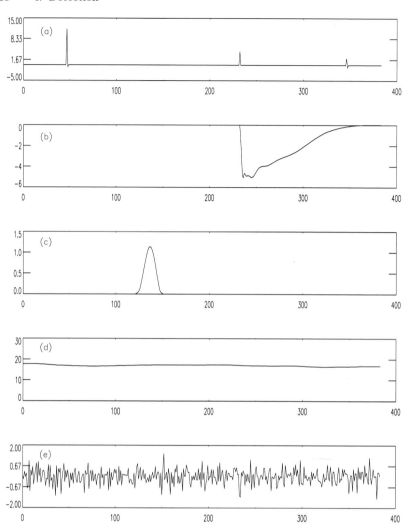

Fig. 4.8. Decomposition of the signal into its main components: (**a**) short glitch, (**b**) glitch negative tail, (**c**) source, (**d**) baseline, (**e**) noise. The simple sum of the five components is exactly equal to the original data (see previous figure). The calibrated background-free data are obtained by addition of signals (**c**) and (**e**).

3. Convolve O^n with the PSF P: $I^n = P * O^n$.
4. Determine the wavelet transform $W(I^n)$ of I^n.
5. Threshold all wavelet coefficients in $W(I^n)$ at position and scales where nothing has been detected (i.e. P_w operator). We get $W_t(I^n)$.
6. Determine the residual $w_r = W - W_t(I^n)$.
7. Reconstruct the residual image R^n by applying an inverse wavelet transform.

8. Add the residual to the solution: $O^{n+1} = O^n + R^n$.
9. Threshold negative values in O^{n+1}.
10. If $\sigma(R^n)/\sigma(O^0) < \epsilon$ then $n = n+1$ and go to step 3.
11. O^{n+1} contains the deconvolved reconstructed object.

In practice, convergence is very fast (less than 20 iterations). The reconstructed image (not deconvolved) can also be obtained just by reconvolving the solution with the PSF.

Space-Variant PSF

Deconvolution methods generally do not take into account the case of a space-variant PSF. The standard approach when the PSF varies is to decompose the image into blocks, and to consider the PSF constant inside a given block. Blocks which are too small lead to a problem of computation time (the FFT cannot be used), while blocks which are too large introduce errors due to the use of an incorrect PSF. Blocking artifacts may also appear. Combining source detection and deconvolution opens up an elegant way for deconvolution with a space-variant PSF. Indeed, a straightforward method is derived by just replacing the constant PSF at step 3 of the algorithm with the PSF at the center of the object. This means that it is not the image which is deconvolved, but its constituent objects.

Undersampled Point Spread Function

If the PSF is undersampled, it can be used in the same way, but results may not be optimal due to the fact that the sampled PSF varies depending on the position of the source. If an oversampled PSF is available, resulting from theoretical calculation or from a set of observations, it should be used to improve the solution. In this case, each reconstructed object will be oversampled. Equation (4.15) must be replaced by

$$J(O) = \| W - (P_w \circ \mathcal{W} \circ \mathcal{D}_l) P * O \| \qquad (4.17)$$

where $\mathcal{D}_l$ is the averaging-decimation operator, consisting of averaging the data in the window of size $l \times l$, and keeping only one average pixel for each $l \times l$ block.

Example: Application to Abell 1689 ISOCAM Data

Fig. 4.9 (left) shows the detections (isophotes) obtained using the MVM method without deconvolution on ISOCAM data. The data were collected using the 6 arcsecond lens at 6.75 μm. This was a raster observation with 10s integration time, 16 raster positions, and 25 frames per raster position. The noise is non-stationary, and the detection of the significant wavelet coefficients was carried out using the root mean square error map $R_\sigma(x,y)$ by the method

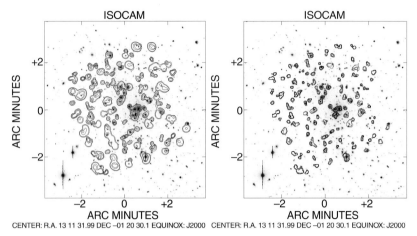

Fig. 4.9. Abell 1689: *left:* ISOCAM source detection (isophotes) overplotted on an optical image (NTT, band V). The ISOCAM image is a raster observation at 7 µm. *Right:* ISOCAM source detection using the PSF (isophotes) overplotted on the optical image. Compared to the left panel, it is clearly easier to identify the detected infrared sources in the optical image.

described in Starck et al. (1999b). The isophotes are overplotted on an optical image (NTT, band V) in order to identify the infrared source. Fig. 4.9 (right) shows the same treatment but using the MVM method with deconvolution. The objects are the same, but the photometry is improved, and it is clearly easier to identify the optical counterpart of the infrared sources.

4.5 Detection in the Cosmological Microwave Background

4.5.1 Introduction

The Cosmic Microwave Background (CMB), discovered in 1965 by Penzias and Wilson (1965), is a relic of radiation emitted some 13 billion years ago, when the Universe was about 370,000 years old. This radiation exhibits characteristics of an almost perfect blackbody at a temperature of 2.726 Kelvin as measured by the FIRAS experiment on board the COBE satellite (Fixsen et al., 1996). The DMR experiment, again on board COBE, detected and measured small angular fluctuations of this temperature, at the level of a few tens of micro Kelvin, and at angular scales of about 10 degrees (Smoot et al., 1992). These so-called temperature anisotropies were predicted as the imprints of the initial density perturbations which gave rise to the present large-scale structures as galaxies and clusters of galaxies. This relation between the present-day universe and its initial conditions has made the CMB radiation one of the most preferred tools of cosmologists to understand the

4.5 Detection in the Cosmological Microwave Background

history of the universe, the formation and evolution of the cosmic structures and physical processes responsible for them, and for their clustering.

As a consequence, the past few years have been a particularly exciting period for observational cosmology focussing on the CMB. With CMB balloon-borne and ground-based experiments such as TOCO (Miller et al., 1999), BOOMERanG (de Bernardis et al., 2000), MAXIMA (Hanany et al., 2000), DASI (Halverson et al., 2002) and Archeops (Benoît et al., 2003), a firm detection of the so-called "first peak" in the CMB anisotropy angular power spectrum at the degree scale was obtained. This detection was very recently confirmed by the WMAP satellite (Bennett et al., 2003), which detected also the second and third peaks. The WMAP satellite mapped the CMB temperature fluctuations (see Fig. 4.10) with a resolution better than 15 arc-minutes, and very good accuracy, marking the starting point of a new era of precision cosmology that enables us to use the CMB anisotropy measurements to constrain the cosmological parameters and the underlying theoretical models.

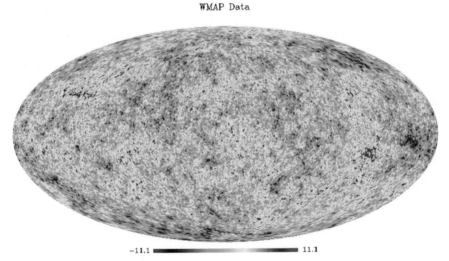

Fig. 4.10. WMAP Data.

CMB data are different from other astronomical data sets in the sense that they are not sparse (typical sparse data are stars or/and galaxies on top of a smooth background). After a component separation processing (see section 6.7), the CMB data are not completely free of contaminations. Point sources still need to be detected and removed. Once we believe the data are clean enough, we want to check if the distribution of CMB temperature fluctuations is Gaussian by using robust statistical Gaussianity tests.

4.5.2 Point Sources on a Gaussian Background

Several methods have been proposed in the last years for point source detection in the CMB such as the the Mexican Hat wavelet (Cayón et al., 2000; Cayón et al., 2001), the pseudo-filter (Sanz et al., 2001), or the biparametric scale-adaptive filter (López-Caniego et al., 2005). A simple and robust technique, which maximizes the signal-to-noise ratio is the Matched Filter (Vio et al., 2002). Assuming an isotropic point spread function (PSF) with known power sprectum $\tau(q)$ and the CMB with power spectrum $P(q)$, the Matched Filter is (Vio et al., 2002):

$$\widehat{\psi}_{MF}(q) = \frac{1}{2\pi\alpha} \frac{\tau(q)}{P(q)}, \qquad \alpha \equiv \int_0^{+\infty} q\frac{\tau^2}{P} \, dq, \qquad (4.18)$$

with minimum variance

$$\sigma^2 = \frac{1}{2\pi\alpha}. \qquad (4.19)$$

If the PSF is unknown (or space-variant), the Mexican Hat wavelet may be a good alternative. It consists of convolving the data with the wavelet function $\psi_{a,b}(x) = \psi(\frac{x-b}{a})$, where $\psi(x) = \frac{1}{\sqrt{2\pi}}(1-x^2)e^{-x^2/2}$. a is the scale parameter and b the position parameter. A fast implementation is obtained by using the Fourier transform (or the spherical harmonic transform for data on the sphere) to perform the convolution products $(\widehat{\psi}_a(q) = \frac{2}{\sqrt{\pi}}(qa)^2 e^{-\frac{1}{2}(qa)^2})$ (López-Caniego et al., 2005).

4.5.3 Non-Gaussianity

The search for non-Gaussian signatures in the CMB temperature fluctuation maps furnished by MAP[1] (Komatsu et al., 2003), and to be furnished by PLANCK[2], is of great interest for cosmologists. Indeed, the non-Gaussian signatures in the CMB can be related to very fundamental questions such as the global topology of the universe (Riazuelo et al., 2002), superstring theory, topological defects such as cosmic strings (Bouchet et al., 1988), and multi-field inflation (Bernardeau and Uzan, 2002). The non-Gaussian signatures can, however, have a different but still cosmological origin. They can be associated with the Sunyaev-Zel'dovich (SZ) effect (Sunyaev and Zeldovich, 1980) (inverse Compton effect) of the hot and ionized intra-cluster gas of galaxy clusters (Aghanim and Forni, 1999; Cooray, 2001), with the gravitational lensing by large scale structures (Bernardeau et al., 2003), or with the reionization of the universe (Aghanim and Forni, 1999; Castro, 2003). They may also be simply due to foreground emission (Jewell, 2001), or to non-Gaussian instrumental noise and systematics (Banday et al., 2000).

---

[1] http://map.gsfc.nasa.gov/
[2] http://astro.estec.esa.nl/SA-general/Projects/Planck/

4.5 Detection in the Cosmological Microwave Background

All these sources of non-Gaussian signatures might have different origins and thus different statistical and morphological characteristics. It is therefore not surprising that a large number of studies have recently been devoted to the subject of the detection of non-Gaussian signatures. Many approaches have been investigated: Minkowski functionals and the morphological statistics (Novikov et al., 2000; Shandarin, 2002), the bispectrum (3-point estimator in the Fourier domain) (Bromley and Tegmark, 1999; Verde et al., 2000; Phillips and Kogut, 2001), the trispectrum (4-point estimator in the Fourier domain) (Kunz et al., 2001), wavelet transforms (Aghanim and Forni, 1999; Forni and Aghanim, 1999; Hobson et al., 1999; Barreiro and Hobson, 2001; Cayón et al., 2001; Jewell, 2001; Starck et al., 2004), and the curvelet transform (Starck et al., 2004). In (Aghanim et al., 2003; Starck et al., 2004), it was shown that the wavelet transform was a very powerful tool to detect the non-Gaussian signatures. Indeed, the excess kurtosis (4th moment) of the wavelet coefficients outperformed all the other methods (when the signal is characterized by a non-zero 4th moment). Based on kurtosis of wavelet coefficients, recent studies have reported non-Gaussian signatures in the WMAP data (Vielva et al., 2004; Mukherjee and Wang, 2004; Cruz et al., 2005). The excess kurtosis is a widely used statistic, based on the 4th moment. For any (symmetrical) random variable X, the kurtosis is:

$$\kappa(X) = \frac{EX^4}{(EX^2)^2} - 3.$$

The kurtosis measures a kind of departure of X from Gaussianity. The non-Gaussianity detector consists of first applying a multiscale transform (e.g., wavelet, or curvelet), and then calculating at each scale the kurtosis. In practice, missing data and instrumental effects may create an artificial kurtosis and it is very important to produce realistic simulations which present the same characteristics as the observed data (e.g., missing data, noise, etc.). Then the kurtosis obtained from the data is compared to the kurtosis level expected from the simulations.

Finally, a major issue of the non-Gaussian studies in CMB remains our ability to disentangle all the sources of non-Gaussianity from one another. Recent progress has been made on the discrimination between different possible origins of non-Gaussianity. Namely, it was possible to separate the non-Gaussian signatures associated with topological defects (cosmic strings) from those due to the Doppler effect of moving clusters of galaxies (both dominated by a Gaussian CMB field) by combining the excess kurtosis derived from both the wavelet and the curvelet transforms (Starck et al., 2004).

The wavelet transform is suited to spherical-like sources of non-Gaussianity, and a curvelet transform is suited to structures representing sharp and elongated structures such as cosmic strings. Each provides an adapted non-Gaussian estimator, namely the normalized mean excess kurtosis. The combination of these transforms through the product of the normalized mean excess kurtosis of wavelet transforms by normalized mean excess kurtosis of curvelet transforms highlights the presence of the cosmic strings in a mixture CMB+SZ+CS. Such a combination gives information about the nature of the non-Gaussian signals. The sensitivity of each transform to a particular shape makes it a very strong discriminating tool (Starck et al., 2004; Jin et al., 2005).

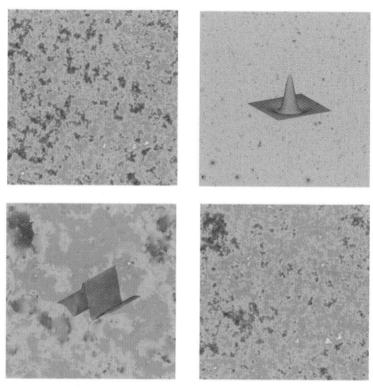

Fig. 4.11. *Top:* primary Cosmic Microwave Background anisotropies (*left*) and kinetic Sunyaev-Zel'dovich fluctuations (*right*). *Bottom:* cosmic string simulated map (*left*) and simulated observation containing the previous three components (*right*). The wavelet function is overplotted on the Sunyaev-Zel'dovich map and the curvelet function is overplotted on the cosmic string map.

In order to illustrate this, we show in Fig. 4.11 a set of simulated maps. Primary CMB, kinetic SZ and cosmic string maps are shown respectively in Fig. 4.11 top left, top right and bottom left. The "simulated observed map", containing the three previous components, is displayed in Fig. 4.11 bottom right. The primary CMB anisotropies dominate all the signals except at very high multipoles (very small angular scales). The wavelet function is overplotted on the kinetic Sunyaev-Zel'dovich map and the curvelet function is overplotted on cosmic string map.

4.6 Conclusion

The multiscale vision model allows us to analyze very complex data sets, and the main advantages of this approach are:

- Faint extended objects can be detected as well as point sources.
- The analysis does not require background estimation. (We know that if the background varies spatially, its estimation becomes a non-trivial task and may produce large errors in object photometry.)

We have shown that the source detection can be combined with a deconvolution when using wavelet based methods such as the Multiscale Vision Model. This leads to the reconstruction of deconvolved objects. The main advantages of this approach are:

- Objects are easier to identify in the deconvolved map.
- Morphological parameters (galaxy ellipticity and so on) are more accurate.
- Since each object is deconvolved separately, a spatially variable point spread function can easily be taken into account.
- Very large images can be deconvolved.

Many images contain stars and small galaxies, with a relatively flat background. In such cases, MVM may not be so attractive, especially if the images to be analyzed are very large. Object by object reconstruction requires much more computation time than a standard approach, especially with well-optimized code like SExtractor. A hybrid wavelet-SExtractor solution, as proposed in (Valtchanov et al., 2000), could then be a solution. Wavelet filtering is first applied to the data, taking into account the correct noise model in wavelet space, and the detection program SExtractor is then used on the noise-free image to find the sources. This combined solution has the advantage of fast and robust detection, as in the standard method, while keeping the ability to detect the faintest objects, which is only possible in wavelet space.

4.7 Chapter Summary

Object and feature detection, as discussed in this chapter, builds on the earlier work of (i) filtering, and (ii) deconvolution. The latter are important, and they may be necessary, in conjunction with the work of object and feature finding and measurement.

Formally, we discussed the issues facing us, as analyst, in terms of a vision model. Computer vision, and signal and data analysis, presuppose models at all stages of the processing path. A vision model is a high-level model.

We also showed how data imperfections, using the case of signal glitch analysis, came under the heading of object and feature detection.

5. Image Compression

5.1 Introduction

From year to year, the quantity of astronomical data increases at an ever growing rate. In part this is due to very large digitized sky surveys in the optical and near infrared, which in turn is due to the development of digital imaging arrays such as CCDs (charge-coupled devices). The size of digital arrays is also continually increasing, pushed by the demands of astronomical research for ever larger quantities of data in ever shorter time periods. Currently, projects such as the European DENIS and American 2MASS infrared sky surveys, or the Franco-Canadian MegaCam Survey and the American Sloan Digital Sky Survey, will each produce of the order of 10 TBytes of image data. The routine and massive digitization of photographic plates has been made possible by the advent of automatic plate scanning machines (MAMA, APM, COSMOS, SuperCOSMOS, APS, PMM, PDS) (Richter, 1998). These machines allow for digitization of the truly enormous amount of useful astronomical data represented in a photograph of the sky, and they have allowed the full potential of large area photographic sky surveys to be brought to fruition. The storage of astronomical data requires the latest innovations in archiving technology (12" or $5\frac{1}{4}$" WORM in the past, CD WORMS or magnetic disks with RAID technology now, DVD in the very near future). The straightforward transfer of such amounts of data over computer networks becomes cumbersome and in some cases practically impossible. Transmission of a high resolution Schmidt plate image over the Internet would take many hours. Facing this enormous increase in pixel volume, and taking into account the fact that catalogs produced by extraction of information from the pixels can always be locally wrong or incomplete, the needs of the astronomer follow two very different paths:

– The development of web technology creates the need for fast access to informative pixel maps, which are more intuitively understandable than the derived catalogs.
– Quantitative work often requires accurate refinement of astrometry and photometry, or effective redetection of missed objects.

In both cases and for different reasons, the astronomical community is confronted with a rather desperate need for data compression techniques.

Several techniques have in fact been used, or even developed, in the field of astronomy. Véran (1994) studied lossless techniques. White et al. (1992) developed HCOMPRESS, based on the Haar wavelet transform, and Press et al. (1992) developed FITSPRESS based on the Daubechies wavelet transform. In addition, the scientist must of course consider JPEG, a general purpose standard. Somewhat counter-intuitively (because the median is by no means the most computationally efficient of operations) effective and efficient compression based on the multiresolution Pyramidal Median Transform (PMT) algorithm was developed by Starck et al. (1996). Huang and Bijaoui (1991) used mathematical morphology in MathMorph for astronomical image processing.

Based on image type and application, different strategies can be used:

1. Lossy compression: in this case compression ratio is relatively low (< 5).
2. Compression without visual loss. This means that one cannot see the difference between the original image and the decompressed one.
3. Good quality compression: the decompressed image does not contain any artifact, but some information is lost.
4. Fixed compression ratio: for some technical reason or other, one may decide to compress all images with a compression ratio higher than a given value, whatever the effect on the decompressed image quality.
5. Signal/noise separation: if noise is present in the data, noise modeling can allow for very high compression ratios just by including filtering in wavelet space during the compression.

According to the image type and the selected strategy the optimal compression method may vary. A major interest in using a multiresolution framework is to get, in a natural way, the possibility for progressive information transfer.

According to Shannon's theorem, the number of bits we need to code an image I without distortion (losslessly) is given by its entropy H. If the image (with N pixels) is coded with L intensity levels, each level having a probability of appearance p_i, the entropy H is

$$H(I) = \sum_{i=1}^{L} -p_i \log_2 p_i \tag{5.1}$$

The probabilities p_i can be easily derived from the image histogram. The compression ratio is given by:

$$\mathcal{C}(I) = \frac{\text{number of bits per pixel in the raw data}}{H(I)} \tag{5.2}$$

and the distortion is measured by

$$R = \parallel I - \tilde{I} \parallel^2 = \sum_{k=1}^{N} (I_k - \tilde{I}_k)^2 \tag{5.3}$$

where $\tilde{I}$ is the decompressed image.

A Huffman, or an arithmetic, coder is generally used to transform the set of integer values into the new set of values, in a reversible way

Compression methods use redundancy in the raw data in order to reduce the number of bits. Efficient methods mostly belong to the transform coding family, where the image is first transformed into another set of data where the information is more compact (i.e. the entropy of the new set is lower than the original image entropy). The typical steps are:

1. transform the image (for example using a discrete cosine transform, or a wavelet transform),
2. quantize the obtained coefficients, and
3. code the values by a Huffman or an arithmetic coder.

The first and third points are reversible, while the second is not. The distortion depends on the way the coefficients are quantized. We may want to minimize the distortion with the minimum of bits, and a trade-off is then necessary in order to also have "acceptable" quality. "Acceptable" is subjective, and depends on the application. Sometimes, any loss is unacceptable, and the price to be paid for this is a very low compression ratio (often between one and two).

In this chapter, we describe the compression method based on the Pyramidal Median Transform (PMT), and we report our findings using five other compression algorithms (HCOMPRESS, FITSPRESS, JPEG, PMT, and MathMorph) from astrometry and photometry. We used a sample of nearly 2000 stars from an ESO Schmidt plate centered on the globular cluster M5. The results indicate that PMT can give compression ratios of up to 5 times the maximum ratio obtained from the other methods, when the regions are not too dense.

The next section contains a brief and general description of image compression techniques, and of the four compression software packages, FITSPRESS, HCOMPRESS, JPEG and PMT. This is followed by a presentation of data and calibrations used for our study (and a discussion of our approach to testing the astronomical quality assessment of the compressed images), and a presentation of our results.

5.2 Lossy Image Compression Methods

5.2.1 The Principle

Numerical image information is coded as an array of intensity values, reproducing the geometry of the detectors used for the observation or the densitometer used for plate digitization. The object signal is stored with noise, background variations, and so on. The relevant information depends on the application domain, and represents what the astronomer wants to study. The information of relevance reflects the limits of the observing instrument and of the digitization process. Reducing the amount of data to be coded requires that the relevant information be selected in the image and that the coding process be reorganized so that we emphasize the relevant information and drop noise and non-meaningful data. For this, we can focus on the region of interest, filter out noise, and quantize coarsely to take into account the limits of our human visual system if the images are only used for browsing.

Furthermore, the usual pixel array representation associated with images stores a lot of redundant information due to correlation of intensity values between nearby pixels and between different scales for large image structures or slowly varying background. A good compression scheme should aim at concentrating on the meaningful information in relation to the scientific purpose of the imaging (survey) project and code it efficiently, thereby limiting as much as possible the redundancy.

For this study, we examined the major available image compression packages – with relevance to astronomy – and compared their strategies with respect to these goals.

5.2.2 Compression with Pyramidal Median Transform

Multiscale Median Transform. The median transform is nonlinear, and offers advantages for robust smoothing (i.e. the effects of outlier pixel values are mitigated). Define the median transform of image f, with square kernel of dimensions $n \times n$, as $med(f, n)$. Let $n = 2s + 1$; initially $s = 1$. The iteration counter will be denoted by j, and J is the user-specified number of resolution scales.

1. Let $c_j = f$ with $j = 1$
2. Determine $c_{j+1} = med(f, 2s + 1)$.
3. The multiresolution coefficients w_{j+1} are defined as: $w_{j+1} = c_j - c_{j+1}$. Image w_1 has zero values.
4. Let $j \longleftarrow j+1$; $s \longleftarrow 2s$. Return to Step 2 if $j \leq J$.

A straightforward expansion formula for the original image is given by:

$$f = c_J + \sum_{j=1}^{J} w_j \qquad (5.4)$$

where c_J is the residual image.

The multiscale coefficient values, w_j, are evidently not necessarily of zero mean, and so the potential artifact-creation difficulties related to this aspect of wavelet transforms do not arise. Note of course that values of w can be negative.

For integer image input values, this transform can be carried out in integer arithmetic only which may lead to computational savings.

Computational requirements of the multiresolution transform are high, and these can be reduced by decimation: one pixel out of two is retained at each scale. Here the transform kernel does not change from one iteration to the next, but the image to which this transform is applied does. This pyramidal algorithm is looked at next.

Pyramidal Median Transform. The Pyramidal Median Transform (PMT) is obtained by the following algorithm:

1. Let $c_j = f$ with $j = 1$.
2. Determine $c^*_{j+1} = med(c_j, 2s+1)$ with $s = 1$.
3. The pyramidal multiresolution coefficients w_{j+1} are defined as: $w_{j+1} = c_j - c^*_{j+1}$.
4. Let $c_{j+1} = dec(c^*_{j+1})$ where the decimation operation, dec, entails 1 pixel replacing each 2×2 subimage.
5. Let $j \longleftarrow j+1$. Return to Step 2 so long as $j < J$.

Here the kernel or mask of dimensions $(2s+1) \times (2s+1)$ remains the same during the iterations. The image itself, to which this kernel is applied, becomes smaller.

While this algorithm aids computationally, the reconstruction formula (equation 5.4 above) is no longer valid. Instead we use the following algorithm based on B-spline interpolation:

1. Take the lowest scale image, c_j.
2. Interpolate c_j to determine the next resolution image (of twice the dimensionality in x and y). Call the interpolated image c'_j.
3. Calculate $c_{j-1} \longleftarrow c'_j + w_j$.
4. Set $j \longleftarrow j-1$. Go to Step 2 if $j > 0$.

This reconstruction procedure takes account of the pyramidal sequence of images containing the multiresolution transform coefficients, w_j. It presupposes, though, that good reconstruction is possible. We ensure that by use of the following refined version of the Pyramidal Median Transform. Using iteration, the coefficients, $w_{j+1} = c_j - c_{j+1}$, are improved relative to their potential for reconstructing the input image.

Iterative Pyramidal Median Transform. An iterative scheme can be proposed for reconstructing an image, based on pyramidal multi-median transform coefficients. Alternatively, the PMT algorithm, itself, can be enhanced to allow for better estimates of coefficient values. The following is an iterative algorithm for this objective:

1. $i \longleftarrow 0$. Initialize f^i with the given image, f. Initialize the multiresolution coefficients at each scale j, w^f_j, to 0.
2. Using the Pyramidal Median Transform, determine the set of transform coefficients, $w^{f_i}_j$.
3. $w^f_j \longleftarrow w^f_j + w^{f_i}_j$.
4. Reconstruct image f_{i+1} from w^f_j (using the interpolation algorithm described in the previous section).

5. Determine the image component which is still not reconstructible from the wavelet coefficients: $f_{i+1} \longleftarrow f - f_{i+1}$.
6. Set $i \longleftarrow i + 1$, and return to Step 2.

The number of iterations is governed by when f_{i+1} in Step 5 approaches a null (zero) image. Normally 4 or 5 iterations suffice. Note that the additivity of the wavelet coefficients in Step 3 is justified by additivity of the image decomposition in Step 5 and the reconstruction formula used in Step 4, both of which are based on additive operations.

Non-Iterative Pyramidal Transform with Exact Reconstruction. A non-iterative version of the pyramidal median transform can be performed by decimating and interpolating the median images during the transform:

1. Let $c_j = f$ with $j = 1$.
2. Determine $c_{j+1} = dec[med(c_j, 2s + 1)]$
3. Determine c^*_{j+1} = interpolation of c_{j+1} to size of c_j
4. The pyramidal multiresolution coefficients w_{j+1} are defined as: $w_{j+1} = c_j - c^*_{j+1}$.
5. Let $j \longleftarrow j + 1$. Return to Step 2 so long as $j < J$.

This saves computation time in two ways. First, there is no need to iterate. Secondly, in step 2 one does not calculate the median for all pixels and then decimate it; rather, one just calculates the median for the pixels to be left after decimation. Thus the median calculations are 4 times fewer. This algorithm is very close to the Laplacian pyramid developed by Burt and Adelson (Burt and Adelson, 1983). The reconstruction algorithm is the same as before, but the reconstructed image has no error. In the following, we will indicate this version by referring to PMT.

5.2.3 PMT and Image Compression

The principle of the method is to select the information we want to keep, by using the PMT, and to code this information without any loss. Thus the first phase searches for the minimum set of quantized multiresolution coefficients which produce an image of "high quality". The quality is evidently subjective, and we will define by this term an image such as the following:

– There is no visual artifact in the decompressed image.
– The residual (original image – decompressed image) does not contain any structure.

Lost information cannot be recovered, so if we do not accept any loss, we have to compress what we take as noise too, and the compression ration will be low (a ratio of 3 or 4 only).

The method employed involves the following sequence of operations:

1. Determination of the multiresolution support.

2. Determination of the quantized multiresolution coefficients which gives the filtered image. (Details of the iterative implementation of this algorithm are dealt with below.)
3. Coding of each resolution level using the Huang-Bijaoui method (1991). This consists of quadtree-coding each image, followed by Huffman-coding (with fixed codes) the quadtree representation. There is no information loss during this phase.
4. Compression of the noise if this is wanted.
5. Decompression consists of reconstituting the noise-filtered image (+ the compressed noise if this was specified).

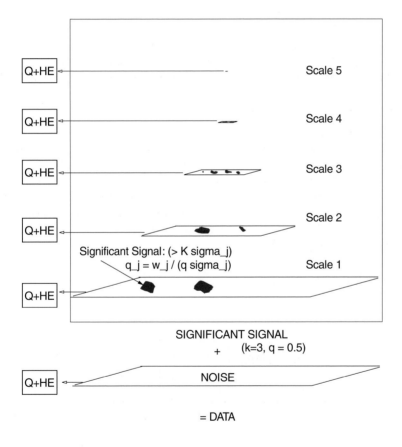

Fig. 5.1. Graphic depiction of the PMT compression method.

Fig. 5.1 shows graphically the PMT compression method. Note that we can reconstruct an image at a given resolution without having to decode the entire compressed file.

The first two phases were described in earlier chapters and we will now describe the last three phases.

Quantized Coefficients. We define the set $\mathcal{Q} = \{q_1, ...q_n\}$ of quantized coefficients, q_j corresponding to the quantized multiresolution coefficients w_j. We have:

- $q_j(x,y) = 0$ if $M(j,x,y) = 0$
- $q_j(x,y) = int(w_j(x,y)/(k_{signal}\sigma_j))$ if $M(j,x,y) = 1$

Here, int denotes integer part. The image reconstructed from $\mathcal{Q}$ gives the decompressed image D. Good compression should produce D such that the image $R = I - D$ contains only noise. Due to the thresholding and to the quantization, this may not be the case. It can be useful to iterate if we want to compress the quantized coefficients such that the best image is rendered possible. The final algorithm which allows us to compute both the quantized coefficients and the multiresolution support is:

1. Set $i = 0$, $R^i = I$
2. Set $M(j,x,y) = 0$ and $q_j^i(x,y) = 0 \quad \forall x,y,j$
3. Compute the PMT of R^i: we obtain w_j
4. If $i = 0$, estimate at each scale j the standard deviation of the noise σ_j.
5. New estimation of the multiresolution support:
   for all j,x,y, if $\mid w_j(x,y) \mid > k\sigma_j$, $M(j,x,y) = 1$
6. New estimation of the set $\mathcal{Q}$:
   for all j,x,y, if $\mid M(j,x,y) \mid = 1$,
   $q_j(x,y) = q_j(x,y) + int(w_j(x,y)/(k_{signal}\sigma_j))$
7. Reconstruction of D^i from $\mathcal{Q}$
8. $i = i + 1$, $R^i = I - D^{i-1}$ and go to 3

In step 6, $k_{signal} = 1.5$. (This value was fixed experimentally and seems to be a good trade-off between quality and efficiency.) After a few iterations, the set $\mathcal{Q}$ contains the multiresolution coefficients. This allows a considerable compression ratio and the filtered image can be reconstructed without artifacts. The residual image R is our reconstruction error ($rec(\mathcal{Q}) + R = I$). The results without iterating are satisfactory too, and are sufficient in most cases. If we are not limited by time computation during the compression, we can carry out a few iterations in order to have subsequently the best quality reconstruction.

Quadtree and Huffman Encoding. To push compression further, we code the multiresolution coefficients using a quadtree (Samet, 1984) followed by Huffman (Held and Marshall, 1987) encoding.

The particular quadtree form we are using was suggested by Huang and Bijaoui (1991) for image compression.

- Divide the bitplane into 4 quadrants. For each quadrant, code as "1" if there are any 1-bits in the quadrant, else code as "0".
- Subdivide each quadrant that is not all zero into 4 more sub-quadrants and code them similarly. Continue until one is down to the level of individual pixels.

Noise Compression. If we want exact compression, we have to compress the noise too. There is no transform which allows better representation of the noise, and the noise compression ratio will be defined by the entropy. Generally, we do not need all the dynamic range of the noise, and we do not compress the residual map R but rather the image $R_q = int(R/(k_{\mathrm{noise}}\sigma_R))$ with $k_{\mathrm{noise}} = \frac{1}{2}$ in the applications below.

Lossless compression can be performed too, but this makes sense only if the input data are integers, and furthermore the compression ratio will be very low.

Image Decompression. Decompression is carried out scale by scale, starting from low resolution, so it is not necessary to decompress the entire file if one is just interested in having a look at the image. Noise is decompressed and added at the end, if this is wanted. (The examples discussed below suppress the noise entirely.)

5.2.4 Compression Packages

Methods used in astronomy include HCOMPRESS (White et al., 1992), FITSPRESS (Press, 1992), and JPEG (Furht, 1995). These are all based on linear transforms, which in principle help to reduce the redundancy of pixel values in a block and decorrelate spatial frequencies or scales. In addition to the PMT, other methods have been proposed for astronomical image such as compression using mathematical morphology (Huang and Bijaoui, 1991). A specific decompression method was also developed in (Bijaoui et al., 1996) in order to reduce artifacts relative to the HCOMPRESS method. In the signal processing domain, two other recent approaches are worthy of mention. The first is based on fractals, and the second uses a bi-orthogonal wavelet transform.

We first briefly review all of these methods, and then compare them in the framework of astronomical images.

HCOMPRESS. HCOMPRESS (White et al., 1992) was developed at the Space Telescope Science Institute (STScI, Baltimore), and is commonly used to distribute archive images from the Digital Sky Survey DSS1 and DSS2. It is based on the Haar wavelet transform. The algorithm consists of

1. applying a Haar wavelet transform to the data,
2. quantizing the wavelet coefficients linearly as integer values,
3. applying a quadtree to the quantized value, and
4. using a Huffman coder.

Sources are available at http://www.stsci.edu/software/hcompress.html.

HCOMPRESS with Iterative Decompression. Iterative decompression was proposed in (Bijaoui et al., 1996) to decompress files which were compressed using HCOMPRESS. The idea is to consider the decompression problem as a restoration problem, and to add constraints on the solution in order to reduce the artifacts.

FITSPRESS. FITSPRESS (Press, 1992) uses a threshold on very bright pixels and applies a linear wavelet transform using the Daubechies-4 filters. The wavelet coefficients are thresholded based on a noise threshold, quantized linearly and runlength encoded. This was developed at the Center for Astrophysics, Harvard. Sources are available at
http://www.eia.brad.ac.uk/rti/guide/fits.html.

JPEG. JPEG is the standard video compression software for single frame images (Furht, 1995). It decorrelates pixel coefficients within 8×8 pixel blocks using the discrete cosine transform (DCT) and uniform quantization.

Wavelet Transform. Various wavelet packages exist which support image compression, leading to more sophisticated compression methods. The wavelet transform we used is based on a bi-orthogonal wavelet transform (using Antonini-Daubechies 7/9 coefficients) with non-uniform coding (Taubman and Zakhor, 1994), and arithmetic encoding. Source code is available at http://www.geoffdavis.net/dartmouth/wavelet/wavelet.html.

Fractal. The image is decomposed into blocks, and each block is represented by a fractal. See (Fisher, 1994) for more explanation.

Mathematical Morphology. This method (Huang and Bijaoui, 1991), denoted MathMorph in this chapter, is based on mathematical morphology (erosion and dilation). It consists of detecting structures above a given level, the level being equal to the background plus three times the noise standard deviation. Then, all structures are compressed by using erosion and dilation, followed by quadtree and Huffman coding. This method relies on a first step of object detection, and leads to high compression ratios if the image does not contain many objects (the image is nearly black, cf. Chapter 3), as is often the case in astronomy.

5.2.5 Remarks on these Methods

The pyramidal median transform (PMT) is similar to the mathematical morphology (MathMorph) method in the sense that both try to understand what is represented in the image, and to compress only what is considered as significant. The PMT uses a multiresolution approach, which allows more powerful separation of signal and noise. The latter two methods are both implemented in the MR/1 package (see http://www.multiresolution.com).

Each of these methods belongs to a general scheme where the following steps can be distinguished:

1. Decorrelation of pixel values inside a block, between wavelength, scales or shape, using orthogonal or nonlinear transforms.
2. Selection and quantization of relevant coefficients.
3. Coding improvement: geometrical redundancy reduction of the coefficients, using the fact that pixels are contiguous in an array.
4. Reducing the statistical redundancy of the code.

Table 5.1. Description and comparison of the different steps in the compression packages tested.

Software	Transform	Coefficient Quantization	Coefficient Organisation	Geometrical Redundancy reduction	Statistical Redundancy reduction
JPEG	DCT 8×8 pixels	Linear	Zigzag sequence	Runlength coding	Huffmann
HCOMPRESS	Haar 2×2 pixels	Linear	Pyramidal	Quadtree on bitplanes	Huffmann
FITSPRESS	Wavelets Daubechies-4	Linear	Increasing resolution	Runlength coding	Huffmann
MR/1 PMT	Pyramidal Median Transform	Linear / Noise estimation	Decreasing resolution	Quadtree on bitplanes	Huffmann
MR/1 Math.Morph.	Erosion/ Dilation	Linear / Noise estimation	–	Quadtree on bitplanes	Huffmann

How each method realizes these different steps is indicated in Table 5.1.

Clearly these methods combine many strategies to reduce geometrical and statistical redundancy. The best results are obtained if appropriate selection of relevant information has been performed before applying these schemes.

For astronomical images, bright or extended objects are sought, as well as faint structures, all showing good spatial correlation of pixel values and within a wide range of greylevels. Noise background, on the contrary, shows no spatial correlation and fewer greylevels. The removal of noisy background helps in regard to data compression of course. This can be done with filtering, greylevel thresholding, or coarse quantization of background pixels. This is used by FITSPRESS, PMT and MathMorph which divide information into a noise part, estimated as a Gaussian process, and a highly correlated signal part. MathMorph simply thresholds the background noise estimated by a 3-sigma clipping, and quantizes the signal as a multiple of sigma (Huang and Bijaoui, 1991). FITSPRESS thresholds background pixels and allows for coarse background reconstruction, but also keeps the highest pixel values in a separate list. PMT uses a multiscale noise filtering and selection approach based on noise standard deviation estimation. JPEG and HCOMPRESS do not carry out noise separation before the transform stage.

Identifying the Information Loss. Apart from signal-to-noise discrimination, information losses may appear after the transforms at two steps: coefficient selection and coefficient quantization. The interpretable resolution of the decompressed images clearly depends upon these two steps.

If the spectral bandwidth is limited, then the more it is shortened, the better the compression rate. The coefficients generally associated with the high spatial frequencies related to small structures (point objects) may be suppressed and lost. Quantization also introduces information loss, but can be optimized using a Lloyd-Max quantizer for example (Proakis, 1995).

All other steps, shown in Table 5.1, such as reorganizing the quantized coefficients, hierarchical and statistical redundancy coding, and so on, will not compromise data integrity. This statement can be made for all packages. The main improvement clearly comes from an appropriate noise/signal discrimination and the choice of a transform appropriate to the objects' properties.

5.2.6 Other Lossy Compression Methods

The number of bits that we need for an accurate representation of CCD (charge coupled device) raw data was discussed in (Watson, 2002). The integer image I in the FITS file is related to the real data D by: $D = B_{zero} + Ib_{scale}$. Lossy compression can then be obtained by (i) quantization of I, and (ii) application of a lossless compression technique such as bzip2 (Seward, 1998), gzip (Gailly, 1993), lzop (Oberhumer, 1998) or HCOMPRESS (White et al., 1992). The image I should be quantized with a quantization step Q:

$$I = I\left[\frac{I}{Q}\right] \tag{5.5}$$

where $[x]$ is the nearest integer to x. The quantization step should be derived from the read-out noise of the detector by (Watson, 2002):

$$Q = \begin{cases} \text{INT}\left(\frac{q\sigma_b}{b_{scale}}\right) & \text{if } q\sigma_b > b_{scale} \\ 1 & \text{if } q\sigma_b \leq b_{scale} \end{cases} \tag{5.6}$$

where $\text{INT}(x)$ is the largest integer no greater than x, q is a user parameter (suitable values for q are 0.5 to 2), and σ_b is the noise standard deviation.

In a recent paper, (Shamir and Nemiroff, 2005), a similar method is proposed, but only pixels in the background are quantized in order to not affect bright pixels. This approach is in the same spirit as the PMT and the MathMorph method, i.e. it searches first as to where the information is, and then compresses as a function of the result of the search. But the problem is that the source/background classification is done on a pixel by pixel basis which leads to a misclassification of all faint extended sources. Such a problem does not occur with the PMT. The second drawback is the need to estimate the background (as in the MathMorph method), which is not trivial, especially when the image contains very extended sources.

5.3 Comparison

5.3.1 Quality Assessment

In order to compare the different compression methods, we can use several characteristics, with constraints on these characteristics depending on the type of applications (see Table 5.2).

Table 5.2. List of criteria for comparison of compression methods for various types of astronomical image-based application.

Application type Comparison criterion	Quick view	Catalog overlay: cross-correl.	Source extract., cross-ident.	Deep detection	Recalibration
quality: visual	medium	high	medium	indifferent	high
quality: precision	low	medium	high	very high	high
Transfer + comput. speed	very fast	fast	medium	slow	medium
Progressive vision	yes	yes	no	no	no

The progressive vision aspect is very useful in the context of quick views (for example on the web) and catalog overlays, where the user can decide when the quality of a displayed image is sufficient. Overall latency, i.e. speed of display (transfer + processing time) is usually critical for applications related to progressive vision. For good quality quick views of a given area, catalog and database overlays, and cross-correlation of sources at various wavelengths, the required quality will be essentially qualitative: good geometry of the objects, no visual artifacts, good contrast, etc.

More quantitative tasks have different requirements. For cross-identification processes, and any situation where recalibration to improve astrometry and photometry is needed, or reprocessing of object detections where some were obviously missed, star/galaxy discrimination or separation of distinct objects falsely merged, the quality estimation must be a quantitative process. The loss of information can be measured by the evolution of "relevant parameters" varying according to compression rate and method.

Quality criteria for estimating the merits and performances of a compression method fall under these headings:

150   5. Image Compression

1. Visual aspect
2. Signal-to-noise ratio
3. Detection of real and faint objects
4. Object morphology
5. Astrometry
6. Photometry

Very few quantitative studies have been carried out up to now in astronomy in order to define which compression method should be used. Two studies were carried out in the framework of the Aladin Interactive Sky Atlas project (Bonnarel et al., 2001). One was in 1993–94, when JPEG, FITSPRESS, and HCOMPRESS were evaluated (Carlsohn et al., 1993; Dubaj, 1994), and another in 1996-1997 (Murtagh et al., 1998; Louys et al., 1999), when JPEG and PMT were compared.

5.3.2 Visual Quality

A quick overview was obtained of each method produced by running all compression algorithms on two images. The first was a 256×256 image of the Coma cluster from an STScI POSS-I digitized plate, and the second was a 1024×1024 image, extracted from the ESO 7992V plate digitized by CAI-MAMA (described in more detail in the next section). The visual quality was estimated from the visual aspect of the decompressed image, and the quality of the residual (original image – decompressed image). Conclusions relative to this study are:

– FITSPRESS leads to cross-like artifacts in the residual image, a loss of faint objects and a decrease in object brightness.
– JPEG cannot be used at compression ratios higher than 40. Above this, artifacts become significant, and furthermore astrometry and photometry become very bad.
– The fractal method cannot be used for astronomical data compression. There are boxy artifacts, but the main problem is that object fluxes are modified after decompression, and the residual contains a lot of information (stars or galaxies can be easily identified in the residual map).
– MathMorph leads to good compression ratios, but background estimation is delicate. For the Coma cluster, the result was relatively bad, due to the difficulty of finding the background. More sophisticated algorithms can certainly be used to do this task. Another drawback of this method is the bad recovery of the contours of the object, which leads also to a loss of flux.
– HCOMPRESS produces artifacts. Iterative reconstruction allows them to be suppressed, but in this case reconstruction takes time. However this approach should be considered when the archived data are already compressed with HCOMPRESS (e.g. HST archive).

- The wavelet method produces very good results for the Coma cluster (compression ratio of 40). For the second image, where a compression ratio of more than 200 is obtained with the PMT or by mathematic morphology, artifacts appear if we try to achieve the same high performances. This method can be used, but not for very high compression ratios.
- PMT produces good quality results for both images. The compression ratio, similarly to the mathematical morphology method, depends on the content of the image. The fewer the pixels of objects in the image, the higher the compression ratio.

An interesting feature of the wavelet method is that the compression ratio is a user parameter. For PMT, and MathMorph, the compression ratio is determined from noise modeling. For other methods, a user parameter allows the compression ratio to be changed, and consequently the image quality, but only iterations can lead to a given compression ratio, or to a given quality.

Comparison between PMT and HCOMPRESS

Fig. 5.2, upper left, shows the Coma cluster from a Space Telescope Science Institute POSS-I digitized plate. Fig. 5.2, upper right, shows the decompressed image using HCOMPRESS (30:1) and bottom left, the decompressed image using the PMT (30:1). Fig. 5.2, bottom right, shows the difference between the original image and the PMT decompressed one.

5.3.3 First Aladin Project Study

Two quantitative studies were conducted at CDS (Strasbourg Data Centre), within the scope of the Aladin Interactive Sky Atlas project, focusing on a small number of methods. The effects of compression for a Schmidt photographic plate in the region of M5 (numbered ESO 7992v), scanned with the CAI-MAMA facility, were examined. The digitized image is a mosaic of 28 × 28 subimages, each of 1024 × 1024 pixels. Sampling is 0.666 arcseconds per pixel. This region was chosen because of the availability of a catalog (Ojha et al., 1994) obtained from the same plate digitization, where positions and blue magnitudes had been estimated for 20,000 stars or galaxies of magnitude 10–19. The position of each object was ascertained by Ojha et al. by marginal Gaussian fitting to the intensity distribution. Magnitude was determined using 120 photometric standards, which allowed the magnitude-integrated density calibration curve to be specified.

To carry out our tests in a reasonable time and to avoid plate boundary effects, we analyzed 25 adjacent subimages, located at the center of the photographic plate. We stress that these test images are real and not simulated. They are representative of the images distributed by the CDS's reference image service, Aladin. The central region used for the astrometry and photometry measurements contains about 2000 objects whose magnitude distribution

152   5. Image Compression

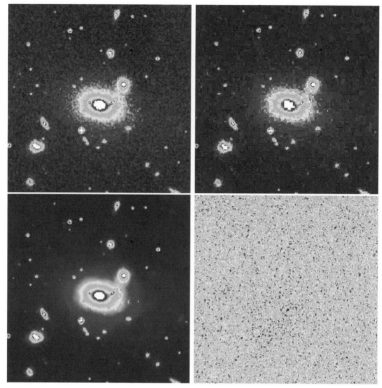

Fig. 5.2. *Upper left:* Coma cluster from a Space Telescope Science Institute POSS-I digitized plate. *Upper right:* decompressed image using HCOMPRESS (30:1). *Bottom left:* decompressed image using Pyramidal Median Transform (30:1). *Bottom right:* difference between the original image and the PMT decompressed one.

(from 14 for the brightest objects, to 19 for the faintest objects) is indicative of that of the global population of the catalog (Dubaj, 1994).

Detection experiments (Carlsohn et al., 1993) were performed to study the effect of compression on the preservation of faint objects. This was done on a test region, where 16 sources of estimated magnitude close to 21 were identified. Results are described in Table 5.3. Detection errors (loss and false detections) clearly increase with the compression rate. FITSPRESS loses the most objects, HCOMPRESS creates the most false objects. JPEG slightly better preserves faint objects but only below compression rate of 40:1. In (Carlsohn et al., 1993), the three methods were compared with respect to the signal-to-noise ratio, and positional and brightness error of known objects. It results that JPEG is better than HCOMPRESS at low signal-to-noise ratios, and is relatively similar at higher levels. Concerning the signal-to-noise ratio, the astrometry, and photometry, JPEG and HCOMPRESS produce images of equivalent quality, but FITSPRESS is again worse than the other two

Table 5.3. Detection of faint objects in a digitized patch of a Schmidt plate image, using the MIDAS detection routines SEARCH/INVENTORY on original and compressed/decompressed images at different compression rates: 4:1, 10:1, 20:1 and 40:1. With comparable detection parameters, and depending on compression method, faint objects can be lost or spurious objects corresponding to local maxima can be found. Visual inspection is necessary to confirm real detected objects.

Method	Compress. ratio	Real objects detected	Lost objects	False objects detected	Number detection errors	Percentage detection errors
JPEG	4	15	1	2	3	19
	10	14	2	2	4	25
	20	14	2	4	6	38
	40	13	3	5	8	50
Hcompress	4	14	2	3	5	31
	10	14	2	5	7	44
	20	11	5	3	8	50
	40	11	5	5	10	53
Fitspress	4	15	1	1	2	13
	10	13	3	0	3	19
	20	10	6	0	6	38
	40	5	11	0	11	69

methods. The first astrometrical tests were undertaken by Dubaj and are summarized in Fig. 5.3.

Star/galaxy discrimination was assessed by measuring the mean density by pixel, and considering the deviation of this quantity relative to its mean value for stars with the same integrated density as the object. Sufficiently low values are considered associated with galaxies. Applying this criterion to a subsample of around 1000 objects known a priori as stars or galaxies led to a contamination rate of 18% on the original image and 21% to 25% with compressed/uncompressed images (compression factor 40, for the three methods). This shows at least that morphological studies can be made on compressed/uncompressed images without substantial degradation.

The general conclusion of this first study was that none of these methods could provide good visual quality above compression rates of 40:1 and that the standard JPEG method was ultimately not so bad, even if block artifacts appear. The available software (i.e., HCOMPRESS and FITSPRESS) developed in astronomy did not score very well in the framework of the Aladin project. When the PMT method was proposed (Starck et al., 1996; Starck et al., 1998a), a second study was carried out in order to compare JPEG and PMT. In the meantime, MathMorph was implemented and underwent the same tests.

154   5. Image Compression

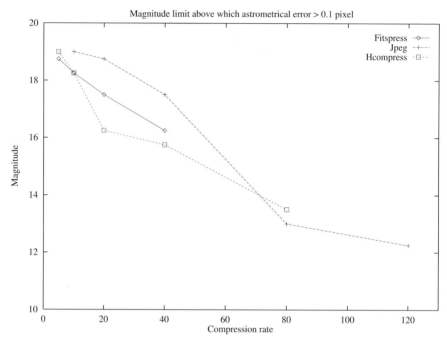

Fig. 5.3. Comparison of the ability of the different packages to recover the position of objects according to object magnitude: astrometrical error increases with magnitude. We recorded the limit of magnitude above which the position error exceeds the catalog precision: 0.1 pixel.

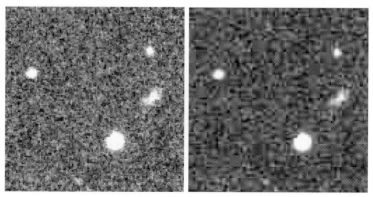

Fig. 5.4. *Left:* Original image, subimage extracted from 1024×1024 patch, extracted in turn from the central region of ESO7992v. *Right:* JPEG compressed image at 40:1 compression rate.

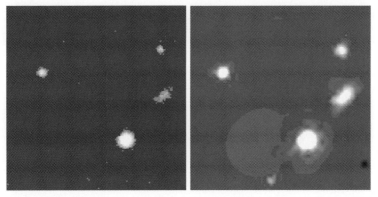

Fig. 5.5. *Left:* MathMorph compressed image of the same patch, at 203:1 compression rate. *Right:* PMT-compressed image at 260:1 compression rate.

5.3.4 Second Aladin Project Study

Visual Quality. For the two compression methods studied here (JPEG and PMT), each implying loss of information, a good compromise between compression rate and visual quality has to be found. In the case of JPEG, various studies (Carlsohn et al., 1993; Dubaj, 1994) confirm that beyond a compression rate of 40:1 this method of compression, when used on 12 bit/pixel images, gives rise to "blocky" artifacts. For PMT, as described in this chapter, the reconstruction artifacts appear at higher compression rates, beyond a rate of 260 in the particular case of our images. Figs. 5.4 and 5.5 allow the visual quality of the two methods to be compared, for test image 325. A subimage of the original image is shown in Fig. 5.4 (left).

Astrometry Quality. To estimate the influence of compression algorithms on astrometrical precision of the objects, the error in the position of the object in the original image compared to the position in the compressed/uncompressed image, was studied. This was done for each object in the catalog. Object position determination was carried out using the MIDAS (ESO, 1995) software (Munich Image Data Analsyis System) routines based on fitting of two marginal Gaussian profiles as used originally (Ojha et al., 1994) for creating the catalog. Knowing the catalog magnitude of the objects, the mean positional error as a function of the object magnitude can be represented.

This was done for magnitude intervals of 0.25 for the 2000 objects of the dataset used. Fig. 5.6 allows the performances of JPEG and PMT to be compared. Note that for the two methods, the error is below the systematic error of the catalog, in particular in the interval from object magnitudes 13 to 19 where sufficient objects warrant asserting a significant result. Outside that interval, the dataset does not contain enough objects to establish a mean error in the astrometry.

156   5. Image Compression

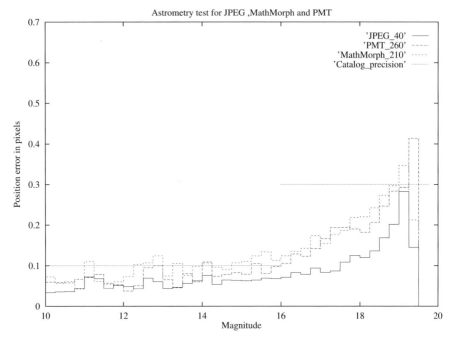

Fig. 5.6. Mean error in astrometry, by interval of 0.25 magnitude, for images compressed 40 times by JPEG, 260 times by PMT, and 210 times for MathMorph.

Photometry Quality. Conservation of photometric properties is also a fundamental criterion for comparison of compression algorithms. The integrated densities of the objects in the 25 original images were compared with the corresponding integrated densities from the images compressed/uncompressed with PMT and with JPEG. This study was carried out in three stages:

- Detection of objects in the original image, and in the reconstructed image, and calculation of the integrated densities. This stage of the processing therefore gives a list of objects characterized by (x_o, y_o, d_o), with (x_o, y_o) the coordinates of the barycenter, and d_o the logarithm of the integrated density. Similarly, (x_r, y_r, d_r) represents the list of objects detected under similar conditions in the reconstructed image.
- Magnitude calibration of the original image and of the reconstructed image.
- Calculation of the error in the logarithm of the integrated density, by magnitude interval.

Each detected object is associated with its nearest neighbor in the catalog, according to the following rule, and we assign the corresponding catalog magnitude, M_c, to the detected object: a detected object, (x, y), is associated with the closest catalog object (x_c, y_c) subject to their distance being less than or equal to 3 pixels. This finally provides two object lists: (x_o, y_o, d_o, Mc_o), for the original image, and (x_r, y_r, d_r, Mc_r) for the reconstructed image. In a sim-

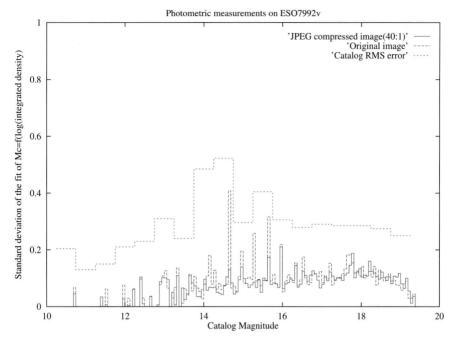

Fig. 5.7. Comparison of the calibration error, by 0.0625 magnitude intervals, between the uncompressed image using JPEG, the original image, and the reference catalog.

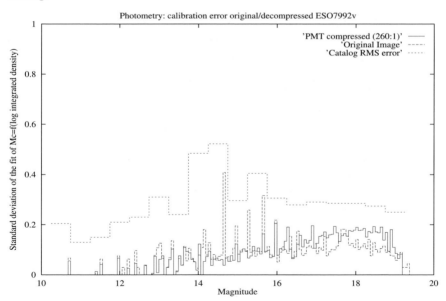

Fig. 5.8. Comparison of the calibration error, by 0.0625 magnitude intervals, between the uncompressed image using PMT, the original image, and the reference catalog.

158   5. Image Compression

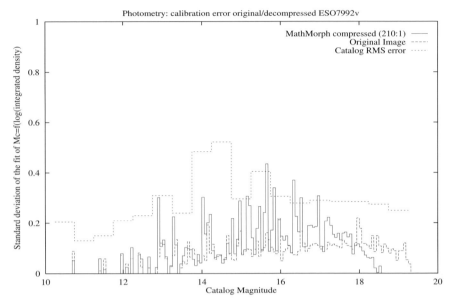

Fig. 5.9. Comparison of the calibration error, by 0.0625 magnitude intervals, measured on the uncompressed image using MathMorph, the original image, and the reference catalog.

ilar manner, the magnitude and logarithm of the integrated density association curves, $Mc_r = f(d_r)$, are studied for the JPEG- and PMT-reconstructed images. To verify the stability of the photometric values in spite of the compression, we hope to obtain curves, and thus to calibrate the reconstructed images, and to find dispersion around an average position which stays close to the dispersion obtained on the calibration curve of the original image. In fact, for varied compression methods, a systematic lowering of integrated densities of images can be noted (Dubaj, 1994), which results in the average calibration function (fitted by an order 3 polynomial) being slightly translated relative to the calibration function of the original image. To estimate the behavior of the dispersion of the calibration curve for both compression methods, we proceeded thus:

– Approximation by polynomial (degree 3) regression of the calibration function. $Mc = f(d)$.
– Calculation of the mean calibration error by magnitude interval, for the set of objects detected in the 25 subimages, i.e. about 2000 objects in all.

Thus we measure the photometric stability of the objects following compression, relative to their representation in the original image. The corresponding error curves are shown in Figs. 5.7, 5.8 and 5.9. The JPEG curve shows a slight increase for magnitudes above 18, and a smoothing effect for brighter objects between 14 and 16. For PMT, an increase in dispersion is noticed for

Table 5.4. Compression of a 1024 × 1024 integer-2 image. Platform: Sun Ultra-Enterprise; 250 MHz and 1 processor. Artifact asks whether or not prominent artifacts are produced. Progressive transmissions points to availability of code.

	Comp. time (sec)	Decomp. time (sec)	Artifact	Comp. ratio	Progressive transmission
JPEG	1.17	4.7	Y	<40	Y (in C)
Wavelet	45	7.1	Y	270	Y
Fractal	18.3	9	Y	< 30	N
Math. Morpho.	13	7.86	N	< 210	N
Hcompress	3.29	2.82	Y	270	Y (in C)
Hcompress + iter rec	3.29	77	N	270	N
PMT	7.8	3.1	N	270	Y (in Java)

high magnitudes, which corresponds to the problem of the detection of faint objects. Lowering the detection threshold from 4σ to 3σ does not change this. We note that the number of intervals below 14 is too small to allow for interpretation of the behavior of very bright objects. Even if PMT brings about greater degradation in the photometry of objects, especially when the objects are faint, the errors stay close to that of the catalog, and as such are entirely acceptable. Of course we recall also that the compression rate used with PMT is 260:1, compared to 40:1 for JPEG.

5.3.5 Computation Time

Table 5.4 presents the computation time required for compression and decompression on a specific platform (Sun Ultra-Enterprise, 250 MHz and 1 processor). With the JPEG, wavelet, and fractal methods, the time to convert our integer-2 FITS format to a one-byte image is not taken into account. Depending on the applications, the constraints are not the same, and this table can help in choosing a method for a given project. The last column indicates if software already exists for progressive image transmission.

Thinking from the point of view of managing a web-based image server, we would like to compare the performances of the different packages considering two scenarios:

160   5. Image Compression

– Archive original and compressed images and distribute both on demand.
– Compress the data before transferring them and let the end-user decompress them at the client side.

This latter situation has been studied and is illustrated in Fig. 5.10. Considering a network rate of 10kbits/second and an image of 2 MBytes, we measured the time necessary to compress, transmit and decompress the image. Methods are ordered from top to bottom according to increasing visual quality of the decompressed image. If we consider 20 seconds to be the maximum delay the end-user can wait for an image to be delivered, only HCOMPRESS and PMT succeed, with less artifacts for PMT.

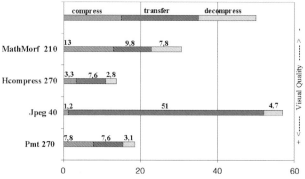

Fig. 5.10. Comparison of the overall time for compression, transmission and decompression for distribution of astronomical images using the web: the network rate is supposed to be 10kbits/second, and the image size is 2MBytes (1024 × 1024 × 2 bytes). The best-preserving codecs with respect to visual quality are shown at the bottom of the graph.

5.3.6 Conclusion

The MathMorph method reproduces the image pixels up to a given threshold. The quality of the image depends on the estimate of the noise standard deviation before the application of MathMorph transformations. The method has good performance on uncrowded astronomical fields. When a crowded field or an extended object is present in the image, the compression rate becomes much lower than the one obtained with the pyramidal median transform and, with traditional estimation of noise standard deviation, the faint extensions of objects and faint objects are lost in the compression.

The PMT method provides impressive compression rates, coupled with acceptable visual quality. This is due to the progressive noise suppression at successive scales. Nevertheless, on some crowded regions the PMT cannot compress more than 50:1, because much object information is to be coded in few image scales.

This method is robust and can allow for certain image imperfections. On a Sun Ultra-Enterprise (250 Mhz, 1 processor), compressing a 1024×1024 image takes about 8 seconds (CPU time), with subsequent very fast decompression.

The decomposition of the image into a set of resolution scales, and furthermore the fact they are in a pyramidal data structure, can be used for effective transmission of image data (Percival and White, 1996). Some work on web progressive image transmission capability has used bit-plane decomposition (Lalich-Petrich et al., 1995). Using resolution-based and pyramidal transfer and display with web-based information transfer is a further step in this direction.

5.4 Lossless Image Compression

5.4.1 Introduction

The compression methods looked at above involve filtering of information which is not considered to be of use. This includes what can demonstrably (or graphically) be shown to be noise. Noise is the part of information which is non-compressible, so that the residual signal is usually very highly compressible. It may be feasible to store such noise which has been removed from images on backing data store, but clearly the access to such information is less than straightforward.

If instead we seek a compression method which is guaranteed not to destroy information, what compression ratios can be expected?

We note firstly that quantization of floating (real) values necessarily involves some loss, which is avoided if we work in integer arithmetic only. This is not a restrictive assumption. Appropriate rescaling of image values may be availed of.

Next we note that the lifting scheme (Sweldens and Schröder, 1996) provides a convenient algorithmic framework for many wavelet transforms. The low-pass and band-pass operations are replaced by predictor and update operators at each resolution level, in the construction of the wavelet transform. When the input data consist of integer values, the wavelet transform no longer consists of integer values, and so we redefine the wavelet transform algorithm to face this problem. The predictor and update operators use, where necessary, a floor truncation function. The lifting scheme formulas for prediction and updating allow this to be carried out with no loss of information. If this had been done in the usual algorithmic framework, truncation would entail some small loss of information.

5.4.2 The Lifting Scheme

The lifting scheme (Sweldens and Schröder, 1996) is a flexible technique that has been used in several different settings, for easy construction and imple-

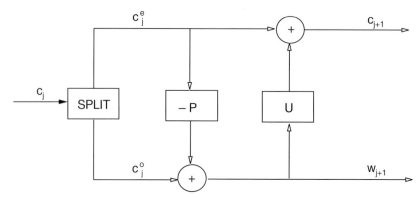

Fig. 5.11. The lifting scheme – forward direction.

mentation of traditional wavelets (Sweldens and Schröder, 1996), and of second generation wavelets (Sweldens, 1997) such as spherical wavelets (Schröder and Sweldens, 1995).

Its principle is to compute the difference between a true coefficient and its prediction:

$$w_{j+1,l} = c_{j,2l+1} - \mathcal{P}(c_{j,2l-2L}, ..., c_{j,2l-2}, c_{j,2l}, c_{j,2l+2}, ..., c_{j,2l+2L}) \quad (5.7)$$

A pixel at an odd location $2l + 1$ is then predicted using pixels at even locations.

The transformation is done in three steps:

1. **Split:** Split the signal into even and odd number samples:
$$\begin{aligned} c^e_{j,l} &= c_{j,2l} \\ c^o_{j,l} &= c_{j,2l+1} \end{aligned} \quad (5.8)$$

2. **Predict:** Calculate the wavelet coefficient $w_{j+1,l}$ as the prediction error of $c^o_{j,l}$ from $c^e_{j,l}$ using the prediction operator $\mathcal{P}$:
$$w_{j+1,l} = c^o_{j,l} - \mathcal{P}(c^e_{j,l}) \quad (5.9)$$

3. **Update:** The coarse approximation c_{j+1} of the signal is obtained by using $c^e_{j,l}$ and $w_{j+1,l}$ and the update operator $\mathcal{U}$:
$$c_{j+1,l} = c^e_{j,l} + \mathcal{U}(w_{j+1,l}) \quad (5.10)$$

The lifting steps are easily inverted by:

$$\begin{aligned} c_{j,2l} &= c^e_{j,l} = c_{j+1,l} - \mathcal{U}(w_{j+1,l}) \\ c_{j,2l+1} &= c^o_{j,l} = w_{j+1,l} + \mathcal{P}(c^e_{j,l}) \end{aligned} \quad (5.11)$$

Some examples of wavelet transforms via the lifting scheme are:

- Haar wavelet via lifting: the Haar transform can be performed via the lifting scheme by taking the predict operator equal to the identity, and an update operator which halves the difference. The transform becomes:

$$w_{j+1,l} = c^o_{j,l} - c^e_{j,l}$$
$$c_{j+1,l} = c^e_{j,l} + \frac{w_{j+1,l}}{2}$$

All computation can be done in place.
- Linear wavelets via lifting: the identity predictor used before is correct when the signal is constant. In the same way, we can use a linear predictor which is correct when the signal is linear. The predictor and update operators are now:

$$\mathcal{P}(c^e_{j,l}) = \frac{1}{2}(c^e_{j,l} + c^e_{j,l+1})$$
$$\mathcal{U}(w_{j+1,l}) = \frac{1}{4}(w_{j+1,l-1} + w_{j+1,l})$$

It is easy to verify that:

$$c_{j+1,l} = -\frac{1}{8}c_{j,2l-2} + \frac{1}{4}c_{j,2l-1} + \frac{3}{4}c_{j,2l} + \frac{1}{4}c_{j,2l+1} - \frac{1}{8}c_{j,2l+2}$$

which is the bi-orthogonal Cohen-Daubechies-Feauveau (1992) wavelet transform.

The lifting factorization of the popular (9/7) filter pair leads to the following implementation (Daubechies and Sweldens, 1998):

$$s^{(0)}_l = c_{j,2l}$$
$$d^{(0)}_l = c_{j,2l+1}$$
$$d^{(1)}_l = d^{(0)}_l + \alpha(s^{(0)}_l + s^{(0)}_{l+1})$$
$$s^{(1)}_l = s^{(0)}_l + \beta(d^{(1)}_l + d^{(1)}_{l-1})$$
$$d^{(2)}_l = d^{(1)}_l + \gamma(s^{(1)}_l + s^{(1)}_{l+1})$$
$$s^{(2)}_l = s^{(1)}_l + \delta(d^{(2)}_l + d^{(2)}_{l-1})$$
$$c_{j+1,l} = u s^{(2)}_l$$
$$c_{j+1,l} = u d^{(2)}_l \tag{5.12}$$

with

$$\alpha = -1.586134342$$
$$\beta = -0.05298011854$$
$$\gamma = 0.8829110762$$
$$\delta = 0.4435068522$$
$$u = 1.149604398 \tag{5.13}$$

Every wavelet transform can be written via lifting.

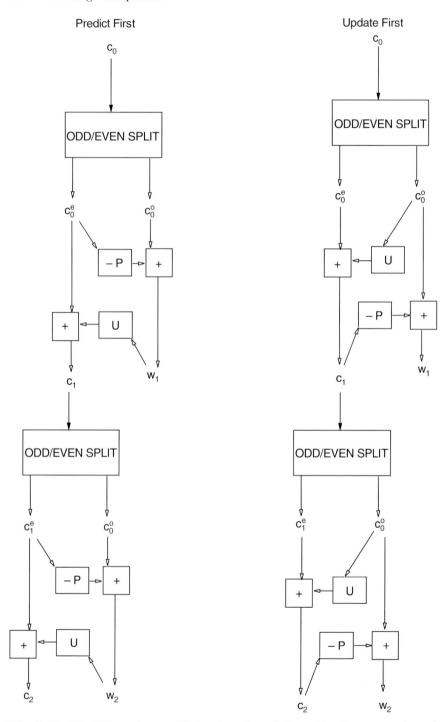

Fig. 5.12. The lifting scheme with two iterations: left, predict-first approach, and right, update-first approach.

Adaptive Wavelet Transform. Adaptivity can be introduced in the wavelet transform by reversing the order of the predict and the update steps in the lifting scheme (Claypoole et al., 2000). When the update operator is first applied, the prediction is based on the low-pass coefficients that are computed as in the standard wavelet transform. In the update-first approach, as illustrated in Fig. 5.12, the detail coefficients are not in the loop for calculating the coefficients at a coarser scale. Hence, we can start the prediction process at the coarser scale, and working from coarse to fine scales. The idea is now to make the predictor data-dependent. The prediction operator is chosen, based on the local properties of the data at a coarser scale. If a feature is detected, the order of the predictor is reduced, while if the data are smooth, a larger order is preferred.

Integer Wavelet Transform. When the input data are integer values, the wavelet transform no longer consists of integers. For lossless coding, it is useful to have a wavelet transform which produces integer values. We can build an integer version of every wavelet transform (Calderbank et al., 1998). For instance, denoting $\lfloor x \rfloor$ as the largest integer not exceeding x, the integer Haar transform (also called "S" transform) can be calculated by:

$$\begin{aligned} w_{j+1,l} &= c_{j,l}^o - c_{j,l}^e \\ c_{j+1,l} &= c_{j,l}^e + \lfloor \frac{w_{j+1,l}}{2} \rfloor \end{aligned} \qquad (5.14)$$

while the reconstruction is

$$\begin{aligned} c_{j,2l} &= c_{j+1,l} - \lfloor \frac{w_{j+1,l}}{2} \rfloor \\ c_{j,2l+1} &= w_{j+1,l} + c_{j,2l} \end{aligned} \qquad (5.15)$$

More generally, the lifting operators for an integer version of the wavelet transform are:

$$\begin{aligned} \mathcal{P}(c_{j,l}^e) &= \lfloor \sum_k p_k c_{j,l-k}^e + \frac{1}{2} \rfloor \\ \mathcal{U}(w_{j+1,l}) &= \lfloor \sum_k u_k w_{j+1,l-k} + \frac{1}{2} \rfloor \end{aligned} \qquad (5.16)$$

The linear integer wavelet transform is

$$\begin{aligned} w_{j+1,l} &= c_{j,l}^o - \lfloor \frac{1}{2}(c_{j,l}^e + c_{j,l+1}^e) + \frac{1}{2} \rfloor \\ c_{j+1,l} &= c_{j,l}^e + \lfloor \frac{1}{4}(w_{j+1,l-1} + w_{j+1,l}) + \frac{1}{2} \rfloor \end{aligned} \qquad (5.17)$$

Even if there is no filter that consistently performs better than all the other filters on all images, the linear integer wavelet transform performs generally better than others (Calderbank et al., 1998). More filters can be found in (Calderbank et al., 1998).

5.4.3 Comparison

For comparison purposes, we use JPEG in lossless mode. We also use the standard Unix gzip command, which implements the Lempel-Ziv run-length encoding scheme, especially widely-used for text data. As before, we base our experiments on image data from a Schmidt photographic plate in the region of M5 (numbered ESO 7992v), scanned with the CAI-MAMA facility. We use subimages of dimensions 1024×1024. Sampling is 0.666 arcseconds per pixel. Ancillary catalog information was available for our tests. The images are representative of those used in Aladin, the reference image service of the CDS (Strasbourg Data Centre, Strasbourg Observatory).

Table 5.5. Compression of a 1024×1024 integer-2 image. Platform: Sun Ultra-Sparc 250 MHz and 2 processors.

Software	CPU time (sec) Compression time	Decompression time	Compression ratio
JPEG lossless	2.0	0.7	1.6
Lifting scheme with Haar	4.3	4.4	1.7
Gzip (Unix)	13.0	1.4	1.4

Table 5.5 shows our findings, where we see that the integer and lifting scheme wavelet transform approach is a little better than the other methods in terms of compression ratio. Not surprisingly, the compression rate is not large for such lossless compression. Routine `mr_lcomp` (MR/1, 2001) was used for this work. Lossless JPEG suffers additionally from rounding errors.

We now turn attention to usage of a lossless wavelet-based compressor, above and beyond the issues of economy of storage space and of transfer time. The lifting scheme implementation of the Haar wavelet transform presents the particularly appealing property that lower resolution versions of an image are *exactly* two-fold rebinned versions of the next higher resolution level. For aperture photometry and other tasks, lower level resolution can be used to provide a partial analysis. A low resolution level image can be used scientifically since its "big" pixels contain the integrated average of flux covered by them.

The availability of efficiently delivered low resolution images can thus be used for certain scientific objectives. This opens up the possibility for an innovative way to analyze distributed image holdings.

5.5 Large Images: Compression and Visualization

5.5.1 Large Image Visualization Environment: LIVE

With new technology developments, images furnished by detectors are larger and larger. For example, current astronomical projects are beginning to deal with images of sizes larger than 8000 by 8000 pixels (VLT: 8k × 8k, MegaCam 16k × 16k, etc.). A digitized mammogram film may lead to images of about 5k × 5k. Analysis of such images is obviously not easy, but the main problem is clearly that of archiving and network access to the data by users.

In order to visualize an image in a reasonable amount of time, transmission has to be based on two concepts:

– data compression
– progressive decompression

With very large images, a third concept is necessary, which is the region of interest. Images are becoming so large it is impossible to display them in a normal window (typically of size 512 × 512), and we need to have the ability to focus on a given area of the image at a given resolution. To move from one area to another, or to increase the resolution of a part of the area is a user task, and is a new active element of the decompression. The goal of LIVE is to furnish this third concept.

An alternative to LIVE for extracting a region of the image would be to let the server extract the selected area, compress it, and send it to the user. This solution is simpler but gives rise to several drawbacks:

– Server load: the server must decompress the full size image and re-compress the selected area on each user request.
– Transfer speed: to improve the resolution of a 256 × 256 image to a 512 × 512 image, the number of bits using the LIVE strategy is of the order of 10 to 100 less than if the full 512 × 512 image is transferred. The lower the compression ratio, the more relevant the LIVE strategy.

The principle of LIVE is to use the previously described technology, and to add the following functionality:

– Full image display at a very low resolution.
– Image navigation: the user can go up (the quality of an area of the image is improved) or down (return to the previous image). Going up or down in resolution implies a four-fold increase or decrease in the size of what is viewed.

Fig. 5.13 illustrates this concept. A large image (say 4000 × 4000), which is compressed by blocks (8 × 8, each block having a size of 500 × 500), is represented at five resolution levels. The visualization window (of size 256 × 256 in our example) covers the whole image at the lowest resolution level (image size 250 × 250), but only one block at the full resolution (in fact between

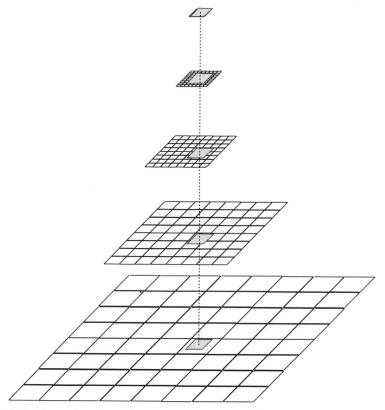

Fig. 5.13. Example of large image, compressed by block, and represented at five resolution levels. At each resolution level, the visualization window is superimposed at a given position. At low resolution, the window covers the whole image, while at the full resolution level, it covers only one block.

one and four, depending on the position in the image). The LIVE concept consists of moving the visualization window into this pyramidal structure, without having to load into memory the large image. The image is first visualized at low resolution, and the user can indicate (using the mouse) which part of the visualized subimage he wishes to zoom on. At each step, only wavelet coefficients of the corresponding blocks and of the new resolution level are decompressed.

5.5.2 Decompression by Scale and by Region

Support of the transfer of very large images in a networked (client-server) setting requires compression and prior noise separation. In addition, progressive transfer may be supported, or delivery by scale and by image region. For such additional functionality, multiscale transform based methods are very attractive because they integrate a multiresolution concept in a natural way. The LIVE prototype (MR/1, 2001), which is Java-based at the client end, allows access to differing resolution levels as well as block-sized regions of the compressed image data.

A prototype has also been developed (Gastaud et al., 2001) allowing the visualization of large images with the SAO DS9 software (available at http://hea-www.harvard.edu/RD/ds9).

Computation Time. We examined compression performance on large numbers of astronomy images. Consider for example a 12451×8268 image from the CFH12K detector at the CFHT (Canada-France-Hawaii Telescope), Hawaii. A single image is 412 MB. As astronomy detectors tend towards 16000×16000 in image dimensions – the case of the UK's Vista telescope now being designed for operation in Chile for instance – it is clear that compression and delivery technologies are very much needed. A typical observing night gives rise to terabytes of data, and image repositories are measured in petabytes.

Using denoising compression, we compressed the CFH12K image to 4.1 MB, i.e. less than 1% of its original size. Compression took 789 seconds on an Ultra-Sparc 10. Decompression to the fifth resolution scale (i.e., dimensions divided by 2^5) took 0.43 seconds. For rigorously lossless compression, compression to 97.8 MB, i.e. 23.75% of the original size, took 224 seconds, and decompression to full resolution took 214 seconds. Decompression to full resolution by block was near real-time.

5.5.3 The SAO-DS9 LIVE Implementation

A user interface was developed (Gastaud et al., 2001) for images compressed by the software package MR/1 (2001), which comes as a plug-in to the widely-used astronomical image viewer SAO-DS9 (Joye and Mandel, 2000). This interface allows the user to load a compressed file and to choose not only the scale, but also the size and the portion of image to be displayed, resulting in reduced memory and processing requirements. Astrometry and all SAO-DS9 functionality are still simultaneously available.

The sources of the interface in Tcl/Tk and the binaries for the decompression (for Unix and Windows operating systems) are available to the astronomical community.

– Compression: The compression and decompression tools are part of the MR1 package (MR/1, 2001). Wavelet, Pyramidal Median, and lifting scheme are implemented, with lossy or lossless options. The final file is stored in a proprietary format (with the extension .fits.MRC). The decompression module is freely available.
– Image viewer: They are many astronomical image viewers. We looked at JSky (because it is written in Java) and SAOImage-DS9. The latter was selected: it is well maintained, and for the programmer it is simpler. DS9

170   5. Image Compression

is a Tk/Tcl application which utilizes the SAOTk widget set. It also incorporates the new X Public Access (XPA) mechanism to allow external processes to access and control its data, and graphical user interface functions.
– Interface: DS9 supports external file formats via an ASCII description file. It worked with the MRC format, but it enables only one scale of the image to be loaded. The selected solution was a Tcl/Tk script file which interacts with XPA. Tcl/Tk is recommended by the SAO team and is free and portable.
This interface enables the user to
– select a file,
– select the maximum size of the displayed window,
– zoom on a selected region (inside the displayed window), and
– unzoom.

The Tcl/Tk script file with DS9 and the decompressed module has been used on Solaris (Sun Microsystems Sparc platform), Linux (Intel PC platform) and Windows NT, 2000 (with some tuning), and can also work on HP-UX, ALPHA-OSF1. On a 3-year old PC, the latency is about one second.

How to Use It. The user first launches DS9, and then xlive.tcl. The XLIVE window (see Fig. 5.14) has four buttons and a text field.

1. **Max Window Size**: this is the maximum size of the displayed window (FITS image given to DS9), and can be changed by the user.
2. **File**: the user selects a compressed image (extension .fits.MRC) The complete image at the lowest resolution appears in the DS9 window (see Fig. 5.14)
3. **Up Resol**: this button zooms the image. If the new image is bigger than the Max Window Size, select the center of the Region of Interest in the DS9 window. Validate by the "OK" button.
4. **Down Resol**: this button unzooms the image.
5. **Exit**: this button quits the program, deleting the temporary files.

Simultaneously all DS9 functionality is availability.

5.6 Hyperspectral Compression for Planetary Space Missions

In recent years, hyperspectral data have come very much to the fore, and here we discuss compression of such data, based on a multiresolution transform.

Recent planetary and future space missions carry or will carry imaging spectrometers, e.g. OMEGA on board Mars-Express, VIMS on board Cassini, VIRTIS on board Rosetta, and Venus-Express and Simbio-Sys on

5.6 Hyperspectral Compression for Planetary Space Missions 171

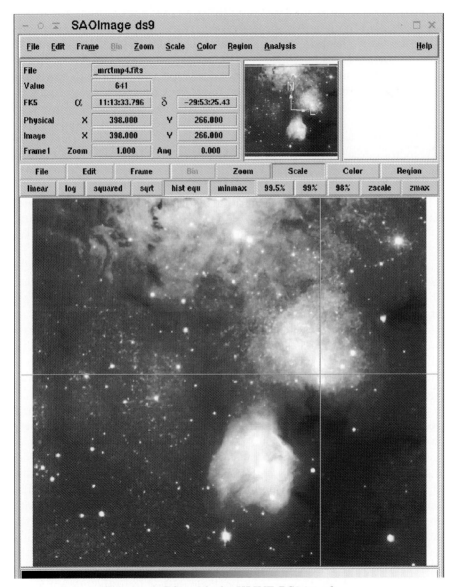

Fig. 5.14. DS9 with the XLIVE-DS9 interface.

board BepiColombo. These types of instruments, working mainly in the visible and infrared domain, furnish hyperspectral cubes of data, two of the dimensions representing the image and the third the spectral response of the imaged scene. For example the VIRTIS imaging spectrometer on board Rosetta (a European Space Agency cornerstone aiming to study the comet Churyomov-Gerasimenko) will acquire cubes of dimensions 128×800 for the spatial directions by 128 spectral channels. Since the on-board memory and computing capabilities are limited on this kind of spatial mission, the hyperspectral cube is obtained by successive acquisition of two-dimensional frames. Moreover since the data downlink rate is generally low and the amount of data to transmit becomes higher with the increasing performances of the instruments, efficient data compression techniques in terms of both speed and SNR are necessary. Lossy compression based on the wavelet transform meets these two requirements and the efficient SPIHT (set partition in hierarchical trees) algorithm (Said and Pearlman, 1996) has become very popular.

The data cube should be compressed frame by frame. In this context two possibilities exist: either the compression is achieved from the spatial images and repeated along the spectral direction, or the compression is applied to the frames having one spectral direction and one spatial direction, i.e. by slice. The latter corresponds to the acquisition mode of grating spectrometers such as OMEGA and VIRTIS.

It has been shown (Langevin and Forni, 2000) on AVIRIS $128 \times 128 \times 128$ 12-bit cube data from the ESA compression database, http://www.estec.esa.nl/tech/datacwg, that the second solution gives far better results. Applying a straightforward wavelet transform on each slice, the signal to noise ratio (SNR) gain was typically 2dB for a compression rate of 1 bit per data value.

However this approach is not satisfactory in terms of spatial/spectral crosstalk. Two contiguous spectra present a contrast of nearly 30% in overall brightness. As a result, the decompressed spectrum at 1 bit/datum of a given pixel is systematically lower than the uncompressed spectrum, with significant distortions of the spectral signatures.

A major improvement on this issue was obtained by (Langevin and Forni, 2000) using a specific wavelet transformation. Instead of performing the wavelet transform in the X and Y directions on the 128×128 unit, and then repeating three times the transform on the low-pass upper left region, a 4-level wavelet transform is performed on each of the 128 spatial lines, and then the same procedure is performed on each column of the result. The same tree coding search (Said and Pearlman, 1996) is implemented as with the nominal transform, notwithstanding the rectangular shape of most of the sub-bands. When performed on a standard image, this approach is slightly less efficient than the nominal transform but on spectral-spatial slices, the SNR increases by typically 4 to 5dB. Even more importantly, the bias on the decompressed spectrum cannot be distinguished from the uncompressed spectrum. The wavelet transform applied in this way is indeed an efficient decorrelator between the spatial and spectral components.

5.7 Chapter Summary

Storage and also delivery have been the main themes of this chapter. Integer wavelet transforms, using the lifting scheme, have been shown to be important for lossless compression. For lossy compression, perhaps surprisingly in view of the anticipated computation time, the Pyramidal Median Transform performs very well.

For large image delivery, a plug-in was provided for the SAO-DS9 image display system to support scale- and region of interest based on-the-fly decompression. This system decompresses and displays very large images – for example, images which are not far short of 2 GB in size – in near real-time.

For some projects, we need to achieve huge compression ratios, which cannot be obtained by current methods without introducing unacceptable distortions. For instance, it was shown (Dollet et al., 2004) that if we wish to extend the GAIA mission in order to make a high-spatial resolution all-sky survey in the visible based on a scanning satellite, then the main limitation is the amount of collected data to be transmitted. A solution is to introduce all our knowledge of both the sky and the instrument in order to compress only the difference between what we know and what we observe. Data simulations become a very important task and errors on the point spread functions, positions of stars, etc., must be under control (Dollet et al., 2004). We imagine that this concept could be extended to a new generation of astronomical image coders which exploit all astronomical ressources available (hence availing of the virtual observatory) and which apply this strategy (viz., sky simulation + difference coding).

Finally, we would like to point out that, to our knowledge, thorough study has yet to be carried out on the use of the new JPEG2000 image compression standard for astronomical data compression.

6. Multichannel Data

6.1 Introduction

A new generation of detectors produce multichannel data, i.e. a set of images taken with different filters.

The challenge for multichannel data filtering and restoration is to have a data representation which takes into account at the same time both the spatial and the spectral (or temporal) correlation. A three-dimensional transform based coding technique was proposed in (Saghri et al., 1995), consisting of a one-dimensional spectral Karhunen-Loève transform, KLT (Karhunen, 1947)) and a two-dimensional spatial discrete cosine transform (DCT). The KLT is used to decorrelate the spectral domain and the DCT is used to decorrelate the spatial domain. All images are first decomposed into blocks, and each block uses its own Karhunen-Loève transform instead of one single matrix for the whole image. Lee (1999) improved on this approach by introducing a varying block size. The block size is made adaptive using a quadtree and bit allocation for each block. The DCT transform can also be replaced by a wavelet transform (WT) (Epstein et al., 1992; Tretter and Bouman, 1995).

We introduce in section 6.2 the Wavelet-Karhunen-Loève transform (WT-KLT) (Starck and Querre, 2001) and show how to use it for noise removal. Decorrelating first the data in the spatial domain using the WT, and afterwards in the spectral domain using the KLT, allows us to derive robust noise modeling in the WT-KLT space, and hence to filter the transformed data in an efficient way. We show also that the correlation matrix can be computed by different methods taking noise modeling into account.

Multichannel data gives us also the opportunity to better understand the physical properties of an observed source. However, in many cases, we cannot isolate the source of interest, and each channel contains the superposition of the flux emitted at a given wavelength by all sources along the line of sight. If the source gives rise to different known spectral behaviors, it is possible to separate them very easily. When the spectral behaviors are unknown, it becomes a problem of *blind source separation* (BSS), which can be solved using independent component analysis (ICA). This consists of recovering unobserved signals or "sources" from several observed mixtures (Cardoso, 1998a). Assuming that n statistically independent signals $s_1(t), ..., s_n(t)$ are mixed by an unknown $n \times n$ mixing matrix $A = [a_{ij}]$, we have:

$$X(t) = AS(t) \tag{6.1}$$

where $S(t) = [s_1(t), \ldots, s_n(t)]^t$, and $X(t) = [x_1(t), \ldots, x_n(t)]^t$ represents the observed signals, with $x_i(t) = \sum_{j=1}^{n} a_{ij} s_j(t)$. The challenge is to find how to achieve separation using as our only assumption that the source signals are statistically independent. The solution consists of finding an $n \times n$ separating matrix B, $Y(t) = BS(t)$, such that Y is an estimate of S. This is achieved by minimizing contrast functions Φ, which are defined in terms of the Kullback-Leibler divergence K:

$$\Phi(Y) = \int p_Y(u) \log \frac{p_Y(u)}{\prod p_{Y_i}(u_i)} du \tag{6.2}$$

The mutual information, expressed by the Kullback-Leibler divergence, vanishes if and only if the variables Y_i are mutually independent, and is strictly positive otherwise. ICA has been used in astronomy to analyze multispectral images (Nuzillard and Bijaoui, 2000) of the galaxy 3C 120, to separate the Cosmic Microwave Background from other sky components (Baccigalupi et al., 2000a; Maino et al., 2002; Delabrouille et al., 2003; Patanchon et al., 2004a), and to study MARS data obtained using the OMEGA spectra (Bibring and OMEGA, 2004) on board Mars Express (Forni et al., 2005). As for PCA, it has been shown (Zibulevsky and Pearlmutter, 2001; Zibulevsky and Zeevi, 2001) that applying ICA to wavelet transformed signals leads to better quality results, especially in the presence of noise. Section 6.6 describes the ICA concept and the recent developments (SMICA,WSMICA methods) for the specific case of the CMB.

6.2 The Wavelet-Karhunen-Loève Transform

6.2.1 Definition

The Karhunen-Loève transform, also often referred to as eigenvector, Hotelling transform, or principal component analysis (PCA) (Karhunen, 1947; Levy, 1948; Hotelling, 1933) allows us to transform discrete signals into a sequence of uncorrelated coefficients. Considering a vector $D = d_1, ..., d_L$ of L signals or images of dimension N (i.e. N pixels per image), we denote $M = \{m_1, ..., m_L\}$ the mean vector of the population (m_i is the mean of the ith signal d_i). The covariance matrix C of D is defined by $C = (D - M)(D - M)^t$, and is of order $L \times L$. Each element $c_{i,i}$ of C is the variance of d_i, and each element $c_{i,j}$ is the covariance between d_i and d_j. The KLT method consists of applying the following transform to all vectors $x_i = \{d_1(i), ..., d_L(i)\}$ ($i = 1..N$):

$$y_i = \Lambda^{-\frac{1}{2}} A(x_i - M) \tag{6.3}$$

6.2 The Wavelet-Karhunen-Loève Transform

where Λ is the diagonal matrix of eigenvalues of the covariance matrix C, and A is a matrix whose rows are formed from the eigenvectors of C (Gonzalez and Woods, 1992), ordered following the monotonic decreasing order of eigenvalues.

Because the rows of A are orthonormal vectors, $A^{-1} = A^t$, and any vector x_i can be recovered from its corresponding y_i by:

$$x_i = \Lambda^{\frac{1}{2}} A^t y_i + M \tag{6.4}$$

The Λ matrix multiplication can be seen as a normalization. Building A from the correlation matrix instead of the covariance matrix leads to another kind of normalization, and the Λ matrix can be suppressed ($y_i = A(x_i - M)$ and $x_i = A^t y_i + M$). Then the norm of y will be equal to the norm of x.

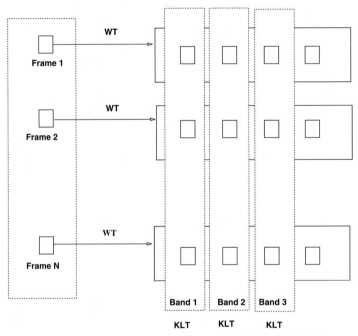

Fig. 6.1. WT-KLT transform data flow. Each frame (or band) of the input data set is first wavelet transformed, and a principal component analysis is applied at each resolution level.

We suppose now that we have L observations of the same view, e.g. at different wavelengths (or at different epochs, etc.), and denote as d_l one observation, $W^{(l)}$ its wavelet transform, and $w_{l,j,k}$ one wavelet coefficient at scale j and at position k. The standard approach would be to use an orthogonal wavelet transform, and to calculate the correlation matrix C from the wavelet coefficients instead of the pixel values:

178   6. Multichannel Data

$$C_{m,n} = \frac{\sum_{j=1}^{J} \sum_{k=1}^{N_j} w_{m,j,k} w_{n,j,k}}{\sqrt{\sum_{j=1}^{J} \sum_{k=1}^{N_j} w_{m,j,k}^2} \sqrt{\sum_{j=1}^{J} \sum_{k=1}^{N_j} w_{n,j,k}^2}} \quad (6.5)$$

where J is the number of bands, and N_j is the number of coefficients in the band j. In (Lee, 1999), a more complex approach was proposed, which is to decompose the images into N_b blocks and apply a KLT for each block separately. We investigate here different approaches for data restoration.

6.2.2 Correlation Matrix and Noise Modeling

We introduce a noise model into our calculation of the correlation matrix. If the input sequence D contains noise, then the wavelet coefficients are noisy too. Eigenvalues at the high scales are computed with noisy WT coefficients and we may lose the true underlying correlation that exists between the input images d_l. The expression of the correlation matrix has to be modified in order to allow us to take noise into account. We add a weighting term to each wavelet coefficient which depends on the signal-to-noise ratio. The correlation matrix is calculated by:

$$C_{m,n} = \frac{\sum_{j=1}^{J} \sum_{k=1}^{N_j} p_j(w_{m,j,k}) w_{m,j,k} p_j(w_{n,j,k}) w_{n,j,k}}{\sqrt{\sum_{j=1}^{J} \sum_{k=1}^{N_j} p_j^2(w_{m,j,k}) w_{m,j,k}^2} \sqrt{\sum_{j=1}^{J} \sum_{k=1}^{N_j} p_j^2(w_{n,j,k}) w_{n,j,k}^2}} \quad (6.6)$$

where p_j is a weighting function. The standard approach corresponds to the specific case where $p_j(w_m) = 1$ (no weighting). By considering that only wavelet coefficients with high signal-to-noise ratio should be used for the correlation matrix calculation, p_j can be defined by:

$$p_j(w) = \begin{cases} 1 & \text{if } w \text{ is significant} \\ 0 & \text{if } w \text{ is not significant} \end{cases} \quad (6.7)$$

and a wavelet coefficient w is said to be "significant" if its probability of being due to noise is smaller than a given ϵ value. In the case of Gaussian noise, it suffices to compare the wavelet coefficient w to a threshold level t_j. t_j is generally taken as $\lambda \sigma_j$, where σ_j is the noise standard deviation at scale j, and λ is chosen between 3 and 5. The value of $\lambda = 3$ corresponds to a probability of false detection of 0.27%, for a Gaussian statistic.

This hard weighting scheme may lead to problems if only a few coefficients are significant. This weighting scheme can be replaced by a soft weighting scheme, by defining $p_j(w)$ as:

$$p_j(w) = 1 - \text{Prob}(W > | w |) \quad (6.8)$$

where $\text{Prob}(W > | w |)$ is the probability that a wavelet coefficient is larger than w due to the noise. For Gaussian noise, we have:

$$p_j(w) = 1 - \frac{2}{\sqrt{2\pi}\sigma_j} \int_{|w|}^{+\infty} \exp(-W^2/2\sigma_j^2)dW$$

$$= \mathrm{erf}\left(\frac{|w|}{\sqrt{2}\sigma_j}\right) \tag{6.9}$$

6.2.3 Scale and Karhunen-Loève Transform

We can also analyze separately each band of the wavelet transform, and then apply one KLT per resolution level. This implies calculating a correlation matrix $C^{(j)}$ for each band j.

$$C_{m,n}^{(j)} = \frac{\sum_{k=1}^{N_j} p_j(w_{m,j,k})w_{m,j,k}p_j(w_{n,j,k})w_{n,j,k}}{\sqrt{\sum_{k=1}^{N_j} p_j^2(w_{m,j,k})w_{m,j,k}^2}\sqrt{\sum_{k=1}^{N_j} p_j^2(w_{n,j,k})w_{n,j,k}^2}} \tag{6.10}$$

This has the advantage of taking into account more complex behavior of the signal. Indeed, structures of different sizes may have a different spectral behavior (for example, stars and galaxies in astronomical images), and a band-by-band independent analysis allows us to better represent this kind of data.

6.2.4 The WT-KLT Transform

The final WT-KLT algorithm has these steps:

1. Estimate the noise standard deviation $\sigma^{(l)}$ of each input data set d_l.
2. Calculate the wavelet transform $W^{(l)}$ of each input data set d_l.
3. For each band j of the wavelet transform, calculate the correlation matrix $C^{(j)}$ relative to the vector $x_j = \left\{W_j^{(1)}, W_j^{(2)}, ..., W_j^{(L)}\right\}$, where $W_j^{(l)}$ represents band j of the wavelet transform $W^{(l)}$ of d_l.
4. For each band j, diagonalize the matrix $C^{(j)}$ and build the transform matrix A_j from the eigenvectors of $C^{(j)}$.
5. For each band j and each position k, apply the matrix A_j to the vector $x_{j,k} = \{w_{1,j,k}, w_{2,j,k}, ..., w_{L,j,k}\}$:

$$y_{j,k} = A_j x_{j,k} \tag{6.11}$$

6. The WT-KLT coefficients $c_{l,j,k}$ are derived from $y_{j,k}$ by $c_{l,j,k} = y_{j,k}(l)$. The l index in the transformed coefficients no longer represent the observation number, but the eigenvector number. $l = 1$ indicates the principal eigenvector while $l = L$ indicates the last one.

The mean vector M disappears in this algorithm because the wavelet coefficients are zero mean.

Fig. 6.1 shows the data flow of the WT-KLT transform.

6.2.5 The WT-KLT Reconstruction Algorithm

The reconstruction algorithm has these steps:

1. For each band j and each position k, we apply the matrix A_j^t to the vector $y_{j,k} = \{c_{1,j,k}, c_{2,j,k}, ..., c_{L,j,k}\}$

$$x_{j,k} = A_j^t y_{j,k} \tag{6.12}$$

2. The wavelet coefficients $w_{l,j,k}$ are derived from $x_{j,k}$ by $w_{l,j,k} = x_{j,k}(l)$.
3. An inverse wavelet transform of $W^{(l)}$ furnishes d_l.

6.3 Noise Modeling in the WT-KLT Space

Since a WT-KLT coefficient c is obtained by two successive linear transforms, broadly applicable noise modeling results from this, and as a consequence determining the noise standard deviation associated with the c value is tractable. In the case of Poisson noise, the Anscombe transformation can first be applied to the data (see Chapter 2). This implies that for the filtering of a data set with Poisson noise or a mixture of Poisson and Gaussian noise, we will first pre-transform the data D into another data set $\mathcal{A}(D)$ with Gaussian noise. Then $\mathcal{A}(D)$ will be filtered, and the filtered data will be inverse-transformed. For other kinds of noise, modeling must be performed in order to define the noise probability distribution of the wavelet coefficients (Starck et al., 1998a). In the following, we will consider only stationary Gaussian noise.

Noise Level on WT-KLT Coefficients

Assuming a Gaussian noise standard deviation σ_l for each signal or image d_l, the noise in the wavelet space follows a Gaussian distribution $\sigma_{l,j}$, j being the scale index. For a bi-orthogonal wavelet transform with an L^2 normalization, $\sigma_{l,j} = \sigma_l$ for all j. Since the WT-KLT coefficients are obtained from a linear transform, we can easily derive the noise standard deviation relative to a WT-KLT coefficient from the noise standard deviation relative to the wavelet coefficients. In fact, considering the noise standard deviation vector $s = \{\sigma_1, ..., \sigma_L\}$, we apply the following transformation:

$$y_j = A_j^2 s^2 \tag{6.13}$$

and the noise standard deviation relative to a WT-KLT coefficient $c_l(j,k)$ is $\sqrt{y_j(l)}$.

6.4 Multichannel Data Filtering

6.4.1 Introduction

KLT-based filtering methods were proposed in the past (Andrews and Patterson, 1976; Lee, 1991; Konstantinides et al., 1997) for single images. The idea proposed was to decompose the image I of $M \times N$ pixels into non-overlapping blocks B_s of size $N_b \times N_b$. Typically, N_b takes values from 4 to 16. Let $\lambda_1, \lambda_2, ... \lambda_n$ be the singular values of the matrix I in decreasing order, and assuming that the matrix I is noisy, the rank r of I was defined such that (Konstantinides and Yao, 1988)

$$\lambda_r \geq \epsilon_1 > \lambda_{r+1} \tag{6.14}$$

where ϵ_1 is the norm of the noise. In the case of a Gaussian distribution of zero mean, an upper bound is $\sqrt{MN}\sigma$ (Konstantinides et al., 1997). A filtered version of I can be obtained by reconstructing each block only from its first r eigenvalues. A novel approach was developed in (Natarajan, 1995; Konstantinides et al., 1997) in order to find the optimal ϵ value, based on the ϵ-compression ratio curve, using a lossless compression method like JPEG. It was found that the maximum of the second derivative of the curve furnishes the optimal ϵ value (Natarajan, 1995).

In the case of multichannel data filtering, several different approaches may be considered based on noise modeling. They are presented in this section, and evaluated in the next section.

6.4.2 Reconstruction from a Subset of Eigenvectors

The WT-KLT transform of a data set $D = \{d_1, ..., d_L\}$ consists of applying a KLT on the wavelet scales. Hence, the vector $W_j = \left\{ W_j^{(1)}, ..., W_j^{(L)} \right\}$ of the scales j of the wavelet transforms W^l can be decomposed uniquely as:

$$W_j = U_j \Lambda_j^{\frac{1}{2}} V_j^{-1} = \sum_{i=1}^{L} \sqrt{\lambda_{j,i}} u_{j,i} v_{j,i}^t \tag{6.15}$$

where Λ_j is the diagonal matrix of eigenvalues of the correlation matrix C_j, U_j and V_j are orthogonal matrices with column vectors $u_{j,i}$ and $v_{j,i}$ which are respectively the eigenvectors of $W_j W_j^t$ and $W_j^t W_j$.

The filtered wavelet coefficients of band j can be obtained by:

$$\tilde{W}_j = \sum_{i=1}^{r} \sqrt{\lambda_{j,i}} u_{j,i} v_{j,i}^t \tag{6.16}$$

where r is the rank of the matrix.

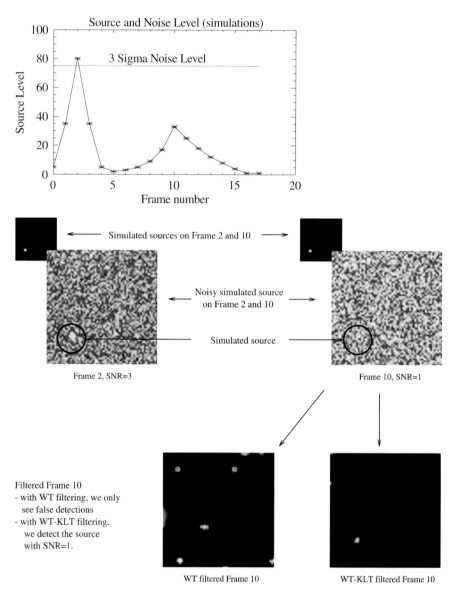

Fig. 6.2. Simulation: the dataset is composed of 18 frames. Each of them contains a source (small Gaussian) at the same position, but at different intensity levels. Top, plot of the source maximum value versus the frame number. Middle, frames 2 and 10, and bottom, filtered version of frame 10 by the wavelet transform and wavelet Karhunen-Loève transform.

6.4.3 WT-KLT Coefficient Thresholding

Hard thresholding can be applied to the WT-KLT coefficients in a fashion analogous to the thresholding of wavelet coefficients.

6.4.4 Example: Astronomical Source Detection

Fig. 6.2 shows a simulation. We created a dataset of 18 frames, each containing a source at the same position, but at different intensity levels. The source is a small Gaussian, and the source SNR is defined as the ratio between the maximum of the source and the noise standard deviation. Fig. 6.2, top, shows the evolution of the SNR in the 18 frames. Frames 2 and 10 are shown in Fig. 6.2, middle, left and right. The source SNR ratio is respectively 3 and 1. Fig. 6.2, bottom, left and right, show respectively the filtered frame 10 using the wavelet transform and the WT-KLT. The WT detects only noise, while the WT-KLT clearly identifies the source.

6.5 The Haar-Multichannel Transform

We have seen in Chapter 2 that the Haar transform has some advantages, especially when the data contains Poisson noise.

In order to decorrelate the information both spatially and in wavelength (or in time), a 2D-wavelet transform must first be performed on each frame of the data cube $D(x, y, z)$. We denote the result $w_{j,k_x,k_y,z}$, where j is the scale index ($j \in [1, J]$, where J is the number of scales), k_x, k_y the spatial position in the scale ($k_x \in [0, N_x - 1]$, $k_y \in [0, N_y - 1]$), and z is the frame number. This set must again be transformed in order to decorrelate the information in the third dimension. We apply a 1D-transform on each vector $w_{j,k_x,k_y,z}$, and we get a new set of data u_{j,j',k_x,k_y,k_z}, where j' and k_z are respectively the scale index and position in the third dimension.

Using the unnormalized Haar transform, a coefficient $u_{j+1,j'+1,k,l,t}$ can be written as:

$$u_{j+1,j'+1,k_x,k_y,k_z} = \sum_{i=2^{j'+1}k_z}^{2^{j'}k_z+2^{j'}-1} w_{j+1,k_x,k_y,i} - \sum_{i=2^{j'+1}k_z+2^{j'}}^{2^{j'+1}(k_z+1)-1} w_{j+1,k_x,k_y,i} \quad (6.17)$$

We assume now a constant background rate λ_i for each frame i of the data cube. Each Haar wavelet coefficient $w_{j+1,k_x,k_y,i}$ is the difference between two random variables X_i and Y_i, which follows a Poisson distribution of parameter $\lambda_{j,i}$, where $\lambda_{j,i}$ represents the rate per pixel over 2^{j+1} pixels of the ith frame, and is equal to $2^{2j}\lambda - i$. Then $u_{j+1,j'+1,k_x,k_y,k_z}$ is the difference of two variables, $X = \sum_i X_i$ and $Y = \sum_i Y_i$, and both follow a Poisson distribution of parameter $\sum_i \lambda_{j,i}$. The thresholding method described in section 2.6 can therefore be used, using the correct λ value.

Filtering. The multichannel filtering algorithm is:

1. For each frame $D(*,*,z)$, apply the 2D-Haar transform: we obtain $w_{j,k,l}(z)$.
2. For each scale j, and at each position k_x, k_y, extract the 1D vector $w_{j,k_x,k_y,*}$ and compute its wavelet transform. We get a new data set.
3. Threshold the coefficients using the correct λ value.
4. Inverse 1D transform.
5. Inverse 2D transform.

For non-constant background, a coarse-to-fine approach, as explained in section 2.6 in Chapter 2, can be used.

6.6 Independent Component Analysis

6.6.1 Definition

In multidimensional data processing, Blind Source Separation (BSS) may be necessary. The overall goal is to recover unobserved signals, images or *sources* S from mixtures of these sources X observed typically at the output of an array of sensors. The simplest mixture model takes the form:

$$X = AS \qquad (6.18)$$

where X and S are random vectors of respective sizes $m \times 1$, $n \times 1$, and A is an $m \times n$ matrix. The entries of S are assumed to be independent random variables. Multiplying S by A linearly mixes the n sources into m observed processes. Independent component analysis methods were developed to solve the BSS problem, i.e. given a batch of T observed samples of X, estimate the mixing matrix A and reconstruct the corresponding T samples of the source vector S, relying mostly on the statistical independence of the source processes. Independence is a strong, yet in many cases plausible, assumption that goes beyond the simple second order decorrelation obtained using PCA for instance. Due to obvious intrinsic indeterminacies in model (6.18), decorrelation is not enough to recover the source processes.

The probability distributions of the individual sources also need to be modeled but coarse assumptions can be made (Cardoso, 1998b; Hyvärinen et al., 2001). Still, algorithms for blind component separation and mixing matrix estimation do depend on the model used for the probability distribution of the sources (Cardoso, 2001). In a first set of techniques, source separation is achieved in a noiseless setting, based on the non-Gaussianity of all but possibly one of the components. Most mainstream ICA techniques belong to this category: Jade (Cardoso, 1999), FastICA, Infomax (Hyvärinen et al., 2001). In a second set of blind techniques, the components are modeled as Gaussian processes and, in a given representation, separation requires that the sources have diverse, i.e. non-proportional, variance profiles. For instance, the Spectral Matching ICA method (SMICA) (Delabrouille et al., 2003; Moudden et al., 2005) considers in this sense the case of mixed stationary Gaussian components and goes further than the above model (Eqn. 6.18) by taking into account additive *instrumental* noise N:

$$X = AS + N \tag{6.19}$$

Moving to a Fourier representation, the idea is that colored components can be separated based on the diversity of their power spectra.

In the case where the main component of interest cannot be modeled as a stationary Gaussian distribution, methods from the first set are expected to yield better results. A brief overview of the non-Gaussian ICA method, JADE, and FastICA (Hyvärinen et al., 2001), are presented next. The SMICA method is described in section 6.7. This is followed by a description of ways to combine wavelets and ICA techniques. Some useful properties of wavelet transforms can indeed enhance the performance of ICA methods in various situations.

6.6.2 JADE

The Joint Approximate Diagonalization of Eigenmatrices method (JADE) assumes the observed data X follows the noiseless mixture model (6.18) where the independent sources S are non-Gaussian i.i.d. [1] random processes. The mixing matrix is assumed to be square and invertible so that (de)mixing is actually just a change of basis. In the case we have more channels than sources, the data can be preprocessed using PCA, and only the first n eigenvectors are retained, n being the number of sources.

As mentioned above, second order statistics do not retain enough information: finding a change of basis in which the data covariance matrix is diagonal cannot lead to source recovery. However, decorrelation is *half the job* and one may seek the basis in which the data is represented by maximally independent processes among those bases in which the data is decorrelated. This leads to so-called orthogonal algorithms: after a proper whitening of the data by multiplication with W, one is then seeking a rotation R so that $\widehat{S}$ defined by

$$\widehat{S} = W^{-1} Y = W^{-1} R X_w = W^{-1} R W X \tag{6.20}$$

and $\widehat{B} = \widehat{A^{-1}} = W^{-1} R W$ are estimates of the sources and of the inverse of the mixing matrix.

JADE is an orthogonal ICA method. As for most mainstream ICA techniques, it exploits higher order statistics and tries to achieve some sort of

---

[1] I.i.d. stands for independently and identically distributed meaning that each entry of X at a given time t is independent of X at any other time t', and that the distribution of X does not depend on time.

nonlinear decorrelation. More precisely, approximate independence is measured using fourth order cross-cumulants:

$$\begin{aligned}
F_{ijkl} &= \text{cum}(y_i, y_j, y_k, y_l) \\
&= \text{E}(y_i, y_j, y_k, y_l) - \text{E}(y_i, y_j)\text{E}(y_k, y_l) \\
&\quad - \text{E}(y_i, y_l)\text{E}(y_j, y_k) - \text{E}(y_i, y_k)\text{E}(y_j, y_l)
\end{aligned} \quad (6.21)$$

where E represents statistical expectation, and where the y_is are the entries of vector Y modeled as random variables, and the correct change of basis (i.e. rotation) is found by somehow *diagonalizing* the fourth order cumulant tensor. Indeed, if the y_is were independent, all the cumulants with at least two different indices would be zero. Following the independence assumption of the source processes S and the decorrelation (*whiteness*) of Y for all rotations R, the fourth order tensor F is well-structured: JADE was in fact devised to take advantage of the algebraic properties of F. JADE's objective function is given by:

$$\begin{aligned}
\mathcal{J}_{\text{jade}}(R) &= \sum_{ijkl \neq ijkk} \text{cum}(y_i, y_j, y_k, y_l)^2 \\
&= \sum_{ij} \sum_{k \neq l} \text{cum}(y_i, y_j, y_k, y_l)^2
\end{aligned} \quad (6.22)$$

which can be interpreted as a joint diagonalization criterion. Fast and robust algorithms are available for the minimization of $\mathcal{J}_{\text{jade}}(R)$ with respect to R based on Jacobi's method for matrix diagonalization (Pham, 2001). More details on JADE can be found in (Cardoso, 1999; Cardoso, 1998b; Hyvärinen et al., 2001), and an interesting application of JADE for Mars Express data analysis (OMEGA instrument) can be found in (Forni et al., 2005).

JADE for Spherical Maps. Applying JADE on multichannel data mapped to the sphere does not require any particular modification of the algorithm. Indeed, JADE estimates the fourth order cumulant tensor from the available data samples assuming an i.i.d. random field. Hence, given a pixelization scheme on the sphere such as provided by the Healpix package, JADE can be directly applied to the multichannel spherical data pixels.

Example. Fig. 6.3 shows three simulated signals. Six observed data sets were derived from the three sources, by a linear combination (see Fig. 6.4). Without any knowledge of the mixing matrix, the JADE-ICA (Cardoso, 1998a) method was applied and the three signals were reconstructed from the six observed data sets (see Fig. 6.5).

6.6.3 FastICA

FastICA is by now a standard technique in ICA. Like JADE, it is meant for the analysis of mixtures of independent non-Gaussian sources in a noise-less setting. A complete description of this method can be found in (Hyvärinen

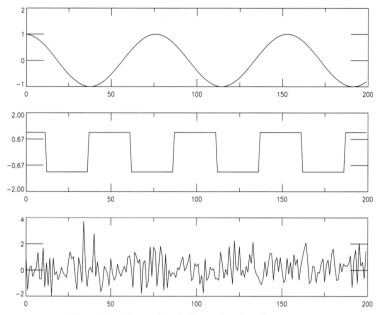

Fig. 6.3. Example of three simulated sources.

et al., 2001) and references therein. Many papers on this algorithm are available at http://www.cs.helsinki.fi/u/ahyvarin/papers/fastica.shtml. We give here a brief and simplified account of the algorithm. FastICA, again like JADE, is a so-called orthogonal ICA method: the independent components are sought by maximizing a measure of non-Gaussianity under the constraint that they are decorrelated. Intuitively, one should understand that mixtures of independent non-Gaussian random variables tend to *look more Gaussian*. An enlightening view of the relation between mutual information, which is a natural measure of independence, decorrelation and non-Gaussianity, can be found in (Cardoso, 2001; Cardoso, 2003). Non-Gaussianity is assessed in FastICA using a contrast function G based on a non-linear approximation to *negentropy* (Hyvärinen et al., 2001). In practice, depending on the application, different approximations or nonlinear (non-quadratic) functions should be experimented with. In a simple deflation scheme, for sphered data, the directions are found sequentially: a direction r of maximal non-Gaussianity is sought by maximizing

$$J_G(r) = \left(\mathrm{E}\{G(r^T x_{\text{white}})\} - \mathrm{E}\{G(\nu)\} \right)^2 \tag{6.23}$$

where ν stands for centered unit variance Gaussian variable, under the constraint that r has unit norm and that r is orthogonal to the directions found previously. The contrast function G can for instance be chosen among the following (Hyvärinen et al., 2001):

188   6. Multichannel Data

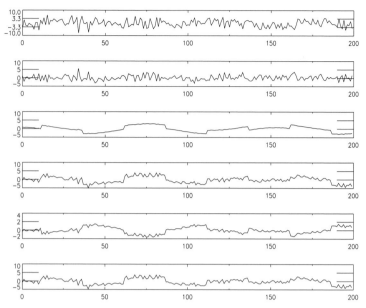

Fig. 6.4. Mixed sources. Each of these six signals is a linear combination of the three simulated sources.

$$G_0(u) = \frac{1}{a}\log\cosh(au)$$
$$G_1(u) = -\frac{1}{a}\exp(-au^2/2)$$
$$G_2(u) = \frac{1}{4}u^4$$

(6.24)

where a is a constant to be determined depending on the application. It can be shown that the maxima of J_G occur at certain maxima of $E\{G(r^T x_{\text{white}})\}$. These are obtained for r solution to:

$$E\{x_{\text{white}} g(r^T x_{\text{white}})\} - \lambda r = 0 \tag{6.25}$$

where λ is a constant easily expressed in terms of the optimal direction r_0, and g is the derivative of G. Solving this equation using Newton's method, and a few approximations, a *fixed-point* algorithm is derived which consists of repeating the following two steps until convergence:

$$r \leftarrow E\{x_{\text{white}} g(r^T x_{\text{white}})\} - E\{g'(r^T x_{\text{white}})\} r$$
$$r \leftarrow \frac{r}{\|r\|}$$

(6.26)

A simple implementation of this algorithm, largely based on Matlab^{TM}, is available at www.cis.hut.fi/projects/ica/fastica/.

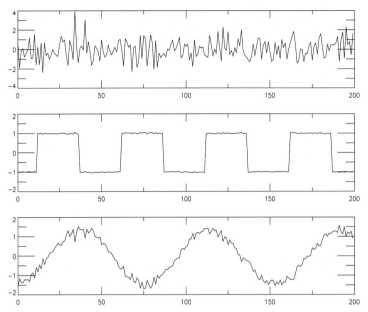

Fig. 6.5. Reconstructed sources from the mixed signals.

6.7 CMB Data and the SMICA ICA Method

6.7.1 The CMB Mixture Problem

In the CMB, the precise measurement of these fluctuations is of utmost importance to cosmology. Their statistical properties (spatial power spectrum, Gaussianity) strongly depend on the cosmological scenarios describing the properties and evolution of our Universe as a whole, and thus permit us to constrain these models as well as to measure the cosmological parameters describing the matter content, the geometry, and the evolution of our Universe.

Accessing this information, however, requires disentangling in the data the contributions of several distinct astrophysical sources, all of which emit radiation in the frequency range used for CMB observations. This problem of component separation, in the field of CMB studies, has been the object of many dedicated studies in the past.

To first order, the total sky emission can be modeled as a linear superposition of a few independent processes. The observation of the sky in direction (θ, φ) with detector d is a noisy linear mixture of N_c components:

$$x_d(\vartheta, \varphi) = \sum_{j=1}^{N_c} A_{dj} s_j(\vartheta, \varphi) + n_d(\vartheta, \varphi) \qquad (6.27)$$

where s_j is the emission template for the jth astrophysical process, here referred to as a *source* or a *component*. The coefficients A_{dj} reflect emission

190   6. Multichannel Data

distributions while n_d accounts for noise. When N_d detectors provide independent observations, this equation can be put in vector-matrix form:

$$X(\vartheta,\varphi) = AS(\vartheta,\varphi) + N(\vartheta,\varphi) \tag{6.28}$$

where X and N are vectors of length N_d, S is a vector of length N_c, and A is the $N_d \times N_c$ mixing matrix.

Blind source separation or independent component analysis (ICA) methods have been implemented specifically for CMB studies. The work of Baccigalupi et al. (2000b), further extended by Maino et al. (2002) implements a blind source separation method exploiting the non-Gaussianity of the sources for their separation, which permits us to recover the mixing matrix A and the maps of the sources. Accounting for spatially varying instrumental noise in the observation model is investigated by Kuruoglu et al. (2003), as well as the possible inclusion of prior information about the distributions of the components using a generic Gaussian mixture model.

Snoussi et al. (2004) propose a Bayesian approach in the Fourier domain assuming known spectra for the components as well as possibly non-Gaussian priors for the Fourier coefficients of the components. A fully blind, maximum likelihood approach is developed in (Cardoso and et al., 2002) and (Delabrouille et al., 2003), with the new point of view that spatial power spectra are actually the main unknown parameters of interest for CMB observations. This method, called Spectral Matching ICA (SMICA) is a source separation method based on spectral matching in Fourier space A key benefit is that parameter estimation can then be based on a set of band-averaged spectral covariance matrices, considerably compressing the data size.

6.7.2 SMICA

Spectral matching ICA (SMICA) is a blind source separation technique which, unlike most standard ICA methods, is able to recover Gaussian sources in noisy contexts. It operates in the spectral domain and is based on *spectral diversity*: it is able to separate sources provided they have different power spectra. This section gives a brief account of SMICA. More details can be found in (Delabrouille et al., 2003; Moudden et al., 2005); first applications to CMB analysis are in (Delabrouille et al., 2003; Patanchon et al., 2004a).

Model and Cost Function. For a second-order stationary N_d-dimensional process, we denote by $R_X(\nu)$ the $N_d \times N_d$ spectral covariance matrix at frequency ν, that is, the (i,i)th entry of $R_X(\nu)$ is the power spectrum of the ith coordinate of X while the off-diagonal entries of $R_X(\nu)$ contain the cross-spectra between the entries of X. If X follows the linear model of equation (6.28) with independent additive noise, then its spectral covariance matrix is structured as

$$R_X(\nu) = AR_S(\nu)A^t + R_N(\nu) \tag{6.29}$$

with $R_S(\nu)$ and $R_N(\nu)$ being the spectral covariance matrices of S and N respectively. The assumption of independence between the underlying components implies that $R_S(\nu)$ is a diagonal matrix. We shall also assume independence of the noise processes between detectors: matrix $R_N(\nu)$ also is a diagonal matrix.

In the definition of $R_X(\nu)$, we have not explicitly defined the frequency ν. This is because SMICA can be applied to the separation of components in many contexts: each observation X_d can be a time series (one-dimensional), an image (two-dimensional random field), or a random field on the sphere (as in full-sky CMB studies). In each case, the appropriate notions of frequency, stationarity and power spectrum should be used.

SMICA estimates all (or a subset of) the model parameters:

$$\theta = \{R_S(\nu_q), R_N(\nu_q), A\}$$

by minimizing a measure of "spectral mismatch" between sample estimates $\widehat{R}_X(\nu)$ (defined below) of the spectral covariance matrices and their ensemble averages which depend on the parameters according to equation (6.29). More specifically, an estimate $\widehat{\theta} = \{\widehat{R}_S(\nu_q), \widehat{R}_N(\nu_q), \widehat{A}\}$ is obtained as $\widehat{\theta} = \arg\min_\theta \phi(\theta)$ where the measure of spectral mismatch $\phi(\theta)$ is defined by

$$\phi(\theta) = \sum_{q=1}^{Q} \alpha_q \mathcal{D}\Big(\widehat{R}_X(\nu_q), AR_S(\nu_q)A^\dagger + R_N(\nu_q)\Big) \tag{6.30}$$

Here, $\{\nu_q | 1 \leq q \leq Q\}$ is a set of frequencies, $\{\alpha_q | 1 \leq q \leq Q\}$ is a set of positive weights, and $\mathcal{D}(\cdot, \cdot)$ is a measure of mismatch between two positive matrices.

This approach is a particular instance of moment matching. As such, if consistent estimates $\widehat{R}_X(\nu_q)$ of the spectral covariance matrices $R_X(\nu_q)$ are available and if the model is identifiable, then any reasonable choice of the weights α_q and of the divergence measure $\mathcal{D}(\cdot, \cdot)$ should lead to *consistent* estimates of the parameters. However, this does not mean that these choices should be arbitrary: in our standard implementation, we make specific choices (described next) in such a way that minimizing $\phi(\theta)$ is identical to maximizing the likelihood of θ in a model of Gaussian stationary processes. Hence, these choices guarantee good statistical efficiency when the underlying processes are well-modeled as Gaussian stationary processes. When this is not the case, though, the performance of SMICA may not be as good as (but not necessarily worse than) the performance of other methods designed to capture other aspects of the statistical distribution of the data, such as non-Gaussian features, for instance.

Given a data set, denote $\widetilde{X}(\nu)$ as its discrete Fourier transform at frequency ν and denote by $\{F_q | 1 \leq q \leq Q\}$ a set of Q frequency domains with F_q centered around frequency ν_q. Spectral covariance matrices are estimated non-parametrically by

$$\widehat{R}_X(\nu_q) = \frac{1}{n_q} \sum_{\nu \in F_q} \widetilde{X}(\nu)\widetilde{X}(\nu)^\dagger \qquad (6.31)$$

where n_q denotes the number of Fourier points $\widetilde{X}(\nu)$ in the spectral domain F_q. We always use symmetric domains in the sense that frequency ν belongs to F_q if and only if $-\nu$ also does. This symmetry guarantees that $\widehat{R}_X(\nu_q)$ is always a real valued matrix when X itself is a real valued process.

In its standard form, the SMICA technique uses positive weights $\alpha_q = n_q$ and a divergence $\mathcal{D}$ defined as

$$\mathcal{D}_{KL}(R_1, R_2) = \frac{1}{2}\left(\mathrm{trace}(R_1 R_2^{-1}) - \log\det(R_1 R_2^{-1}) - m\right) \qquad (6.32)$$

which is the Kullback-Leibler divergence between two m-variate zero-mean Gaussian distributions with covariance matrices R_1 and R_2. These choices stem from the Whittle approximation according to which each $\widetilde{X}(\nu)$ has a zero-mean normal distribution with covariance matrix $R_X(\nu)$ and is uncorrelated with $\widetilde{X}(\nu')$ for $\nu \neq \nu'$. In this case, it is easily checked that $-\phi(\theta)$ evaluated with $\alpha_q = n_q$ and $\mathcal{D} = \mathcal{D}_{KL}$ is (up to a constant) the log-likelihood for T data samples. This is actually true when the spectral domains are shrunk to just one DFT point ($n_q=1$ for all q); when the spectral domains F_q are chosen to contain several (usually many) DFT points, then $-\phi(\theta)$ is the log-likelihood, in the Whittle approximation, of the Gaussian stationary model with constant power spectrum over each domain F_q. This approximation is at small statistical loss when the spectrum is smooth enough to show little variation over each spectral domain.

The major gain of assuming constant spectrum over each F_q is the resulting reduction of the data set to a small number of covariance matrices. This may be a crucial benefit in applications like astronomical imaging where very large data sets are frequent.

Regarding our application to CMB analysis, the hypothesized isotropy of the distribution of the sources leads us to integrate over spectral domains with the corresponding symmetry. For sky maps small enough to be considered as flat, the spectral decomposition is the two-dimensional Fourier transform and the "natural" spectral domains are rings centered on the null frequency. For larger maps where curvature cannot be neglected, the spectral decomposition is over spherical harmonics and the natural spectral domains contain all the modes associated with a set of scales (Patanchon et al., 2004b).

Component Map Estimation. When running SMICA, power spectral densities for the sources and detector noise are obtained along with the estimated mixing matrix. They are used in reconstructing the source maps via Wiener filtering in Fourier space: a Fourier mode $X(\nu)$ in frequency band $\nu \in F_q$ is used to reconstruct the maps according to

$$\widehat{S}(\nu) = (\widehat{A}^t \widehat{R}_N(\nu)^{-1} \widehat{A} + \widehat{R}_S(\nu)^{-1})^{-1} \widehat{A}^t \widehat{R}_N(\nu)^{-1} X(\nu) \qquad (6.33)$$

In the limiting case where noise is small compared to signal components, the Wiener filter reduces to

$$\widehat{S}(\nu) = (\widehat{A}^t \widehat{R}_N(\nu)^{-1} \widehat{A})^{-1} \widehat{A}^t \widehat{R}_N(\nu)^{-1} X(\nu) \qquad (6.34)$$

Note however that the above Wiener filter is optimal only for stationary Gaussian processes. For weak, point-like sources such as galaxy clusters seen via the Sunyaev–Zel'dovich effect much better reconstruction can be expected from nonlinear methods.

6.8 ICA and Wavelets

Several properties of wavelets have been recognized as particularly useful in multichannel data processing: bringing wavelets and independent component analysis together has proven quite profitable. Extensions WJADE and WS-MICA of the two ICA methods described previously are discussed in this section.

Wavelets are remarkable at data compression, meaning that data that are structured in the initial representation require fewer significant coefficients in a wavelet representation. In imprecise and general terms, wavelets grab the coherence between coefficients of the structured data and produce a smaller set of significant coefficients which are then less coherent and which have a sparser statistical distribution. Then, the super-Gaussian[2] i.i.d. statistical model which appears in most standard ICA methods may better suit the wavelet coefficients of the data than the data samples in the initial representation.

Wavelets have been developed for the analysis of non-stationary and singular data in order to overcome certain difficulties associated with the Fourier transform. Wavelets are widely used to reveal variations in the spectral content of time series or images as they permit the singling out of regions in direct space while retaining localization in the frequency domain. Astrophysical data analysis has much to gain in avoiding the assumption of stationarity underlying Fourier analysis. Moreover, observed data maps are commonly imperfectly shaped and incomplete with missing or masked patches due to experimental settings, scanning strategies, etc. This will impair direct application of the Spectral Matching ICA method described previously. One might consider resorting to wavelets.

6.8.1 WJADE

Wavelets come into play as a sparsifying transform. Applying a wavelet transform on both sides of (6.18) does not affect the mixing matrix and the model

---
[2] A super-Gaussian distribution is also called a lepto-kurtic distribution, referring to a distribution with a narrow central peak and heavy tails. A typical example is the Laplacian distribution.

structure is preserved. Also, moving the data to a wavelet representation does not affect its information content. However, the statistical distribution of the data coefficients in the new representation is different: wavelets are known to lead to sparse i.i.d. representations of structured data. Further, the *local* (coefficientwise) signal to noise ratio depends on the choice of a representation. A wavelet transform tends to grab the informative coherence between pixels while averaging the noise contributions, thus enhancing structures in the data. Although the standard ICA model (6.18) is for a noiseless setting, the derived methods can be applied to real data. Performance will depend on the detectability of significant coefficients i.e. on the sparsity of the statistical distribution of the coefficients. Moving to a wavelet representation will often lead to more robustness to noise.

Once the data has been transformed to a proper representation (e.g. wavelets but also ridgelets and curvelets in the case of strongly anisotropic 2D or 3D data), WJADE consists of applying the standard JADE method to the new multichannel coefficients. Once the mixing matrix is estimated, the initial source maps are obtained using the appropriate inverse transform after some nonlinear denoising or thresholding of the coefficients if necessary.

6.8.2 Covariance Matching in Wavelet Space: WSMICA

SMICA, which was designed to address some of the general problems raised by Cosmic Microwave Background data analysis, has already shown significant success for CMB spectral estimation in multidetector experiments (Delabrouille et al., 2003; Patanchon et al., 2004a). However, SMICA suffers from the non-locality of the Fourier transform which has undesired effects when dealing with non-stationary components or noise, or with incomplete data maps. The latter is a common issue in astrophysical data analysis: either the instrument scanned only a fraction of the sky or some regions of the sky were masked due to localized strong astrophysical sources of contamination (compact radio-sources or galaxies, strong emitting regions in the galactic plane). A simple way to overcome these effects is to move instead to a wavelet representation so as to benefit from the localization property of wavelet filters, which leads to wSMICA (Moudden et al., 2005). The wSMICA method uses an isotropic undecimated wavelet transform. The wavelet and scaling functions have small compact supports which allows missing patches in the data maps to be handled easily.

Using this wavelet transform algorithm, the multichannel data X is decomposed into J detail maps X_j^w and a smooth approximation map X_{J+1}^w over a dyadic resolution scale. Since applying a wavelet transform on (6.28) does not affect the mixing matrix A, the covariance matrix of the observations at scale j, is still structured as

$$R_w^X(j) = A R_w^S(j) A^t + R_w^N(j) \tag{6.35}$$

where $R_w^S(j)$ and $R_w^N(j)$ are the diagonal spectral covariance matrices in the wavelet representation of S and N respectively. It was shown (Moudden et al., 2005) that replacing in the SMICA method the covariance matrices derived from the Fourier coefficients, by the covariance matrices derived from wavelet coefficients, leads to much better results when the data are incomplete. This is due to the fact that the wavelet filter response on scale j is short enough compared to data size and gap widths and most of the samples in the filtered signal remain then unaffected by the presence of gaps. Using exclusively these samples yields an estimated covariance matrix $\widehat{R}_w^X(j)$ which is not biased by the missing data.

6.8.3 Numerical Experiments

The application of wSMICA to synthetic mixtures of CMB, galactic dust and Sunyaev Zel'dovich (SZ) maps is considered here. Dust emission is the greybody emission of small dust particles in our own galaxy. The intensity of this emission is strongly concentrated towards the galactic plane, although cirrus clouds at high galactic latitudes are present as well (Schlegel et al., 1998). The SZ effect is a small distortion of the CMB blackbody emission that can be modeled, to first order, as a small additive emission, negative at frequencies below 217 GHz, and positive at frequencies above (Birkinshaw, 1999).

The component maps used, shown in Fig. 6.6, were obtained as described in (Delabrouille et al., 2003). The problem of instrumental point spread functions is not addressed here, and all maps are assumed to have the same resolution. The high level foreground emissions from the galactic plane region were discarded using the $Kp2$ mask from the WMAP team website[3]. These three *incomplete* maps were mixed using the matrix in Table 6.1 to simulate observations in the six channels of the Planck high frequency instrument (HFI).

Table 6.1. Entries of A, the mixing matrix used in our simulations.

CMB	DUST	SZ	Channel
1.0	1.0	−1.51	100 GHz
1.0	2.20	−1.05	143 GHz
1.0	7.16	0.0	217 GHz
1.0	56.96	2.22	353 GHz
1.0	1.1×10^3	5.56	545 GHz
1.0	1.47×10^5	11.03	857 GHz

Gaussian *instrumental* noise was added in each channel according to model (6.28). The relative noise standard deviations between channels were

---
[3] http://lambda.gsfc.nasa.gov/product/map/intensity_mask.cfm

196   6. Multichannel Data

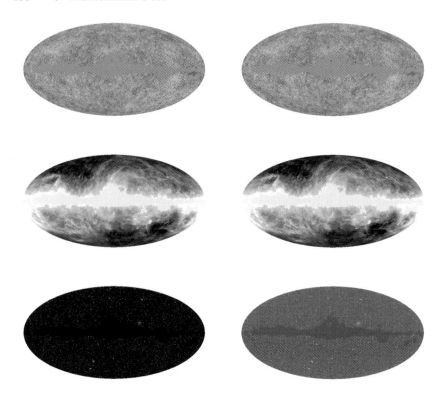

Fig. 6.6. The maps on the left are the templates for CMB, galactic dust and SZ used in the experiment described in section 6.8.3. The maps on the *right* were estimated using wSMICA and scalewise Wiener filtering. (The different maps are drawn here in different color scales in order to enhance structures and ease visual comparisons).

set according to the nominal values of the Planck HFI given in Table 6.2 and we experimented with five *global* noise levels at -6, -3, 0, $+3$ and $+6$ dB from nominal values.

The synthetic observations were decomposed into six scales using our wavelet transform on the sphere and wSMICA was used to obtain estimates of the initial source templates. For the sake of comparison, a separation with SMICA was also performed based on Fourier statistics computed in the same six dyadic bands imposed by our choice of wavelet transform.

The resulting component maps estimated using wSMICA, for nominal noise levels, are shown in Fig. 6.6 where the quality of reconstruction can be visually assessed by comparison with the initial components. Fig. 6.7 gives more quantitative results in the particular case of CMB, comparing the performance of SMICA and wSMICA in terms of reconstruction error MQE

Table 6.2. Nominal noise standard deviations in the six channels of the Planck HFI.

100 GHz	143 GHz	217 GHz	Channel
2.65×10^{-6}	2.33×10^{-6}	3.44×10^{-6}	Noise std
353 GHz	545 GHz	857 GHz	Channel
1.05×10^{-5}	1.07×10^{-4}	4.84×10^{-3}	Noise std

which we defined by

$$MQE = \frac{\mathbf{std}(CMB(\vartheta,\varphi) - \alpha \times \widehat{CMB}(\vartheta,\varphi))}{\mathbf{std}(CMB(\vartheta,\varphi))} \tag{6.36}$$

where **std** stands for empirical standard deviation (obviously computed outside the masked regions), and α is a linear regression coefficient estimated in the least squares sense. These results clearly show that using wavelet-based covariance matrices provides a simple and effective way to cancel the bad impact that gaps actually have on the performance of estimation using Fourier based statistics. Another way in which the effect of the gap on the performance of SMICA could probably be reduced is by applying a proper apodizing window on the data prior to estimating the spectral covariance, which is standard practice in harmonic analysis. With the mask used, building such a window is not straightforward so that, in the present experiments, SMICA was applied without correction for the gaps. The results given in Fig. 6.7 should be interpreted considering this aspect.

It may be argued that the proposed wavelet based approach offers little flexibility in the spectral bands available for wSMICA while the Fourier approach gives complete flexibility in this respect. But actually it is possible to use other transforms on the sphere (e.g. wavelet packet transform, *continuous wavelet transform*) or in fact any set of linear filters, preferably well-localized both on the sphere and in the spherical harmonics domain. In this way gaps are dealt with well, and spectral information is preserved, to achieve the source separation objective.

We have considered here a linear mixing model, which does not reflect the reality. Some components have a spectrum which varies spatially and this spectral index variation is not considered by the previously described methods. A solution could be to separate the wavelet coefficients of the data into different subsets, each subset corresponding to a given spatial area, and to apply successively the wSMICA method on each subset. As long as the

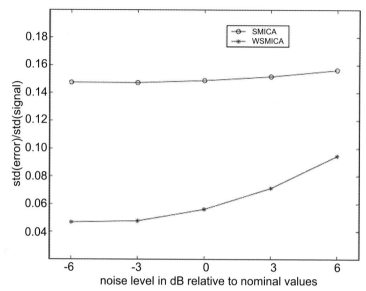

Fig. 6.7. Relative reconstruction error defined by (6.36) of the CMB component map using SMICA and wSMICA as a function of the instrumental noise level.

spectral index variation is relatively smooth, this solution may be acceptable. But this locally invariant mixture model may also not be good enough for an exact representation of the reality. Another, and maybe more impotant, problem is the fact that the ICA methods make the same statistical assumption for all components (i.e. they are all non-Gaussian in JADE and FastICA and they are all Gaussian in SMICA and wSMICA), while we know that one component is Gaussian (i.e. the CMB) and all others are not. An ICA method allowing us to consider different statistical models for the different components would certainly improve dramatically the quality of the component separation in the CMB mixture problem.

6.9 Chapter Summary

It has been seen that applying analysis methods on wavelet noise filtered data may lead to better results: this was noted for principal components analysis (or Karhunen-Loève transform), and for independent component analysis.

ICA methods are powerful techniques for recovering the original sky emissions from a mixture of different components. The recent developments in combined wavelet–ICA approaches such as wSMICA lead to flexible tools to deal with non-stationary components. Maps with missing patches are a particular example of practical significance. Our numerical experiments, based on realistic simulations of the astrophysical data expected from the Planck mission, clearly show the benefits of correctly processing existing gaps in the data which is not a real surprise. By moving to the wavelet domain, it is possible to easily cope with gaps of any shape in a very simple manner, while still retaining spectral information for component separation. Clearly, other possible types of non-stationarities in the collected data such as spatially varying noise or component variance, etc. could be dealt with very simply in a similar fashion using the wavelet extension of SMICA (Moudden et al., 2005).

The integrated analysis strategies discussed in this chapter cater for the analysis of data "cubes", or more correctly multiple band signals. This includes hyperspectral data also. Given its special role in facilitating analysis of signals with Poisson noise, we paid particular attention to the Haar wavelet transform.

7. An Entropic Tour of Astronomical Data Analysis

7.1 Introduction

The term "entropy" is due to Clausius (1865), and the concept of entropy was introduced by Boltzmann to statistical mechanics in order to measure the number of microscopic ways that a given macroscopic state can be realized. Shannon (1948) founded the mathematical theory of communication when he suggested that the information gained in a measurement depends on the number of possible outcomes from which one is realized. Shannon also suggested that the entropy be used for maximization of the bit transfer rate under a quality constraint. Jaynes (1957) proposed to use the entropy measure for radio interferometric image deconvolution, in order to select between a set of possible solutions that which contains the minimum of information or, following his entropy definition, that which has maximum entropy. In principle, the solution verifying such a condition should be the most reliable. A great deal of work has been carried out in the last 30 years on the use of the entropy for the general problem of data filtering and deconvolution (Ables, 1974; Bontekoe et al., 1994; Burg, 1978; Frieden, 1978a; Gull and Skilling, 1991; Narayan and Nityananda, 1986; Pantin and Starck, 1996; Skilling, 1989; Weir, 1992; Mohammad-Djafari, 1994; Mohammad-Djafari, 1998). Traditionally information and entropy are determined from events and the probability of their occurrence. Signal and noise are basic building-blocks of signal and data analysis in the physical sciences. Instead of the probability of an event, we are led to consider the probabilities of our data being either signal or noise.

Observed data Y in the physical sciences are generally corrupted by noise, which is often additive and which follows in many cases a Gaussian distribution, a Poisson distribution, or a combination of both. Other noise models may also be considered. Using Bayes' theorem to evaluate the probability distribution of the realization of the original signal X, knowing the data Y, we have

$$p(X|Y) = \frac{p(Y|X).p(X)}{p(Y)} \tag{7.1}$$

p($Y|X$) is the conditional probability distribution of getting the data Y given an original signal X, i.e. it represents the distribution of the noise. It is given, in the case of uncorrelated Gaussian noise with variance σ^2, by:

$$p(Y|X) = \exp\left\{-\sum_{pixels} \frac{(Y-X)^2}{2\sigma^2}\right\} \tag{7.2}$$

The denominator in equation 7.1 is independent of X and is considered as a constant (stationary noise). p(X) is the a priori distribution of the solution X. In the absence of any information on the solution X except its positivity, a possible course of action is to derive the probability of X from its entropy, which is defined from information theory.

The main idea of information theory (Shannon, 1948) is to establish a relation between the received information and the probability of the observed event (Bijaoui, 1984). If we denote $\mathcal{I}(E)$ the information related to the event E, and p the probability of this event occurring, then we consider that

$$\mathcal{I}(E) = f(p) \tag{7.3}$$

Thereafter we assume the two following principles:

- The information is a decreasing function of the probability. This implies that the more information we have, the less will be the probability associated with one event.
- Additivity of the information. If we have two independent events E_1 and E_2, the information $\mathcal{I}(E)$ associated with the occurrence of both is equal to the addition of the information of each of them.

$$\mathcal{I}(E) = \mathcal{I}(E_1) + \mathcal{I}(E_2) \tag{7.4}$$

Since E_1 (of probability p_1) and E_2 (of probability p_2) are independent, then the probability of both occurring is equal to the product of p_1 and p_2. Hence

$$f(p_1 p_2) = f(p_1) + f(p_2) \tag{7.5}$$

A good choice for the information measure is

$$\mathcal{I}(E) = k \ln(p) \tag{7.6}$$

where k is a constant. Information must be positive, and k is generally fixed at -1.

Another interesting measure is the mean information which is denoted

$$H = -\sum_i p_i \ln(p_i) \tag{7.7}$$

This quantity is called the entropy of the system and was established by Shannon (1948).

This measure has several properties:

- It is maximal when all events have the same probability $p_i = 1/N_e$ (N_e being the number of events), and is equal to $\ln(N_e)$. It is in this configuration that the system is the most undefined.
- It is minimal when one event is sure. In this case, the system is perfectly known, and no information can be added.
- The entropy is a positive, continuous, and symmetric function.

If we know the entropy H of the solution (the next section describes different ways to calculate it), we derive its probability by

$$p(X) = \exp(-\alpha H(X)) \tag{7.8}$$

Given the data, the most probable image is obtained by maximizing $p(X|Y)$. Taking the logarithm of equation 7.1, we thus need to maximize

$$\ln(p(X|Y)) = -\alpha H(X) + \ln(p(Y|X)) - \ln(p(Y)) \tag{7.9}$$

The last term is a constant and can be omitted. In the case of Gaussian noise, the solution is found by minimizing

$$J(X) = \sum_{pixels} \frac{(Y-X)^2}{2\sigma^2} + \alpha H(X) = \frac{\chi^2}{2} + \alpha H(X) \tag{7.10}$$

which is a linear combination of two terms: the entropy of the signal, and a quantity corresponding to χ^2 in statistics measuring the discrepancy between the data and the predictions of the model. α is a parameter that can be viewed alternatively as a Lagrangian parameter or a value fixing the relative weight between the goodness-of-fit and the entropy H.

For the deconvolution problem, the object-data relation is given by the convolution

$$Y = P * X \tag{7.11}$$

where P is the point spread function, and the solution is found (in the case of Gaussian noise) by minimizing

$$J(X) = \sum_{pixels} \frac{(Y - P*X)^2}{2\sigma^2} + \alpha H(X) \tag{7.12}$$

The way the entropy is defined is fundamental, because from its definition will depend the solution. The next section discusses the different approaches which have been proposed in the past. Multiscale Entropy, presented in section 3, is based on the wavelet transform and noise modeling. It is a means of measuring information in a data set, which takes into account important properties of the data which are related to content. We describe how it can be used for signal and image filtering, and for image deconvolution. The case of multi-channel data is then considered before we proceed to the use of multiscale entropy for description of image content. We pursue two directions of enquiry: determining whether signal is present in the image or not, possibly at or below the image's noise level; and how multiscale entropy is very well correlated with the image's content in the case of astronomical stellar fields. Knowing that multiscale entropy represents well the content of the image, we finally use it to define the optimal compression rate of the image. In all cases, a range of examples illustrate these new results.

7.2 The Concept of Entropy

We wish to estimate an unknown probability density $p(X)$ of the data. Shannon (1948), in the framework of information theory, defined the entropy of an image X by

$$H_s(X) = -\sum_{k=1}^{N_b} p_k \log p_k \quad (7.13)$$

where $X = \{X_1, ...X_N\}$ is an image containing integer values, N_b is the number of possible values of a given pixel X_k (256 for an 8-bit image), and the p_k values are derived from the histogram of X:

$$p_k = \frac{\#X_j = k}{N} \quad (7.14)$$

$\#X_j = k$ gives the number of pixels with value k, i.e., $X_j = k$.

If the image contains floating values, it is possible to build up the histogram L of values L_i using a suitable interval Δ, counting up how many times m_k each interval $(L_k, L_k + \Delta)$ occurs among the N occurrences. The probability that a data value belongs to an interval k is $p_k = \frac{m_k}{N}$, and each data value has a probability p_k.

The entropy is minimum and equal to zero when the signal is flat, and increases when we have some fluctuations. Using this entropy in equation 7.10 leads to minimization of

$$J(X) = \frac{\chi^2}{2} + \alpha H_s(X) \quad (7.15)$$

This is a minimum entropy restoration method.

The trouble with this approach is that, because the number of occurrences is finite, the estimate p_k will be in error by an amount proportional to $m_k^{-\frac{1}{2}}$ (Frieden, 1978b). The error becomes significant when m_k is small. Furthermore this kind of entropy definition is not easy to use for signal restoration, because the gradient of equation 7.15 is not easy to compute. For these reasons, other entropy functions are generally used. The main ones are:

– Burg (1978):

$$H_b(X) = -\sum_{k=1}^{N} \ln(X_k) \quad (7.16)$$

– Frieden (1978a):

$$H_f(X) = -\sum_{k=1}^{N} X_k \ln(X_k) \tag{7.17}$$

– Gull and Skilling (1991):

$$H_g(X) = \sum_{k=1}^{N} X_k - M_k - X_k \ln(\frac{X_k}{M_k}) \tag{7.18}$$

where M is a given model, usually taken as a flat image

In all definitions N is the number of pixels, and k represents an index pixel.

Each of these entropies can be used, and they correspond to different probability distributions that one can associate with an image (Narayan and Nityananda, 1986). See (Frieden, 1978a; Skilling, 1989) for descriptions. The last definition of the entropy above has the advantage of having a zero maximum when X equals the model M. All of these entropy measures are negative (if $X_k > 1$), and maximum when the image is flat. They are negative because an offset term is omitted which has no importance for the minimization of the functional. The fact that we consider that a signal has maximum information value when it is flat is evidently a curious way to measure information. A consequence is that we must now maximize the entropy if we want a smooth solution, and the probability of X must be redefined by:

$$\mathrm{p}(X) = \exp(\alpha H(X)) \tag{7.19}$$

The sign has been inverted (see equation 7.8), which is natural if we want the best solution to be the smoothest. These three entropies, above, lead to the Maximum Entropy Method method, MEM, for which the solution is found by minimizing (for Gaussian noise)

$$J(X) = \sum_{k=1}^{N} \frac{(Y_k - X_k)^2}{2\sigma^2} - \alpha H(X) \tag{7.20}$$

To recapitulate, the different entropy functions which have been proposed for image restoration have the property of being maximal when the image is flat, and of decreasing when we introduce some information. So minimizing the information is equivalent to maximizing the entropy, and this has led to the well-known Maximum Entropy Method, MEM. For the Shannon entropy (which is obtained from the histogram of the data), the opposite is the case. The entropy is null for a flat image, and increases when the data contains some information. So, if the Shannon entropy were used for restoration, this would lead to a Minimum Entropy Method.

In 1986, Narayan and Nityanda (1986) compared several entropy functions, and concluded by saying that all were comparable if they have good properties, i.e. they enforce positivity, and they have a negative second derivative which discourages ripple. They showed also that results varied strongly with the background level, and that these entropy functions produced poor results for negative structures, i.e. structures under the background level (absorption area in an image, absorption band in a spectrum, etc.), and compact structures in the signal. The Gull and Skilling entropy gives rise to the difficulty of estimating a model. Furthermore it has been shown (Bontekoe et al., 1994) that the solution is dependent on this choice.

The determination of the α parameter is also not an easy task and in fact it is a very serious problem facing the maximum entropy method. In the historic MAXENT algorithm of Skilling and Gull, the choice of α is such that it must satisfy the ad hoc constraint $\chi^2 = N$ when the deconvolution is achieved, N being the number of degrees of freedom of the system i.e. the number of pixels in image deconvolution problems. But this choice systematically leads to an under-fitting of the data (Titterington, 1985) which is clearly apparent for imaging problems with little blurring. In reality, the χ^2 statistic is expected to vary in the range $N \pm \sqrt{2N}$ from one data realization to another. In the Quantified Maximum Entropy point of view (Skilling, 1989), the optimum value of α is determined by including its probability $P(\alpha)$ in Bayes' equation and then by maximizing the marginal probability of having α, knowing the data and the model m. In practice, a value of α which is too large gives a resulting image which is too regularized with a large loss of resolution. A value which is too small leads to a poorly regularized solution showing unacceptable artifacts. Taking a flat model of the prior image softens the discontinuities which may appear unacceptable for astronomical images, often containing as they do stars and other point-like objects. Therefore the basic maximum entropy method appears to be not very appropriate for this kind of image which contains high and low spatial frequencies at the same time. Another point to be noted is a ringing effect of the maximum entropy method algorithm, producing artifacts around bright sources.

To solve these problems while still using the maximum entropy concept, some enhancements of the maximum entropy method have been proposed. Noticing that neighboring pixels of reconstructed images with MAXENT could have values differing a lot in expected flat regions (Charter, 1990), Gull and Skilling introduced the concepts of hidden image S and intrinsic correlation function C (Gaussian or cubic spline-like) in the Preblur MAXENT algorithm.

The intrinsic correlation function, ICF, describes a minimum scale length of correlation in the desired image O which is achieved by assuming that

$$O = C * S \tag{7.21}$$

This corresponds to imposing a minimum resolution on the solution O. Since the hidden space image S is not spatially correlated, this can be regularized by the entropy

$$H_g(h) = \sum_{k=1}^{N} S_k - M_k - S_k \ln(\frac{S_k}{M_k}) \tag{7.22}$$

In astronomical images many scale lengths are present, and the *Multichannel Maximum Entropy Method*, developed by Weir (Weir, 1991; Weir, 1992; Bridle et al., 1998; Marshall et al., 2002), uses a set of ICFs having different scale lengths, each defining a channel. The visible-space image is now formed by a weighted sum of the visible-space image channels O_j:

$$O = \sum_{j=1}^{N_c} p_j O_j \qquad (7.23)$$

where N_c is the number of channels. Like in Preblur MAXENT, each solution O_j is supposed to be the result of the convolution between a hidden image S_j with a low-pass filter (ICF) C_j:

$$O_j = C_j * S_j \qquad (7.24)$$

But such a method has several drawbacks:

1. The solution depends on the width of the ICFs (Bontekoe et al., 1994).
2. There is no rigorous way to fix the weights p_j (Bontekoe et al., 1994).
3. The computation time increases linearly with the number of pixels.
4. The solution obtained depends on the choice of the models M_j ($j = 1 \ldots N_c$) which were chosen independently of the channel.

In 1993, Bontekoe et al. (1994) used a special application of this method which they called Pyramid Maximum Entropy on infrared image data. The pyramidal approach allows the user to have constant ICF width, and the computation time is reduced. It is demonstrated (Bontekoe et al., 1994) that all weights can be fixed ($p_j = 1$ for each channel).

This method eliminates the first three drawbacks, and gives better reconstruction of the sharp and smooth structures. But in addition to the two last drawbacks, a new one is added: since the images O_j have different sizes (due to the pyramidal approach), the solution O is built by duplicating the pixels of the subimages O_j of each channel. This procedure is known to produce artifacts due to the appearance of high frequencies which are incompatible with the real spectrum of the true image $\hat{O}$.

However this problem can be easily overcome by duplicating the pixels before convolving with the ICF, or expanding the channels using linear interpolation. Thus the introduction of the "pyramid of resolution" has solved some problems and brought lots of improvements to the classic maximum entropy method, but has also raised other questions. In order to derive the model from a physical value, Pantin and Starck (1996) brought the wavelet transform into this context, and defined entropy as follows:

$$H(O) = \frac{1}{\sigma_I^2} \sum_{j=1}^{l} \sum_{k=1}^{N_j} \sigma_j (w_{j,k} - M_j - |w_{j,k}| \ln \frac{|w_{j,k}|}{M_j}) \qquad (7.25)$$

where l is the number of scales, and N_j is the number of samples in the band j ($N_j = N$ for the à trous algorithm). The multiscale entropy is the sum of the entropy at each scale.

The coefficients $w_{j,k}$ are wavelet coefficients, and we take the absolute value of $w_{j,k}$ in this definition because the values of $w_{j,k}$ can be positive or negative, and a negative signal contains also some information in the wavelet transform.

The advantage of such a definition of entropy is the fact we can use previous work concerning the wavelet transform and image restoration (Murtagh et al., 1995; Starck and Murtagh, 1994; Starck and Bijaoui, 1994). The noise behavior has already been studied in the case of the wavelet transform and we can estimate the standard deviation of the noise σ_j at scale j. These estimates can be naturally introduced in our models m_j:

$$M_j = k_m \sigma_j \tag{7.26}$$

The model M_j at scale j represents the value taken by a wavelet coefficient in the absence of any relevant signal and, in practice, it must be a small value compared to any significant signal value. Following the Gull and Skilling procedure, we take M_j as a fraction of the noise because the value of σ_j can be considered as a sort of physical limit under which a signal cannot be distinguished from the noise ($k_m = \frac{1}{100}$).

As described above, many studies have been carried out in order to improve the functional to be minimized. But the question which should be raised is: what is a good entropy for signal restoration?

In (Starck et al., 1998b; Starck and Murtagh, 2001; Starck et al., 2001), the benchmark properties for a good "physical" definition of entropy were discussed. Assuming that a signal X is the sum of several components:

$$X = S + B + N \tag{7.27}$$

where S is the signal of interest, B the background, and N the noise, we proposed that the following criteria should be verified:

1. The information in a flat signal is zero ($S = 0$, $N = 0$ and $B =$ Const.).
2. The amount of information in a signal is independent of the background ($H(X)$ is independent of B).
3. The amount of information is dependent on the noise ($H(X)$ is dependent on N). A given signal X does not furnish the same information if the noise N is high or small.
4. The entropy must work in the same way for a pixel which has a value $B + \epsilon$, and for a pixel which has a value $B - \epsilon$. $H(X)$ must be a function of the absolute value of S instead of S.
5. The amount of information is dependent on the correlation in the signal. If the signal S presents large features above the noise, it contains a lot of information. By generating a new set of data from S, by randomly taking the pixel values in S, the large features will evidently disappear, and this

7.2 The Concept of Entropy   209

Fig. 7.1. Saturn image (*left*) and the same data distributed differently (*right*). These two images have the same entropy using any of the standard entropy definitions.

new signal will contain less information. But the pixel values will be the same as in S.

Fig. 7.1 illustrates the last point perfectly. The second image is obtained by distributing randomly the Saturn image pixel values, and the standard entropy definitions produce the same information measurement for both images. The concept of information becomes really subjective, or at least it depends on the application domain. Indeed, for someone who is not involved in image processing, the second image contains *less* information than the first one. For someone working on image transmission, it is clear that the second image will require more bits for lossless transmission, and from this point of view, he or she will consider that the second image contains *more* information. Finally, for data restoration, all fluctuations due to noise are not of interest, and do not contain relevant information. From this physical point of view, the standard definition of entropy seems badly adapted to information measurement in signal restoration.

These points are not axioms, but rather desirable properties that should be respected by the entropy functional in order to characterize well the data. We see that in these properties we are taking account of: (i) the background – very much a relative notion, associated with our understanding of the image or signal; and (ii) the noise – important when handling scientific, medical and other images and signals. The background can also be termed continuum, or DC component, and is often very dependent on the semantics of the image. Our signal generation process could be conceived of in terms of thermodynamics (Ferraro et al., 1999): the rate of variation of entropy is composed of internal heat changes, and heat transfers from external sources. The latter is our noise, N, and the former is signal including background.

It is clear that among all entropy functions proposed in the past, it is the Shannon one (1948) which best respects these criteria. Indeed, if we assume that the histogram bin is defined as a function of the standard deviation of the noise, the first four points are verified, while none of these criteria are verified with other entropy functions (and only one point is verified for the Gull and Skilling entropy by taking the model equal to the background). We reiterate that our critique of information measures is solely in view of our overriding goals, namely to define a demonstrably appropriate measure for image and signal processing in the physical sciences.

7.3 Multiscale Entropy

7.3.1 Definition

Following on from the desirable criteria discussed in the previous section, a possibility is to consider that the entropy of a signal is the sum of the information at each scale of its wavelet transform, and the information of a wavelet coefficient is related to the probability of it being due to noise. Let us look at how this definition holds up in practice. Denoting h the information relative to a single wavelet coefficient, we define

$$H(X) = \sum_{j=1}^{l} \sum_{k=1}^{N_j} h(w_{j,k}) \tag{7.28}$$

with $h(w_{j,k}) = -\ln p(w_{j,k})$. l is the number of scales, and N_j is the number of samples in band (scale) j. For Gaussian noise, we get

$$h(w_{j,k}) = \frac{w_{j,k}^2}{2\sigma_j^2} + \text{Const.} \tag{7.29}$$

where σ_j is the noise at scale j. Below, when we use the information in a functional to be minimized, the constant term has no effect and we will take the liberty of omitting it. We see that the information is proportional to the energy of the wavelet coefficients. We will call this entropy definition, *ENERGY-MSE*, in the following. A similar result has been derived in (Maisinger et al., 2004) by a different method.

The larger the value of a normalized wavelet coefficient, then the lower will be its probability of being noise, and the higher will be the information furnished by this wavelet coefficient. We can see easily that this entropy fulfills all the requirements listed in the previous section. Just as for Shannon entropy, here information increases with entropy. Using such an entropy for optimization purposes will ordinarily lead to a minimum entropy method. If we consider two signals S_1, S_2, derived from a third one S_0 by adding noise:

$$S_1 = S_0 + N_1(\sigma_1)$$
$$S_2 = S_0 + N_2(\sigma_2)$$
(7.30)

then we have:

if $\sigma_1 < \sigma_2$ then $H(S_1) > H(S_2)$ (7.31)

and a flat image has zero entropy.

This entropy definition is completely dependent on the noise modeling. If we consider a signal S, and we assume that the noise is Gaussian, with a standard deviation equal to σ, we won't measure the same information compared to the case when we consider that the noise has another standard deviation value, or if the noise follows another distribution. As for the Shannon entropy, the information increases with the entropy, and using such an entropy leads to a Minimum Entropy Method.

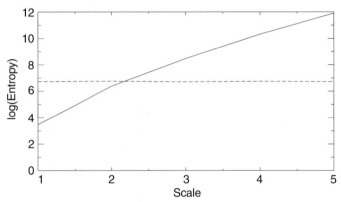

Fig. 7.2. Multiscale entropy of Saturn (*continuous curve*), and multiscale entropy of the scrambled image (*dashed curve*).

Fig. 7.2 shows the information measure at each scale for both the Saturn image and its scrambled version. The global information is the addition of the information at each scale. We see that for the scrambled image (dashed curve), the information-versus-scale curve is flat, while for the unscrambled saturn image, it increases with the scale.

Equation 7.28 holds if the wavelet coefficients are statistically independent, which should imply that our approach is limited to an orthogonal or biorthogonal transform. However, this disadvantage may be addressed through the use of the so-called cycle-spinning algorithm (also named translation-invariant algorithm) (Coifman and Donoho, 1995), which consists of performing the process of "transform," "denoise," and "inverse transform" on every orthogonal basis corresponding to versions of the data obtainable by combinations of circular left-right and upwards-downwards translations. Furthermore Donoho and Coifman (1995) have shown that using a non-decimating wavelet transform is equivalent to performing a set of decimated transforms with shift on the input signal. This means that equation 7.28 remains true for non-decimated wavelet transforms if it is normalized by the number of shifts. We will consider the orthogonal case in the following, knowing it can be generalized to non-decimated transforms.

7.3.2 Signal and Noise Information

Assuming that the signal X is still composed of the three components S, B, N ($X = S+B+N$), H is independent of B but not of N. Hence, our information measure is corrupted by noise, and we decompose our information measure into two components, one (H_s) corresponding to the non-corrupted part, and the other (H_n) to the corrupted part. We have (Starck et al., 1998b)

$$H(X) = H_s(X) + H_n(X) \tag{7.32}$$

We will define in the following H_s as the signal information, and H_n as the noise information. It is clear that noise does not contain any meaningful information, and so H_n describes a semantic component which is usually not informative for us. For each wavelet coefficient $w_{j,k}$, we have to estimate the fractions h_n and h_s of h (with $h(w_{j,k}) = h_n(w_{j,k}) + h_s(w_{j,k})$) which should be assigned to H_n and H_s. Hence signal information and noise information are defined by

$$H_s(X) = \sum_{j=1}^{l} \sum_{k=1}^{N_j} h_s(w_{j,k})$$

$$H_n(X) = \sum_{j=1}^{l} \sum_{k=1}^{N_j} h_n(w_{j,k}) \tag{7.33}$$

If a wavelet coefficient is small, its value can be due to noise, and the information h relative to this single wavelet coefficient should be assigned to h_n. If the wavelet coefficient is high, compared to the noise standard deviation, its value cannot be due to the noise, and h should be assigned to h_s. Note that $H_s(X) + H_n(X)$ is always equal to $H(X)$.

N1-MSE. A first approach for deriving h_s and h_n from p_s and p_n is to just consider p_s and p_n as weights on the information h. Then we have:

$$h_s(w_{j,k}) = p_s(w_{j,k})h(w_{j,k}) \tag{7.34}$$
$$h_n(w_{j,k}) = p_n(w_{j,k})h(w_{j,k})$$

and the noise and signal information in a signal are

$$H_s(X) = \sum_{j=1}^{l} \sum_{k=1}^{N_j} h_s(w_{j,k}) \tag{7.35}$$

$$H_n(X) = \sum_{j=1}^{l} \sum_{k=1}^{N_j} h_n(w_{j,k})$$

which leads for Gaussian noise to:

$$H_s(X) = \sum_{j=1}^{l}\sum_{k=1}^{N_j} \frac{w_{j,k}^2}{2\sigma_j^2} \operatorname{erf}\left(\frac{|w_{j,k}|}{\sqrt{2}\sigma_j}\right) \tag{7.36}$$

$$H_n(X) = \sum_{j=1}^{l}\sum_{k=1}^{N_j} \frac{w_{j,k}^2}{2\sigma_j^2} \operatorname{erfc}\left(\frac{|w_{j,k}|}{\sqrt{2}\sigma_j}\right)$$

We will refer to these functions by the name N1-MSE in the following.

N2-MSE. By the previous entropy measure, information relative to high wavelet coefficients is completely assigned to the signal. For a restoration, this allows us also to exclude wavelet coefficients with high signal-to-noise ratio (SNR) from the regularization. This leads to perfect fit of the solution with the data at scales and positions with high SNR. If we want to consider the information due to noise, even for significant wavelet coefficients, the noise information relative to a wavelet coefficient must be estimated differently.

The idea for deriving h_s and h_n is the following: we imagine that the information h relative to a wavelet coefficient is a sum of small information components dh, each of them having a probability to be noise information. To understand this principle, consider two coefficients u and w ($w > u$) with Gaussian noise ($\sigma = 1$). The information relative to w is $h(w) = w^2$. When u varies from 0 to w with step du, the information $h(u)$ increases until it becomes equal to $h(w)$. When it becomes closer to w, the probability that the difference $w - u$ can be due to the noise increases, and the added information dh is more corrupted by the noise. By weighting the added information by the probability that the difference $w - u$ is due to the noise, we have:

$$h_n(w_{j,k}) = \int_0^{|w_{j,k}|} p_n(|w_{j,k}|-u)\left(\frac{\partial h(x)}{\partial x}\right)_{x=u} du \tag{7.37}$$

is the noise information relative to a single wavelet coefficient, and

$$h_s(w_{j,k}) = \int_0^{|w_{j,k}|} p_s(|w_{j,k}|-u)\left(\frac{\partial h(x)}{\partial x}\right)_{x=u} du \tag{7.38}$$

is the signal information relative to a single wavelet coefficient. For Gaussian noise, we have

$$h_n(w_{j,k}) = \frac{1}{\sigma_j^2}\int_0^{|w_{j,k}|} u\,\operatorname{erfc}\left(\frac{|w_{j,k}|-u}{\sqrt{2}\sigma_j}\right) du$$

$$h_s(w_{j,k}) = \frac{1}{\sigma_j^2}\int_0^{|w_{j,k}|} u\,\operatorname{erf}\left(\frac{|w_{j,k}|-u}{\sqrt{2}\sigma_j}\right) \tag{7.39}$$

Equations 7.35 and 7.39 lead to two different ways to regularize a signal. The first requires that we use all the information which is furnished in high wavelet coefficients, and leads to an exact preservation of the flux in a structure. If the signal gives rise to high discontinuities, artifacts can appear in the solution due to the fact that the wavelet coefficients located at the discontinuities are not noisy, but have been modified like noise. The second equation does not have this drawback, but a part of the flux of a structure (compatible with noise amplitude) can be lost in the restoration process. It is however not as effective as in the standard maximum entropy methods.

LOG-MSE. The multiscale entropy function used in (Pantin and Starck, 1996) – we call it LOG-MSE in the following – can be considered in our framework if h is defined by:

$$h(w_{j,k}) = \frac{\sigma_j}{\sigma_X^2}[w_{j,k} - M_j - \mid w_{j,k} \mid \log\left(\frac{\mid w_{j,k} \mid}{k_m \sigma_j}\right)] \tag{7.40}$$

where σ_X is the noise standard deviation in the data. And h_n is defined by:

$$h_n(w_{j,k}) = A(p_n(w_{j,k}))h(w_{j,k}) \tag{7.41}$$

where A is a function which takes the values 0 or 1 depending on $p_n(w_{j,k})$:

$$A(p_n(w_{j,k})) = \begin{cases} 1 & \text{if } p_n(w_{j,k}) > \epsilon \\ 0 & \text{if } p_n(w_{j,k}) \leq \epsilon \end{cases} \tag{7.42}$$

Wavelet coefficients which are significant will force $A(p_n(w_{j,k}))$ to be equal to 0 (because their probabilities of being due to noise is very small), and do not contribute to H_n. This means that using H_n in a regularization process will have an effect only on scales and positions where no significant wavelet coefficient is detected.

In practice we prefer N1-MSE and N2-MSE for several reasons. First the way the model is used in equation 7.40 is somewhat artificial, and there is an undetermination when the wavelet coefficient is equal to 0. A better choice for the LOG-MSE would be the Herbert and Leaby function (1989):

$$h(w_{j,k,l}) \propto \log\left(1 + \frac{\mid w_{j,k,l} \mid}{\sigma_j}\right) \tag{7.43}$$

LOG-MSE seems difficult to generalize to other classes of noise, which is not the case for N1-MSE and N2-MSE. The ENERGY-MSE is quadratic and leads to a strong penalization even for wavelet coefficients with high signal-to-noise ratio. Such penalization terms are known to oversmooth the strongest peaks and should not be used. N2-MSE has the advantage of estimating the corrupted part in the measured information h, even for large wavelet coefficients. Fig. 3.12 shows the multiscale entropy penalization function.

Multiscale MEM and ICF In (Maisinger et al., 2004), it was argued that the multiscale entropy is merely a special case of the intrinsic correlation function approach, where we replace the ICF kernel by a wavelet function. From the strict mathematical point of view, this is right, but this perspective minimizes completely the improvement related to the wavelets. All the concepts of sparse representation (which is the key to wavelet success in many applications), fast decomposition and reconstruction, zero mean coefficients (which allows us to get wavelet coefficients which are independent of the background and to derive robust noise modeling) do not exist in the ICF-MEM approach. Furthermore, the ICF-MEM approach requires us to estimate accurately the background, which may be a very difficult task, and it has be shown (Bontekoe et al., 1994) that the solution depends strongly on this estimation. On the contrary, Multiscale MEM needs only an estimation of the noise standard deviation, which is easy to determine. We prefer to keep our vision of the multiscale entropy method as a specific case of the generalized wavelet regularization techniques rather than as an extension of the ICF approach.

7.4 Multiscale Entropy Filtering

7.4.1 Filtering

The problem of filtering or restoring data D can be expressed by the following: We search for a solution $\tilde{D}$ such that the difference between D and $\tilde{D}$ minimizes the information due to the signal, and such that $\tilde{D}$ minimizes the information due to the noise.

$$J(\tilde{D}) = H_s(D - \tilde{D}) + H_n(\tilde{D}) \tag{7.44}$$

Furthermore, the smoothness of the solution can be controlled by adding a parameter:

$$J(\tilde{D}) = H_s(D - \tilde{D}) + \alpha H_n(\tilde{D}) \tag{7.45}$$

In practice (Chambolle et al., 1998), we minimize for each wavelet coefficient $w_{j,k}$:

$$j(\tilde{w}_{j,k}) = h_s(w_{j,k} - \tilde{w}_{j,k}) + \alpha h_n(\tilde{w}_{j,k}) \tag{7.46}$$

$j(\tilde{w}_{j,k})$ can be obtained by any minimization routine. In our examples, we have used a simple binary search.

Fig. 7.3 shows the result when minimizing the functional j with different α values, and noise standard deviation equal to 1. The corrected wavelet coefficient is plotted versus the wavelet coefficient. From the top curve to the bottom one, α is respectively equal to 0, 0.1, 0.5, 1, 2, 5, 10. The higher the value of α, the more the corrected wavelet coefficient is reduced. When α is equal to 0, there is no regularization and the data are unchanged.

7.4.2 The Regularization Parameter

The α parameter can be used in different ways:

– It can be fixed to a given value (user parameter): $\alpha = \alpha_u$. This method leads to very fast filtering using the optimization proposed in the following.

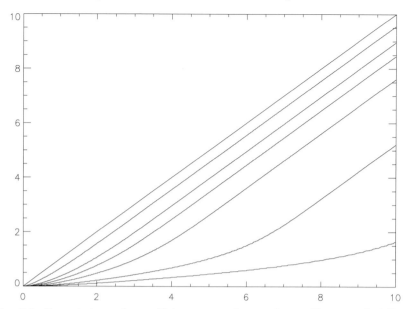

Fig. 7.3. Corrected wavelet coefficient versus the wavelet coefficient with different α values (from the top curve to the bottom one, α is respectively equal to 0, 0.1, 0.5, 1, 2, 5, 10).

- It can be calculated under the constraint that the residual should have some specific characteristic. For instance, in the case of Gaussian noise, we expect a residual with a standard deviation equal to the noise standard deviation. In this case, $\alpha = \alpha_c \alpha_u$. The parameter finally used is taken as the product of a user parameter (defaulted to 1) and the calculated value α_c. This allows the user to keep open the possibility of introducing an under-smoothing, or an over-smoothing. It is clear that such an algorithm is iterative, and will always take more time than a simple hard thresholding approach.
- We can permit more constraints on α by using the fact that we expect a residual with a given standard deviation at each scale j equal to the noise standard deviation σ_j at the same scale. Then rather than a single α we have an α_j per scale.

A more sophisticated way to fix the α value is to introduce a distribution (or a priori knowledge) of how the regularization should work. For instance, in astronomical image restoration, the analyst generally prefers that the flux (total intensity) contained in a star or in a galaxy is not modified by the restoration process. This means that the residual at positions of astronomical objects will approximately be equal to zero. All zero areas in the residual map obviously do not relate to realistic noise behavior, but from the user's point of view they are equally important. For the user, all visible objects in the

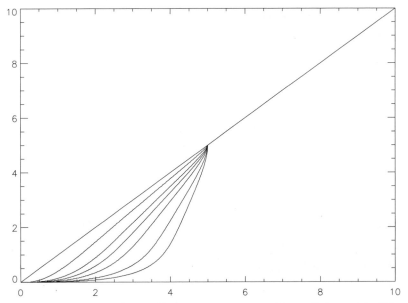

Fig. 7.4. Corrected wavelet coefficient versus the wavelet coefficient with different α values.

filtered map contain the same flux as in the raw data. In order to obtain this kind of regularization, the α parameter is no longer a constant value, but depends on the raw data. Hence we have one α per wavelet coefficient, which will be denoted $\alpha_s(w_{j,k})$, and it can be derived by

$$\alpha_s(w_{j,k}) = \alpha_j \frac{1 - L(w_{j,k})}{L(w_{j,k})} \tag{7.47}$$

with $L(w_{j,k}) = MIN(1, \frac{|w_{j,k}|}{k_s \sigma_j})$, where k_s is a user parameter (typically defaulted to 3).

When $L(w_{j,k})$ is close to 1, $\alpha_s(w_{j,k})$ becomes equal to zero, there is no longer any regularization, and the obvious solution is $\tilde{w}_{j,k} = w_{j,k}$. Hence, the wavelet coefficient is preserved from any regularization. If $L(w_{j,k})$ is close to 0, $\alpha_s(w_{j,k})$ tends toward infinity, the first term in equation (7.46) is negligible, and the solution will be $\tilde{w}_{j,k} = 0$. In practice, this means that all coefficients higher than $k_s \sigma_j$ are untouched as in the hard thresholding approach. We also notice that by considering a distribution $L(w_{j,k})$ equal to 0 or 1 (1 when $\mid w_{j,k} \mid > k_s \sigma_j$ for instance), the solution is then the same as a hard thresholding solution.

7.4.3 Use of a Model

Using a model in wavelet space has been successfully applied for denoising: see for example (Chipman et al., 1997; Crouse et al., 1998; Jansen and Roose,

1998). If we have a model D_m for the data, this can also naturally be inserted into the filtering equation:

$$J_m(\tilde{D}) = H_s(D - \tilde{D}) + \alpha H_n(\tilde{D} - D_m) \tag{7.48}$$

or, for each wavelet coefficient $w_{j,k}$:

$$j_m(\tilde{w}_{j,k}) = h_s(w_{j,k} - \tilde{w}_{j,k}) + \alpha h_n(\tilde{w}_{j,k} - w_{j,k}^m) \tag{7.49}$$

where $w_{j,k}^m$ is the corresponding wavelet coefficient of D_m.

The model can be of quite different types. It can be an image, and in this case, the coefficients $w_{j,k}^m$ are obtained by a simple wavelet transform of the model image. It can also be expressed by a distribution or a given function which furnishes a model wavelet coefficient w^m from the data. For instance, the case where we want to keep intact high wavelet coefficients (see equation 7.47) can also be treated by the use of a model, just by calculating $w_{j,k}^m$ by

$$w_{j,k}^m = p_s(w_{j,k}) w_{j,k} \tag{7.50}$$

When $w_{j,k}$ has a high signal-to-noise ratio, $P_s(w_{j,k})$ is close to 1, and $w_{j,k}^m$ is equal to $w_{j,k}$. Then $\alpha h_n(\tilde{w}_{j,k} - w_{j,k}^m)$ is equal to zero and $\tilde{w}_{j,k} = w_{j,k}$, i.e. no regularization is carried out on $w_{j,k}$.

Other models may also be considered. When the image contains contours, it may be useful to derive the model from the detected edges. Zero-crossing wavelet coefficients indicate where the edges are (Mallat, 1991). By averaging three wavelet coefficients in the direction of the detected edge, we get a value w_a, from which we derive the SNR S_e of the edge ($S_e = 0$ if there is no detected edge). The model value w^m is set to w_a if a contour is detected, and 0 otherwise. This approach has the advantage of filtering the wavelet coefficient, and even if an edge is clearly detected the smoothing operates in the direction of the edge.

There is naturally no restriction on the model. When we have a priori information of the content of an image, we should use it in order to improve the quality of the filtering. It is clear that the way we use knowledge of the presence of edges in an image is not a closed question. The model in the entropy function is an interesting aspect to investigate further in the future.

7.4.4 The Multiscale Entropy Filtering Algorithm

The Multiscale Entropy Filtering algorithm, MEF (Starck and Murtagh, 1999), consists of minimizing for each wavelet coefficient $w_{j,k}$ at scale j

$$j_m(\tilde{w}_{j,k}) = h_s(w_{j,k} - \tilde{w}_{j,k}) + \alpha_j h_n(\tilde{w}_{j,k} - w_{j,k}^m) \tag{7.51}$$

or

$$j_{ms}(\tilde{w}_{j,k}) = h_s(w_{j,k} - \tilde{w}_{j,k}) + \alpha_j \alpha_s(w_{j,k}) h_n(\tilde{w}_{j,k} - w_{j,k}^m) \tag{7.52}$$

if the SNR is used. By default the model $w_{j,k}^m$ is set to 0. There is no user parameter because the α_j are calculated automatically in order to verify the noise properties. If over-smoothing (or under-smoothing) is desired, a user parameter must be introduced. We propose in this case to calculate the α_j in the standard way, and then to multiply the calculated values by a user value α_u defaulted to 1. Increasing α_u will lead to over-smoothing, while decreasing α_u implies under-smoothing.

Using a simple dichotomy, the algorithm becomes:

1. Estimate the noise in the data σ (see chapter 2).
2. Determine the wavelet transform of the data.
3. Calculate from σ the noise standard deviation σ_j at each scale j.
4. Set $\alpha_j^{min} = 0$, $\alpha_j^{max} = 200$.
5. For each scale j do
   a) Set $\alpha_j = \frac{\alpha_j^{min} + \alpha_j^{max}}{2}$
   b) For each wavelet coefficient $w_{j,k}$ of scale j, find $\tilde{w}_{j,k}$ by minimizing $j_m(\tilde{w}_{j,k})$ or $j_{ms}(\tilde{w}_{j,k})$
   c) Calculate the standard deviation of the residual:
   $$\sigma_j^r = \sqrt{\frac{1}{N_j} \sum_{k=1}^{N_j} (w_{j,k} - \tilde{w}_{j,k})^2}$$
   d) If $\sigma_j^r > \sigma_j$ then the regularization is too strong, and we set α_j^{max} to α_j, otherwise we set α_j^{min} to α_j (σ_j is derived from the method described in section 2.4).
6. If $\alpha_j^{max} - \alpha_j^{min} > \epsilon$ then go to 5.
7. Multiply all α_j by the constant α_u.
8. For each scale j and for each wavelet coefficient w find $\tilde{w}_{j,k}$ by minimizing $j_m(\tilde{w}_{j,k})$ or $j_{ms}(\tilde{w}_{j,k})$.
9. Reconstruct the filtered image from $\tilde{w}_{j,k}$ by the inverse wavelet transform.

The minimization of j_m or j_{ms} (step 5b) can be carried out by any method. For instance, a simple dichotomy can be used in order to find $\tilde{w}$ such that

$$\frac{\partial h_s(w - \tilde{w})}{\partial \tilde{w}} = -\alpha_j \frac{\partial h_n(\tilde{w})}{\partial \tilde{w}} \qquad (7.53)$$

The idea to treat the wavelet coefficients such that the residual respects some constraint has also been used in cross-validation (Nason, 1994; Nason, 1996; Amato and Vuza, 1997). However, cross validation appears to overfit the data (Strang and Nguyen, 1996).

See Appendices D and E for the calculation of the derivative of h_s and h_n.

7.4.5 Optimization

In the case of Gaussian noise, the calculation of the *erf* and *erfc* functions could require considerable time compared to a simpler filtering method. This can be easily avoided by precomputing tables, which is possible due to the specific properties of $\frac{\partial h_s}{\partial \tilde{w}}$ and $\frac{\partial h_n}{\partial \tilde{w}}$. h_s and h_n are functions of the standard deviation of the noise, and we denote the reduced functions by h_s^r and h_n^r, i.e. h_s and h_n for noise standard deviation equal to 1. It is easy to verify that

$$\frac{\partial h_s(w_{j,k})}{\partial \tilde{w}} = \sigma_j \frac{\partial h_s^r(\frac{w_{j,k}}{\sigma_j})}{\partial \tilde{w}} \tag{7.54}$$

$$\frac{\partial h_n(w_{j,k})}{\partial \tilde{w}} = \sigma_j \frac{\partial h_n^r(\frac{w_{j,k}}{\sigma_j})}{\partial \tilde{w}} \tag{7.55}$$

Furthermore, $\frac{\partial h_n^r}{\partial \tilde{w}}$ and $\frac{\partial h_s^r}{\partial \tilde{w}}$ are symmetric functions, $\frac{\partial h_n^r}{\partial \tilde{w}}$ converges to a constant value C ($C = 0.798$), and $\frac{\partial h_s^r}{\partial \tilde{w}}$ tends to $C - w$ when w is large enough (> 5). In our implementation, we precomputed the tables using a step-size of 0.01 from 0 to 5. If no model is introduced and if the SNR is not used, the filtered wavelet coefficients are a function of α and $\frac{w_j}{\sigma_j}$, and a second level of optimization can be performed using precomputed tables of solutions for different values of α.

7.4.6 Examples

1D Data Filtering

Figs. 7.5, 7.6 and 7.7 show the results of the multiscale entropy method on simulated data (2048 pixels). From top to bottom, each figure shows simulated data, the noisy data, the filtered data, and both noisy and filtered data overplotted. For the two first filterings, all default parameters were taken (noise standard deviation and α_j automatically calculated, $\alpha_u = 1$, and the chosen wavelet transform algorithm is the à trous one). For the block signal (Fig. 7.5), default parameters were also used, but the multiresolution transform we used is the multiresolution median transform.

Simulations have shown (Starck and Murtagh, 1999) that the MEF method produces a better result than the standard soft or hard thresholding, from both the visual aspect and PSNR (peak signal-to-noise ratio).

7.5 Deconvolution

7.5.1 The Principle

The most realistic solution of the deconvolution problem is that which minimizes the amount of information, but remains compatible with the data. For the MEM method, minimizing the information is equivalent to maximizing the entropy and the functional to minimize is

$$J(O) = \sum_{k=1}^{N} \frac{(I_k - (P * O)_k)^2}{2\sigma_I^2} - \alpha H(O) \tag{7.56}$$

7.5 Deconvolution 221

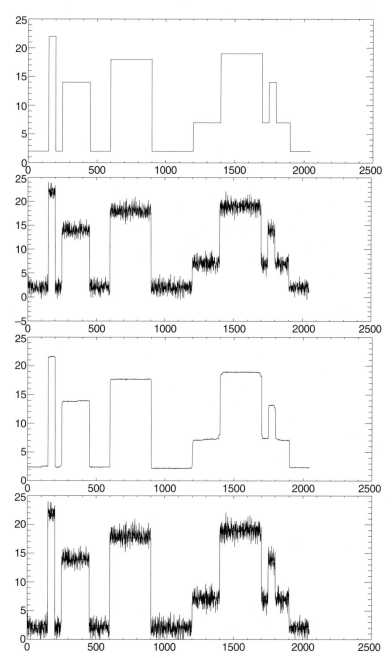

Fig. 7.5. From top to bottom, simulated block data, noise blocks, filtered blocks, and both noisy and filtered blocks overplotted.

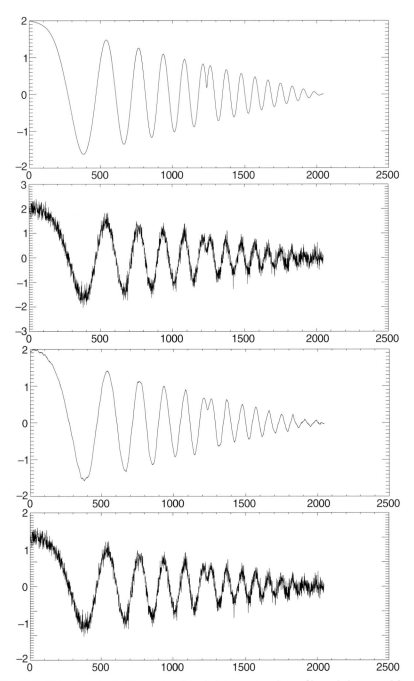

Fig. 7.6. From top to bottom, simulated data, noisy data, filtered data, and both noisy and filtered data overplotted.

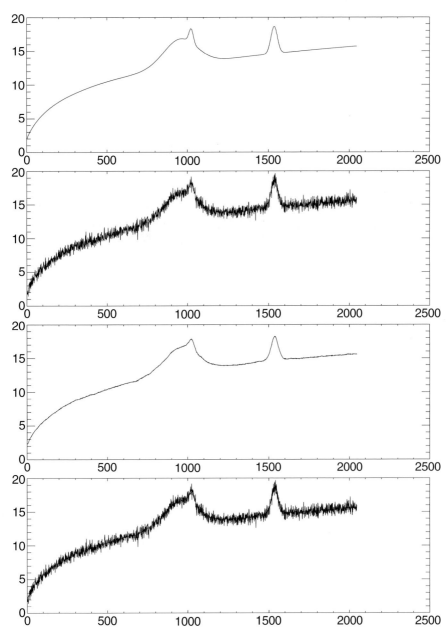

Fig. 7.7. From top to bottom, simulated data, noisy data, filtered data, and both noisy and filtered data overplotted.

where H is either the Frieden or the Gull and Skilling entropy.

Similarly, using the multiscale entropy, minimizing the information is equivalent to minimizing the entropy and the functional to minimize is

$$J(O) = \sum_{k=1}^{N} \frac{(I_k - (P*O)_k)^2}{2\sigma_I^2} + \alpha H(O) \tag{7.57}$$

We have seen that in the case of Gaussian noise, H is given by the energy of the wavelet coefficients. We have

$$J(O) = \sum_{k=1}^{N} \frac{(I_k - (P*O)_k)^2}{2\sigma_I^2} + \alpha \sum_{j=1}^{l} \sum_{k=1}^{N_j} \frac{w_{j,k}^2}{2\sigma_j^2} \tag{7.58}$$

where σ_j is the noise at scale j, N_j the number of pixels at the scale j, σ_I the noise standard deviation in the data, and l the number of scales.

Rather than minimizing the amount of information in the solution, we may prefer to minimize the amount of information which can be due to the noise. The function is now:

$$J(O) = \sum_{k=1}^{N} \frac{(I_k - (P*O)_k)^2}{2\sigma_I^2} + \alpha H_n(O) \tag{7.59}$$

and for Gaussian noise, H_n has been defined by

$$H_n(X) = \sum_{j=1}^{l} \sum_{k=1}^{N_j} \frac{1}{\sigma_j^2} \int_0^{|w_{j,k}|} u \operatorname{erf}\left(\frac{|w_{j,k}| - u}{\sqrt{2}\sigma_j}\right) \tag{7.60}$$

The solution is found by computing the gradient $\nabla(J(O))$ and performing the following iterative scheme:

$$O^{n+1} = O^n - \gamma \nabla(J(O^n)) \tag{7.61}$$

We consider an α_j per scale, and introduce thereby an adaptive regularization which depends on the signal-to-noise ratio of the input data wavelet coefficients.

7.5.2 The Parameters

In order to introduce flexibility in the way we restore the data, we introduce two parameters $\beta_{j,k}$ and $\alpha_{j,k}$ which allow us to weight, respectively, the two terms of the equation to be minimized:

$$J(O) = \frac{1}{2\sigma_I^2} \sum_{k=1}^{N} \left(\sum_j \sum_l \beta_{j,k} w_{j,l}(R) \psi_{j,l}(k) \right)^2 + \sum_{j=1}^{l} \sum_{k=1}^{N_j} \alpha_{j,k} h(w_{j,k}(O))$$

where $R = I - P*O$, and $R = \sum_j \sum_k w_{j,k}(R)\psi_{j,k}$ ($w_{j,k}(R)$ are the wavelet coefficients of R, and $w_{j,k}(O)$ are the wavelet coefficients of O).

We consider three approaches for estimating $\beta_{j,k}$

1. No weighting: $\beta_{j,k} = 1$
2. Soft weighting: $\beta_{j,k} = p_s(w_{j,k}(I))$
   In this case, $\beta_{j,k}$ is equal to the probability that the input data wavelet coefficient is due to signal (and not to noise).
3. Hard weighting: $\beta_{j,k} = 0$ or 1 depending on $p_n(w_{j,k}(I))$ ($p_n(w_{j,k}(I)) = 1 - p_s(w_{j,k}(I))$). This corresponds to using only significant input data wavelet coefficients.

$\alpha_{j,k}$ is the product of two values: $\alpha_{j,k} = \alpha_u \beta'_{j,k}$.

- α_u is a user parameter (defaulted to 1) which allows us to control the smoothness of the solution. Increasing α_u produces a smoother solution.
- $\beta'_{j,k}$ depends on the input data and can take the following value:
  1. No regularization ($\beta'_{j,k} = 0$): only the first term of the functional is minimized.
  2. No protection from regularization ($\beta'_{j,k} = 1$): the regularization is applied at all positions and at all the scales.
  3. Soft protection ($\beta'_{j,k} = p_n(w_{j,k}(I))$): the regularization becomes adaptive, depending on the probability that the input wavelet coefficient is due to noise.
  4. Hard protection ($\beta'_{j,k} = 0$ or 1 depending on $p_n(w_{j,k}(I))$).
  5. Soft + hard protection: ($\beta'_{j,k} = 0$ or $p_n(w_{j,k}(I))$ depending on $p_n(w_{j,k}(I))$).

We easily see that choosing a hard weighting and no regularization leads to deconvolution from the multiresolution support (Starck et al., 1998a).

7.5.3 Examples

Fig. 7.8 shows a simulation. The original image, panel (a), contains stars and galaxies. Fig. 7.8b shows the data (blurred image + Gaussian noise), Fig. 7.8c shows the deconvolved image, and Fig. 7.8d the residual image (i.e. data minus solution reconvolved by the PSF). The blurred image SNR is 12dB, and the deconvolved image SNR is 23.11 dB.

7.6 Multichannel Data Filtering

The multiscale entropy relative to a set of observations $D(1 \ldots M)$ can be written as:

$$H(D) = \sum_{l=1}^{L} \sum_{j=1}^{J} \sum_{k=1}^{N_j} h(c_{l,j,k}) \tag{7.62}$$

where J is the number of scales used in the wavelet transform decomposition, L the number of observations, k a pixel position, c a WT-KLT coefficient, and l denotes the eigenvector number.

226   7. An Entropic Tour of Astronomical Data Analysis

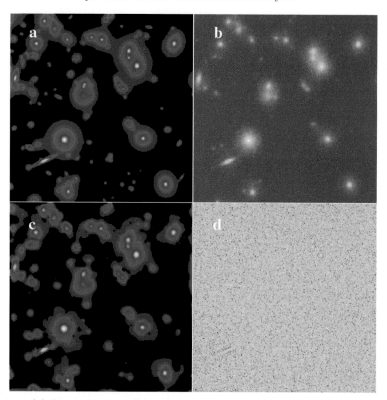

Fig. 7.8. (a) Original image, (b) blurred image + Gaussian noise, (c) deconvolved image, and (d) residual image.

The last scale of the wavelet transform is not used, as previously, so this entropy measurement is background-independent, which is important because the background can vary greatly from one wavelength to another.

As for wavelet coefficients in the case of mono-channel data, we know the noise standard deviation relative to a coefficient, and coefficients are of zero mean. Therefore, we can apply the same filtering method. The filtered WT-KLT coefficients are found by minimizing for each $c_{l,j,k}$:

$$j(\tilde{c}_{l,j,k}) = h_s(c_{l,j,k} - \tilde{c}_{l,j,k}) + \alpha h_n(\tilde{c}_{l,j,k}) \tag{7.63}$$

Example

Fig. 7.9 shows the results of a simulation. We created a dataset of 18 frames, each of them containing a source at the same position, but at different intensity levels. The source is a small Gaussian. The data cannot be coadded because the level of the source varies from one frame to another (variable source). Additive noise was used, and the data were filtered. The root mean square error (RMSE) was calculated on each individual frame on a 5 × 5

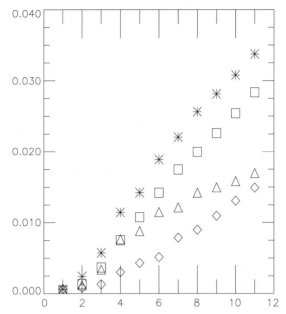

Fig. 7.9. Simulation: Integrated root mean square error (*ordinate*) versus the noise standard deviation (*abscissa*). See text for discussion.

square centered on the source. Hence, the RMSE reflects well the photometric errors, and the addition over the 18 RMSE, which we call IRMSE (Integrated RMSE), provides us with a reliable measurement of the filtering quality. The simulation was repeated with 12 noise levels, and four different filtering methods were compared. Fig. 7.9 shows the IRMSE versus the noise standard deviation plot. The four methods are (i) multiscale entropy applied to the WT-KLT coefficients (diamond), (ii) reconstruction from a subset of eigenvectors of the KLT (triangle), (iii) multiscale entropy applied to each frame independently (square), and (iv) thresholding applied to the wavelet transform of each frame (star). This simulation shows clearly that the approach described here, multiscale entropy applied to the WT-KLT coefficients, outperforms all other methods.

The same experiments were performed using a simulated Planck data set. The data set contains ten images, each a linear combination of 6 sky components images (CMB, SZ, free-free, etc.). As in the previous simulation, noise was added, and the data were filtered by the four methods. The only difference is that the RMSE is calculated on the full frames. Fig. 7.10 shows IRMSE versus the noise standard deviation plot. Diamonds, triangles, square and stars represent the same methods as before. Again, the multiscale entropy applied to the WT-KLT coefficients outperforms the other methods.

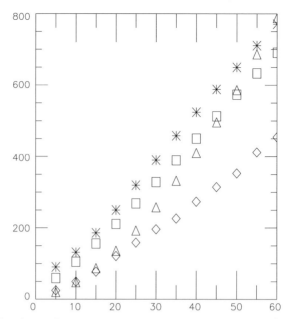

Fig. 7.10. Planck simulation: Integrated root mean square error (*ordinate*) versus the noise standard deviation (*abscissa*). See text for discussion.

7.7 Relevant Information in an Image

Since the multiscale entropy extracts the information from the signal only, it was a challenge to see if the astronomical content of an image was related to its multiscale entropy.

For this purpose, we used the astronomical content of 200 images of 1024 × 1024 pixels extracted from scans of 8 different photographic plates carried out by the MAMA digitization facility (Paris, France) (Guibert, 1992) and stored at CDS (Strasbourg, France) in the Aladin archive (Bonnarel et al., 1999). We estimated the content of these images in three different ways:

1. By counting the number of objects in an astronomical catalog (USNO A2.0 catalog) within the image. The USNO (United States Naval Observatory) catalog was originally obtained by source extraction from the same survey plates as we used in our study.
2. By counting the number of objects estimated in the image by the SExtractor object detection package (Bertin and Arnouts, 1996). As in the case of the USNO catalog, these detections are mainly point sources (i.e. stars, as opposed to spatially extended objects like galaxies).
3. By counting the number of structures detected at several scales using the MR/1 multiresolution analysis package (MR/1, 2001).

Fig. 7.11 shows the results of plotting these numbers for each image against the multiscale signal entropy of the image. The best results are ob-

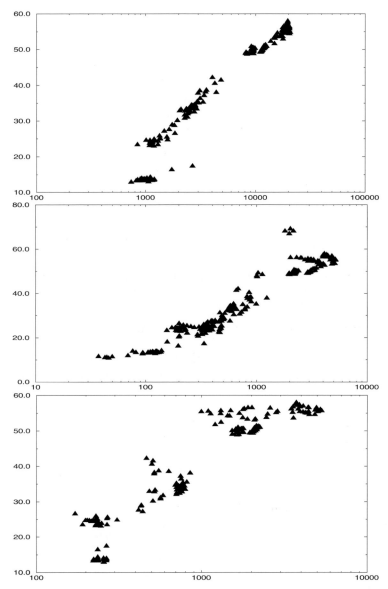

Fig. 7.11. Multiscale entropy versus the number of objects: the number of objects is, respectively, obtained from (*top*) the USNO catalog, (*middle*) the SExtractor package, and (*bottom*) the MR/1 package.

tained using the MR/1 package, followed by SExtractor and then by the number of sources extracted from USNO. The latter two basically miss the content at large scales, which is taken into account by MR/1. Unlike MR/1, SExtractor does not attempt to separate signal from noise.

SExtractor and multiresolution methods were also applied to a set of CCD (charge coupled detector, i.e. digital, as opposed to the digitized photographic plates used previously) images from CFH UH8K, 2MASS and DENIS near infrared surveys. Results obtained were very similar to what was obtained above. This lends support to (i) the quality of the results based on MR/1, which take noise and scale into account, and (ii) multiscale entropy being a good measure of content of such a class of images.

7.8 Multiscale Entropy and Optimal Compressibility

Subsequently we looked for the relation between the multiscale entropy and the optimal compression rate of an image which we can obtain by multiresolution techniques (Starck et al., 1998a). By optimal compression rate we mean a compression rate which allows all the sources to be preserved, and which does not degrade the astrometry (object positions) and photometry (object intensities). Louys et al. (1999) and Couvidat (1999) estimated this optimal compression rate using the compression program of the MR/1 package (2001).

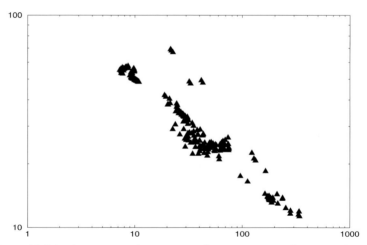

Fig. 7.12. Multiscale entropy of astronomical images versus the optimal compression ratio. Images which contain a high number of sources have a small ratio and a high multiscale entropy value. The relation is almost linear.

Fig. 7.12 shows the relation obtained between the multiscale entropy and the optimal compression rate for all the images used in our previous tests, both digitized plate and CCD images. The power law relation is obvious, thus allowing us to conclude that:

- Compression rate depends strongly on the astronomical content of the image. We can then say that compressibility is also an estimator of the content of the image.
- Multiscale entropy allows us to predict the optimal compression rate of the image.

7.9 Conclusions and Chapter Summary

We have seen that information must be measured from the transformed data, and not from the data itself. This is so that a priori knowledge of physical aspects of the data can be taken into account. We could have used the Shannon entropy, perhaps generalized, cf. Sporring and Weickert (1999), to measure the information at a given scale, and derive the bins of the histogram from the standard deviation of the noise, but for several reasons we thought it better to directly introduce noise probability into our information measure. Firstly, we have seen that this leads, for Gaussian noise, to a very physically meaningful relation between the information and the wavelet coefficients: information is proportional to the energy of the wavelet coefficients normalized by the standard deviation of the noise. Secondly, this can be generalized to many other kinds of noise, including such cases as multiplicative noise, non-stationary noise, or images with few photons/events. We have seen that the equations are easy to manipulate. Finally, experiments have confirmed that this approach gives good results.

For filtering, multiscale entropy has the following advantages:

- It provides a good trade-off between hard and soft thresholding.
- No a priori model on the signal itself is needed as with other wavelet-based Bayesian methods (Chipman et al., 1997; Crouse et al., 1998; Vidakovic, 1998; Timmermann and Nowak, 1999).
- It can be generalized to many kinds of noise.
- The regularization parameter α can be easily fixed automatically. Cross-validation (Nason, 1996) could be an alternative, but with the limitation to Gaussian noise.

Replacing the standard entropy measurements by Multiscale Entropy avoids the main problems in the MEM deconvolution method.

We have seen also how this new information measure allows us to analyze image background fluctuation. In the example discussed, we showed how signal which was below the noise level could be demonstrated to be present. The SNR was 0.25. This innovative analysis leads to our being able to affirm that signal is present, without being able to say where it is.

To study the semantics of a large number of digital and digitized photographic images, we took already prepared – external – results, and we also used two other processing pipelines for detecting astronomical objects within these images. Therefore we had three sets of interpretations of these images. We then used Multiscale Entropy to tell us something about these three sets of results. We found that Multiscale Entropy provided interesting insight into the performances of these different analysis procedures. Based on strength of correlation between Multiscale Entropy and analysis result, we argued that this provided circumstantial evidence of one analysis result being superior to the others.

We finally used Multiscale Entropy to provide a measure of optimal image compressibility. Using previous studies of ours, we had already available to us a set of images with the compression rates which were consistent with the best recoverability of astronomical properties. These astronomical properties were based on positional and intensity information, – astrometry and photometry. Papers cited contain details of these studies. Therefore we had optimal compression ratios, and for the corresponding images, we proceeded to measure the Multiscale Entropy. We found a very good correlation. We conclude that Multiscale Entropy provides a good measure of image or signal compressibility.

The breadth and depth of our applications lend credence to the claim that Multiscale Entropy is a good measure of image or signal content. The image data studied is typical not just of astronomy but other areas of the physical and medical sciences. Compared to previous work, we have built certain aspects of the semantics of such data into our analysis procedures. As we have shown, the outcome is a better ability to understand our data.

Could we go beyond this, and justify this work in the context of, for example, content-based image retrieval? Yes, clearly, if the user's query is for data meeting certain SNR requirements, or with certain evidence (which we can provide) of signal presence in very noisy data. For more general content-based querying, this work opens up another avenue of research. This is simply that in querying large data collections, we can at any time allow greater recall, at the expense of precision. Our semantics-related Multiscale Entropy measure can be used for ranking any large recall set. Therefore it can be employed in an interactive image content-based query environment.

8. Astronomical Catalog Analysis

8.1 Introduction

Galaxies are not uniformly distributed throughout the universe. Voids, filaments, clusters, and walls of galaxies can be observed, and their distribution constrains our cosmological theories. Therefore we need reliable statistical methods to compare the observed galaxy distribution with theoretical models and cosmological simulations.

The standard approach for testing models is to define a point process which can be characterized by statistical descriptors. This could be the distribution of galaxies of a specific type in deep redshift surveys of galaxies (or of clusters of galaxies). In order to compare models of structure formation, the different distributions of dark matter particles in N-body simulations could be analyzed as well, with the same statistics. In this chapter we will be concerned with analysis methods which can be applied to catalogs. We will look at:

– Correlation and clustering in galaxy catalogs.
– The Genus function.
– Fractal analysis methods, and models, of data.
– Graph data structures, and in particular the Voronoi diagram.
– Statistical model-based clustering.
– Catalog data noise filtering.

Usually catalogs of extragalactic objects contain the angular coordinates of the objects (galaxies, groups, clusters, superclusters, voids) on the sky. In the equatorial coordinate system they are right ascension (α) and declination (δ). They are galactic longitude (l) and galactic latitude (b) in galactic coordinates, and they are supergalactic longitude (SL) and supergalactic latitude (SB) in supergalactic coordinates. Recently derived extragalactic catalogs contain a great number of objects with their respective redshifts z, so in principle it is possible to transform the angular coordinates together with redshift to a rectangular coordinate system. To do this one has to assume a cosmological model (H_0, q_0) in order to transform the redshift to the distance in Mpc, and also to choose which distance measure to use (e.g. "luminosity distance", "angular diameter distance", "comoving distance").

The choice for the angular coordinate system depends on the problem, and it is convenient to use the system for which the catalog boundaries can be most easily demarcated.

Methods for estimation of the correlation function can be employed also for the particular case of having only angular positions on the sky. This is the case for example in catalogs from radio observations where there is no information for the distance. Then usually the correlation function, denoted $w(\theta)$, is a function of the angular separation θ.

8.2 Two-Point Correlation Function

8.2.1 Introduction

The two-point correlation function $\xi(r)$ has been the primary tool for quantifying large-scale cosmic structure (Peebles, 1980). Assuming that the galaxy distribution in the Universe is a realization of a stationary and isotropic random process, the two-point correlation function can be defined from the probability δP of finding an object within a volume element δV at distance r from a randomly chosen object or position inside the volume:

$$\delta P = n(1 + \xi(r))\delta v, \tag{8.1}$$

where n is the mean density of objects. The function $\xi(r)$ measures the clustering properties of objects in a given volume. It is zero for a uniform random distribution, and positive (respectively, negative) for a more (respectively, less) clustered distribution. For a hierarchical clustering or fractal process, $1+\xi(r)$ follows a power-law behavior with exponent $D_2 - 3$. Since $\xi(r) \sim r^{-\gamma}$ for the observed galaxy distribution, the correlation dimension for the range where $\xi(r) \gg 1$ is $D_2 \simeq 3 - \gamma$. The Fourier transform of the correlation function is the power spectrum. The direct measurement of the power spectrum from redshift surveys is of major interest because model predictions are made in terms of the power spectral density. It seems clear that the real space power spectrum departs from a single power-law ruling out simple unbounded fractal models (Tegmark et al., 2004).

In an unbounded volume embedded in a three-dimensional Euclidean space, we can compute $\xi(r)$ by considering a large number of points N and calculate the average

$$1 + \xi(r) = \frac{N(r)}{N_p(r)} \tag{8.2}$$

where $N(r)$ is the number of pairs of points with a separation in the interval $[r - \Delta r, r + \Delta r]$, and $N_p(r)$ is the number of pairs for a Poisson distribution in the same volume. Since $N_p(r) = 4\pi r^2 n dr$, we have

$$1 + \xi(r) = \frac{1}{N} \sum_{i=1}^{N} \frac{N_i(r)}{4\pi r^2 n dr} \qquad (8.3)$$

where $N_i(r)$ is the number of points lying in a shell of thickness dr, with radius r, and centered at the point labeled i. However, when the calculation has to be performed in a finite volume, the effect of the edges has to be seriously considered. For this reason, other estimators have been proposed, which consider the estimation of the volume around each data point by means of Monte Carlo random catalog generations. In this section we present a description of the most widely-used estimators, and show the results obtained for some samples with well-known or well-studied clustering properties.

8.2.2 Determining the 2-Point Correlation Function

Standard Method. Given a catalog containing N_d points, we introduce a random catalog with N_R points, and denote

- $DD(r)$ = number of pairs in the interval $(r \pm dr/2)$ in the data catalog.
- $RR(r)$ = number of pairs in the interval $(r \pm dr/2)$ in the random catalog.

The two–point correlation function $\xi(r)$ is derived from

$$\tilde{\xi}(r) = \frac{N_R(N_R - 1)}{N_D(N_D - 1)} \frac{DD(r)}{RR(r)} - 1, \qquad (8.4)$$

where $N_R(N_R - 1)/2$ and $N_D(N_D - 1)/2$ are the number of pairs in the random and data catalogs. The ratio of these two values is a normalization term.

Davis-Peebles Method. Davis and Peebles (1983) proposed a more robust estimator by introducing DR, the number of pairs between the data and the random sample within the same interval.

$$\tilde{\xi}_{DP}(r) = 2 \frac{N_R}{N_D - 1} \frac{DD(r)}{DR(r)} - 1 \qquad (8.5)$$

Hamilton Method. Another possibility is to use the Hamilton (1993) approach, which corrects for a presence in the data of large-scale artificial correlation due to some periodicity caused by the volume boundaries or by some selection effects.

$$\tilde{\xi}_{HAM}(r) = \frac{DD(r) RR(r)}{DR^2(r)} - 1 \qquad (8.6)$$

Landy-Szalay Method. The Landy-Szalay estimator (1993) was introduced with the goal of producing a minimum variance estimator and also like the Hamilton estimator is not affected by large-scale correlations:

$$\tilde{\xi}_{LS}(r) = c_1 \frac{DD(r)}{RR(r)} - c_2 \frac{DR(r)}{RR(r)} + 1 \tag{8.7}$$

with

$$\begin{aligned} c_1 &= \frac{N_R(N_R - 1)}{N_D(N_D - 1)} \\ c_2 &= \frac{2N_R(N_R - 1)}{N_D N_R} \end{aligned} \tag{8.8}$$

From simulations, it has been shown that this estimator is better than the others (Pons-Bordería et al., 1999; Kerscher et al., 2000; Martínez and Saar, 2002).

8.2.3 Error Analysis

Assuming the errors in the correlation function are distributed normally (which is not completely true bearing in mind the cross-correlation in the different separation bins) we can estimate the uncertainty as a Poisson statistic for the corresponding errors in bins:

$$\Delta_P \tilde{\xi}(r) = \frac{1 + \tilde{\xi}(r)}{\sqrt{DD(r)}} \tag{8.9}$$

If C random catalogs $R_1, ..., R_C$ are created instead of one, then $\tilde{\xi}(r)$ can be estimated C times, and our final estimate is:

$$\tilde{\xi}(r) = \frac{1}{C} \sum_{i=1}^{C} \tilde{\xi}_i(r) \tag{8.10}$$

and the error is obtained by

$$\Delta_{STD} \tilde{\xi}(r) = \sqrt{\frac{(\tilde{\xi}_i(r) - \tilde{\xi}(r))^2}{C - 1}} \tag{8.11}$$

Finally, a third approach is also popular, and consists of using a bootstrap method (Efron and Tibshirani, 1986). C bootstrap samples $B_1, ..., B_C$ are created by taking randomly with replacement the same number of points that form the original sample. Then the bias-corrected 68% bootstrap confidence interval is $[\xi_{boot}(0.16), \xi_{boot}(0.84)]$, where

$$\xi_{boot}(t) = G^{-1} \left\{ \Phi \left[\Phi^{-1}(t) + 2\Phi^{-1}[G(\xi_0)] \right] \right\}. \tag{8.12}$$

Here G is the cumulative distribution function (CDF) of the $\xi(r)$ values, for a given bin $r \pm \Delta r$, for all bootstrap resamplings, Φ is the CDF of the normal distribution, Φ^{-1} is its inverse function and ξ_0 is the estimated $\xi(r)$ taken from other than the bootstrap resampling results (e.g. from random catalog generations). Note that $\Phi(0.16) = -1.0$ and $\Phi(0.84) = 1.0$.

This confidence estimator is valid when G is not Gaussian (Efron and Tibshirani, 1986). However, this method requires a large number of bootstrap resamplings (usually more than 100) and for large datasets with tens of thousands of points it becomes quite time-consuming.

Note that we can take also the Δ_{STD}^{boot} for the bootstrap resamplings and then it can be simply written as $[\xi_{boot}(0.16), \xi_{boot}(0.84)] = \xi_0 \pm \sigma_\xi$, where

$$\sigma_\xi = \sqrt{\frac{(\xi_{boot}^i(r) - \xi_0(r))^2}{B - 1}}. \tag{8.13}$$

8.2.4 Correlation Length Determination

The two-point correlation function for the gravitational clustering or fractal distribution can be given as a power law:

$$\xi(r) = Ar^{-\gamma} \tag{8.14}$$

where A is the amplitude and γ is the power law index.

The correlation length r_c is defined by

$$\xi(r) = \left(\frac{r}{r_c}\right)^{-\gamma}, \tag{8.15}$$

and it is the separation at which the correlation is 1. This scale in principle divides the regime of strong, non-linear clustering ($\xi \gg 1$) from linear clustering.

It is easy to connect r_c with A and γ by:

$$r_c = \exp^{-\frac{A}{\gamma}} \tag{8.16}$$

8.2.5 Creation of Random Catalogs

One of the crucial steps in estimation of the correlation function is random catalog creation. The catalogs contain data which is subject to various selection effects and incompleteness. Not taking such properties into account could lead to false correlation. The major effects to be considered are the distance selection function – the number of objects as a function of the distance, and the galactic latitude selection function.

The first effect is caused by the geometry of space-time and the detection of only the strongest objects at great distances. For uniform distribution of points in 3D Euclidean space $N(R) \sim R^{-3}$.

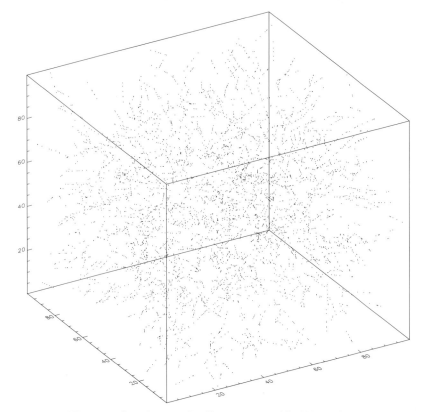

Fig. 8.1. Simulation of a Cox process with 6000 points.

The second effect is caused by light absorption from our Galaxy and it depends on the galactic latitude. Usually it is modeled as a cosec function and in terms of probability density function. It can be given as:

$$P(b) = 10^{\alpha(1-\text{cosec}|b|)}. \tag{8.17}$$

These two effects must be incorporated appropriately in the random catalog generation process.

Since the different catalogs of objects are subject to different selections, it is not possible to have one single procedure for random catalog generation. We will provide, however, versions for some interesting particular cases.

8.2.6 Examples

Simulation of Cox Process. The segment Cox point process (Pons-Bordería et al., 1999) is a clustering process for which an analytical expression of its 2–point correlation function is known and therefore can be used as a test to check the accuracy of the ξ–estimators. Segments of length l are randomly

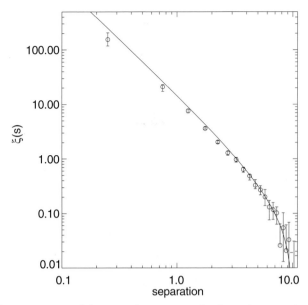

Fig. 8.2. Analytical $\xi_{\text{Cox}}(r)$ curve (*continuous line*), and two–point correlation function of the Cox process with 6000 points overplotted, using the Landy-Szalay method. The error bars are obtained from the minimum and maximum of twenty realizations.

scattered inside a cube W and, on these, segment points are randomly distributed. Let L_V be the length density of the system of segments, $L_V = \lambda_s l$, where λ_s is the mean number of segments per unit volume. If λ_l is the mean number of points on a segment per unit length, then the intensity λ of the resulting point process is

$$\lambda = \lambda_l L_V = \lambda_l \lambda_s l \,. \tag{8.18}$$

For this point field the correlation function can be easily calculated, taking into account that the point field has a driving random measure equal to the random length measure of the system of segments. It has been shown (Stoyan et al., 1995) that

$$\xi_{\text{Cox}}(r) = \frac{1}{2\pi r^2 L_V} - \frac{1}{2\pi r l L_V} \tag{8.19}$$

for $r \leq l$ and vanishes for larger r. The expression is independent of the intensity λ_l. Fig. 8.1 shows the simulation of a Cox process with 6000 points. Fig. 8.2 shows the analytical $\xi_{\text{Cox}}(r)$ curve (continuous line), and the estimated two–point correlation function overplotted. The Landy-Szalay method was used with 10000 random points. The errors are the results from 20 random Cox process realizations.

240   8. Astronomical Catalog Analysis

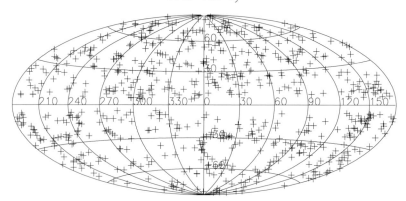

Fig. 8.3. Aitoff equal-area projection in galactic coordinates of IRAS galaxies with $F_{60\mu m} > 1.2$ Jy and distance < 100 Mpc. Their total number is 710.

Application to IRAS Data. We present in this section the two point correlation function analysis of the IRAS 1.2 Jy Redshift Survey (Fisher et al., 1995) for a volume limited subsample.

In order to create a volume limited subsample from the IRAS catalog, we applied the following steps:

- Extract from the catalog the right ascension α (hh,mm,ss,1950), the declination δ (sign,dg,mm,ss,1950), and the velocity $Hvel$ (km/s).
- Convert α, δ to galactic coordinates l, b (in radians) because the catalog boundaries ($|b| > 5$ deg) are most easily defined in this system.
- Convert velocity to redshift ($z = Hvel/c$).
- Assuming $H_0 = 100$ and $\Omega = 1$, calculate the distance d by the luminosity distance formulae proposed in (Pen, 1999):

$$d_L = \frac{c}{H_0}(1+z)[F(1,\Omega_0) - F(\frac{1}{1+z},\Omega_0)]$$

$$F(a,\Omega_0) = 2\sqrt{s^3+1}[\frac{1}{a^4} - 0.1540\frac{s}{a^3} + 0.4302\frac{s^2}{a^2}$$

$$+ 0.19097 * \frac{s^3}{a} + 0.066941 s^4]^{-\frac{1}{8}}$$

$$s^3 = \frac{1-\Omega_0}{\Omega_0} \qquad (8.20)$$

- Select galaxies (statusflag in [O,H,Z,F,B,D,L]) with distance $d < 100$ Mpc, and flux $F_{60\mu m} > 1.2$ Jy in the galaxy rest frame. So the luminosity of a galaxy is given by:

$$L = 4\pi d^2 F_{60\mu m} \qquad (8.21)$$

and the luminosity of a galaxy at the limiting distance (100 Mpc) with the limiting flux (1.2 Jy) is

$$L_{limit} = 4\pi 100^2 (1.2) \tag{8.22}$$

We select all the galaxies with L larger than L_{limit}.
- Calculate the coordinates in a cube:

$$\begin{aligned} X &= d\cos(b)\cos(l) \\ Y &= d\cos(b)\sin(l) \\ Z &= d\sin(b) \end{aligned} \tag{8.23}$$

Fig. 8.3 shows the galaxy positions. The result for the redshift space correlation function for the combined north+south IRAS catalog is presented in Fig. 8.4. The result is consistent with published results for this catalog (Fisher et al., 1995):

$$r_0 = 4.27^{+0.66}_{-0.81} \text{ and } \gamma = 1.68^{+0.36}_{-0.29} \tag{8.24}$$

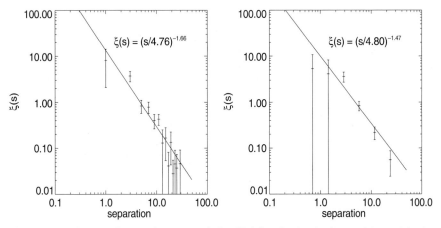

Fig. 8.4. *Left:* correlation function of the IRAS galaxies in linear bins with the corresponding linear least square fit for the data in separation range 1 – 20 Mpc. *Right:* correlation function of the IRAS galaxies in logarithmic bins.

For comparison we present the correlation function for the same data catalog but in logarithmic separation bins. As is clear from Fig. 8.4, right, strong fluctuations for the correlation function at large separations are significantly smoothed.

Application to Numerical Simulations – ΛCDM Model. For cosmological studies it is very important to test the predictions of various cosmological models for the clustering properties of matter, and to put constraints on various parameters by analyzing the results from numerical simulations and their correspondence to what is observed. In simulations we have a large parameter space of the objects (coordinates, velocities, masses, and so on) so it is natural to examine clustering properties by means of various statistical

242   8. Astronomical Catalog Analysis

tools used in the analysis of the observational data: correlation functions, power spectrum analysis, etc.

We will present here the results for the correlation function for one cosmological model (ΛCDM, $h = 0.7$, $\Omega_0 = 0.3$, $\Omega_\Lambda = 0.7$) from a Hubble volume simulation. The data are available at the following address: http://www.physics.lsa.umich.edu/hubble-volume

We extracted a volume limited slice with objects with redshift less than 0.4 (for the cosmological model this corresponds to 1550 Mpc). A view of the data is presented in Fig. 8.5, left, for the XY plane and in Fig. 8.5 right for the XZ plane. All the points represent groups or clusters of galaxies with masses greater than $\sim 6.6 \times 10^{13} M_*$.

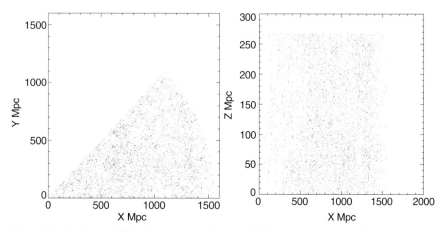

Fig. 8.5. *Left:* the XY plane view of the ΛCDM slice used for the correlation function analysis. The opening angle is 45 deg. and the total number of objects is 6002. *Right:* the XZ plane view of the ΛCDM slice.

The results with the corresponding linear least squares fit are presented in Fig. 8.6, left, for linear separation bins and in Fig. 8.6, right, for logarithmic.

The results are consistent with the normalization used in the simulations – the clustering properties of the simulation should correspond to the observed clustering for redshift of 0 ($r_0 \approx 15$, $\gamma \approx 1.8$).

8.2.7 Limitation of the Two-Point Correlation Function: Toward Higher Moments

In order to illustrate the limitation of the two-point correlation function, we use two simulated data sets. The first one is a simulation from stochastic geometry. It is based on a Voronoi model. The second one is a mock catalog of the galaxy distribution drawn from a Λ-CDM N-body cosmological model

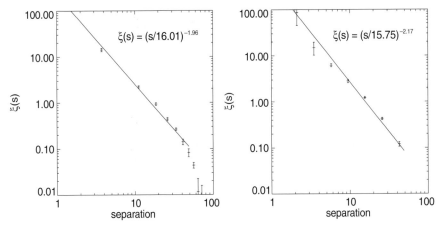

Fig. 8.6. *Left:* the correlation function of the ΛCDM model for linear separation bins. The fit is carried out in 1–50 Mpc separations and the error bars are the standard deviations of 5 random catalog generations (the second method for estimating the uncertainty of the correlation function, i.e. Δ_R). *Right:* the correlation function for logarithmic separation bins.

(Kauffmann et al., 1999). Both processes have very similar two-point correlation functions at small scales, although they look quite different and were generated following completely different algorithms.

- The first comes from Voronoi simulation: We locate a point in each of the vertices of a Voronoi tessellation of 1500 cells defined by 1500 nuclei distributed following a binomial process. There are 10,085 vertices lying within a box of 141.4 h^{-1} Mpc side.
- The second point pattern represents the galaxy positions extracted from a cosmological Λ-CDM N-body simulation. The simulation was carried out by the Virgo consortium and related groups (see http://www.mpa-garching.mpg.de/Virgo). The simulation is a low-density ($\Omega = 0.3$) model with cosmological constant $\Lambda = 0.7$. It is, therefore, an approximation to the real galaxy distribution (Kauffmann et al., 1999). There are 15,445 galaxies within a box with side 141.3 h^{-1} Mpc. Galaxies in this catalog have stellar masses exceeding $2 \times 10^{10}\ M_\odot$.

Fig. 8.7 shows the two simulated data sets, and Fig. 8.8 shows the two-point correlation function curve for the two point processes. The two point fields are different, but as can be seen in Fig. 8.8, both have very similar two-point correlation functions in a huge range of scales (2 decades).

In order to improve the discrimination power, we need to use higher order statistics. The two-point correlation function can been generalized to the N-point correlation function (Szapudi and Szalay, 1998; Peebles, 2001), and the entire hierarchy can be related with the physics responsible for the clustering of matter. Nevertheless they are difficult to measure, and therefore

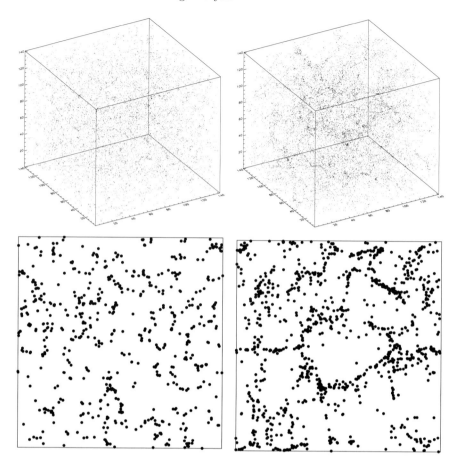

Fig. 8.7. Simulated data sets. Top, the Voronoi vertices point pattern (*left*) and the galaxies of the GIF Λ-CDM N-body simulation (*right*). The bottom panels show one $10\ h^{-1}$ width slice of the each data set.

other related statistical measures have been introduced as a complement in the statistical description of the spatial distribution of galaxies (Martínez and Saar, 2002), such as the void probability function (Maurogordato and Lachieze-Rey, 1987), the multifractal approach (Martínez et al., 1990), the minimal spanning tree (Bhavsar and Splinter, 1996; Krzewina and Saslaw, 1996; Doroshkevich et al., 2001), the Minkowski functionals (Mecke et al., 1994; Kerscher, 2000) or the J function (Lieshout and Baddeley, 1996; Kerscher et al., 1999) which is defined as the ratio

$$J(r) = \frac{1 - G(r)}{1 - F(r)} \qquad (8.25)$$

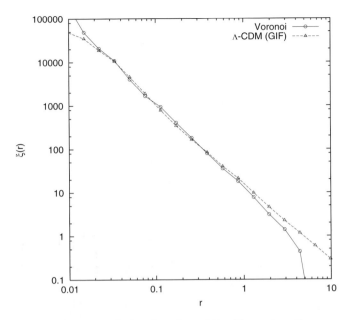

Fig. 8.8. The two-point correlation function of the Voronoi vertices process and the GIF Λ-CDM N-body simulation. They are very similar in the range $[0.02,2]\ h^{-1}$ Mpc.

where F is the distribution function of the distance r of an arbitrary point in $\mathbf{R}^3$ to the nearest object in the catalog, and G is the distribution function of the distance r of an object to the nearest object. Wavelets have also been used for analyzing the projected 2D or the 3D galaxy distribution (Escalera et al., 1992; Slezak et al., 1993; Martínez et al., 1993b; Pagliaro et al., 1999; Kurokawa et al., 2001). Some of these methods are described in the following.

8.3 The Genus Curve

The two-point correlation function is not sensitive to the phase of the Fourier transform of the data. This explains why it describes poorly the spatial distribution of the galaxy. The first morphological descriptor used was the genus (Gott et al., 1986a). The genus $G(S)$ measures the connectivity of a surface, S, with holes and disconnected pieces, by the difference of the number of holes and the number of isolated regions:

$G(S) =$ number of holes $-$ number of isolated regions $+$ 1.

The genus of a sphere is $G = 0$, a torus or a sphere with a handle has the genus $G = +1$, a sphere with N handles has the genus $G = +N$, while the collection of N disjoint spheres has the genus $G = -(N-1)$. The genus describes the topology of the isodensity surfaces. Thus its study is, in the cosmological literature, frequently called "topological analysis".

The genus curve is usually parameterized by two related quantities, the filling factor, f, which is the fraction of the survey volume above the density threshold or, alternatively, by the quantity ν defined by

$$f = \frac{1}{\sqrt{2\pi}} \int_{\nu}^{\infty} e^{-t^2/2} dt. \tag{8.26}$$

In the case of a Gaussian random field, ν is also the number of standard deviations by which the threshold density departs from the mean density, and with this parametrization, the genus per unit volume of a surface, S, corresponding to a given density threshold, $g \equiv (G(S) - 1)/V$, follows the analytical expression

$$g(\nu) = N(1 - \nu^2) \exp\left(-\frac{\nu^2}{2}\right), \tag{8.27}$$

where the amplitude N depends on the power spectrum of the random field (Hamilton et al., 1986).

In practice, the genus is calculated by (i) convolving the data by a kernel, generally a Gaussian, (ii) setting to zero all values under a threshold ν in the obtained distribution, and (iii) taking the difference D between the number of holes and the number of isolated regions. The genus curve $G(\nu)$ is obtained by varying the threshold level ν.

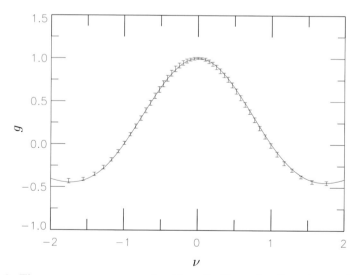

Fig. 8.9. The average genus curve for 50 realizations of a Gaussian random field with $P(k) \sim k^{-1}$ together with the expected analytical result (*solid line*). The error bars show 1 σ deviations.

Using the algorithm described in (Martínez et al., 2005), we have calculated the genus curve for 50 realizations of a Gaussian random field with a power-law power spectrum $P(k) \sim k^{-1}$ in a 128^3 box. The realizations were smoothed with a Gaussian kernel of $\sigma = 3$. The results are shown in Fig. 8.9 and are very close to the theoretical expectations.

When rich clusters dominate the distribution, the genus curves are shifted to the left, and the morphology is referred to as "meat-ball", while the expression "Swiss-cheese" is used for right-shifted genus curves corresponding to distributions with empty bubbles surrounded by a single high density region. As discussed in (Martínez et al., 2005), the first step of the algorithm, the convolution by a Gaussian, may be dramatic for the description of filaments, which are spread out along all directions, and better results are obtained if we replace the Gaussian smoothing by a wavelet denoising.

8.4 Minkowski Functionals

An elegant generalization of the genus statistic is to consider this measure as one of the four Minkowski functionals which describe different morphological aspects of the galaxy distribution (Mecke et al., 1994). These functionals provide a complete family of morphological measures since all additive, motion invariant and conditionally continuous functionals defined for any hypersurface are linear combinations of its Minkowski functionals.

The Minkowski functionals (MF for short) describe the morphology of isodensity surfaces, and depend thus on two factors – the smoothing procedure and the specific density level (see (Sheth and Sahni, 2005) for a recent review). An alternative approach starts from the point field, decorating the points with spheres of the same radius, and studying the morphology of the resulting surface (Schmalzing et al., 1996). These functionals depend only on one parameter (the radius of the spheres), but this approach does not refer to a density; we shall not use that for the present study.

The Minkowski functionals are defined as follows. Consider an excursion set F_ϕ of a field $\phi(\mathbf{x})$ in 3-D: i.e., the set of all points where $\phi(\mathbf{x} \geq \phi)$. Then, the first Minkowski functional (the volume functional) is the volume of the excursion set:

$$V_0(\phi) = \int_{F_\phi} d^3x.$$

The second MF is proportional to the surface area of the boundary δF_ϕ of the excursion set:

$$V_1(\phi) = \frac{1}{6} \int_{\delta F_\phi} dS(\mathbf{x}).$$

The third MF is proportional to the integrated mean curvature of the boundary:

$$V_2(\phi) = \frac{1}{6\pi} \int_{\delta F_\phi} \left(\frac{1}{R_1(\mathbf{x})} + \frac{1}{R_2(\mathbf{x})} \right) dS(\mathbf{x}),$$

where R_1 and R_2 are the principal curvatures of the boundary. The fourth Minkowski functional is proportional to the integrated Gaussian curvature (the Euler characteristic) of the boundary:

$$V_3(\phi) = \frac{1}{4\pi} \int_{\delta F_\phi} \frac{1}{R_1(\mathbf{x}) R_2(\mathbf{x})} dS(\mathbf{x}).$$

The last MF is simply related to the morphological genus g introduced in the previous subsection by

$$V_3 = \chi = \frac{1}{2}(1 - G)$$

(χ is the usual notation for the Euler characteristic). The functional V_3 is somewhat more comfortable to use – it is additive, while G is not, and it gives just twice the number of isolated balls (or holes). Although the genus remains to be widely used, in several recent papers authors have chosen to present the Minkowski functional V_3.

Instead of the functionals, their spatial densities V_i are frequently used:

$$v_i(f) = V_i(f)/V, \quad i = 0, \ldots, 3,$$

where V is the total sample volume.

All the Minkowski functionals have analytic expressions for isodensity slices of realizations of Gaussian random fields. For three-dimensional space they are:

$$\begin{aligned}
v_0 &= \frac{1}{2} - \frac{1}{2} \Phi\left(\frac{\nu}{\sqrt{2}}\right), \\
v_1 &= \frac{2}{3} \frac{\lambda}{\sqrt{2\pi}} \exp\left(-\frac{\nu}{2}\right), \\
v_2 &= \frac{2}{3} \frac{\lambda^2}{\sqrt{2\pi}} \nu \exp\left(-\frac{\nu}{2}\right), \\
v_3 &= \frac{\lambda^3}{\sqrt{2\pi}} (\nu^2 - 1) \exp\left(-\frac{\nu}{2}\right),
\end{aligned}$$

where $\Phi(\cdot)$ is the Gaussian error integral, and λ is determined by the correlation function $\xi(r)$ of the field as:

$$\lambda^2 = \frac{1}{2\pi} \frac{\xi''(0)}{\xi(0)}.$$

Numerical Algorithms

Several algorithms can be used to calculate the Minkowski functionals for a given density field and a given density threshold. We can either try to follow exactly the geometry of the isodensity surface, e.g., using triangulation (Sheth et al., 2003), or to approximate the excursion set on a simple cubic lattice. The algorithm that was proposed first by (Gott et al., 1986b) uses a decomposition of the field into filled and empty cells, and another popular algorithm (Coles et al., 1996) uses a grid-valued density distribution. The lattice-based algorithms are simpler and faster, but not as accurate as the triangulation codes. The main difference is in the edge effects – while surface triangulation algorithms do not suffer from these, edge effects may be rather serious for the lattice algorithms. In (Martínez et al., 2005), a simple grid-based algorithm has been proposed. It consists of finding the density thresholds for given filling fractions by sorting the grid densities, first. Vertices with higher densities than the threshold form the excursion set. This set is characterized by its basic sets of different dimensions – points (vertices), edges formed by two neighboring points, squares (faces) formed by four edges, and cubes formed by six faces. The algorithm counts the numbers of all basic sets, and finds the values of the Minkowski functionals as

$$V_0(f) = a^3 N_3,$$
$$V_1(f) = a^2 \left(\frac{2}{9}N_2(f) - \frac{2}{3}N_3(f)\right),$$
$$V_2(f) = a \left(\frac{2}{9}N_1(f) - \frac{4}{9}N_2(f) + \frac{2}{3}N_3(f)\right),$$
$$V_3(f) = N_0(f) - N_1(f) + N_2(f) - N_3(f),$$

where a is the grid step, f is the filling factor, N_0 is the number of vertices, N_1 is the number of edges, N_2 is the number of squares (faces), and N_3 is the number of basic cubes in the excursion set for a given filling factor (density threshold). This formula was proven by (Adler, 1981) and was first used in cosmological studies by (Coles et al., 1996).

This algorithm is simple to program, and it gives excellent results, provided the grid step is substantially smaller than the characteristic lengths of the isosurfaces (the smoothing length). This is needed to be able to accurately follow the geometry of the surface. It is also very fast, allowing us to use Monte-Carlo simulations for error estimation (Martínez et al., 2005).

8.5 Fractal Analysis

8.5.1 Introduction

The word "fractal" was introduced by Mandelbrot (1983), and comes from the Latin word *fractus* which means "break". According to Mandelbrot, a fractal is an object which has a greater dimension than its topological dimension. A typical fractal is the Cantor set. It is built in the following way: considering a segment of dimension L, we separate it into three equal parts, and suppress the middle part. There remain two segments of size $\frac{L}{3}$. Repeating the process on both segments, we get four segments of size $3^{-2}L$. After n iterations, we have 2^n segments of size $3^{-n}L$. The Cantor set is obtained when n tends to infinity. The set has the following properties:

1. It is self-similar.
2. It has a fine structure, i.e. detail on arbitrary small scales.
3. It is too irregular to be described in traditional geometrical language, both locally and globally.
4. It is obtained by successive iterations.
5. It has an infinite number of points but its length is negligeable.

These properties define in fact a fractal object.

A real fractal does not exist in nature, and we always need to indicate at which scales we are talking about fractality. It is now well-established that the universe is fractal at small scales ($r < 10h^{-1}\mathrm{Mpc}$) (Durrer and Labini, 1998; Sylos Labini, 1999; Joyce et al., 1999; Martínez et al., 2001; Pietronero and Sylos Labini, 2001; Gaite and Manrubia, 2002; Ribeiro, 2005; Seshadri, 2005). A recent review of the fratal approach to large-scale galaxy distribution can be found in (Yadav et al., 2005).

8.5.2 The Hausdorff and Minkowski Measures

Measure. An object dimension describes how an object F fills space. A simple manner of measuring the length of curves, the area of surfaces or the volume of objects is to divide the space into small boxes (segment in one dimension, surface in 2D, and cubes in 3D) of diameter δ. These boxes are chosen so that their diameter is not greater than a given size δ, which corresponds to the measure resolution.

We consider now the quantity:

$$L_\delta^d(F) = \sum diam(B_i)^d \tag{8.28}$$

where d is a real number, $diam(B_i)$ is the diameter of the box i. $L_\delta^s(F)$ represents an estimation of the size of F at the resolution δ. Depending on the choice of the boxes, the measure is more or less correct. Generally, it is easier to manipulate the Minkowski-Bouligand measure which fixes all the boxes to the same size δ. Using this measure, the size of F is given by:

$$M^s(F) = \lim_{\delta \to 0} \sum diam(B_i)^s = \delta^s N_B(\delta) \tag{8.29}$$

where N_B is the number of boxes needed to cover the object F. Then the curves of length L_* can be measured by finding the number $N_B(\delta)$ of line segments (respectively squares and cubes for the second and third object) of length δ needed to cover the object. The three sizes are:

$$L = M^1(F) = N_B(\delta)\delta^1 \underset{\delta \to 0}{\to} L_*\delta^0 \tag{8.30}$$

$$A = M^2(F) = N_B(\delta)\delta^2 \underset{\delta \to 0}{\to} L_*\delta^1 \tag{8.31}$$

$$V = M^3(F) = N_B(\delta)\delta^3 \underset{\delta \to 0}{\to} L_*\delta^2 \tag{8.32}$$

8.5.3 The Hausdorff and Minkowski Dimensions

The Hausdorff dimension d_H of the set F is the *critical dimension* for which the measure $H^d(F)$ jumps from infinity to zero:

$$H^d(F) = \begin{cases} 0, & d > d_H, \\ \infty, & d < d_H. \end{cases} \tag{8.33}$$

But $H^{d_H}(F)$ can be finite or infinite. For a simple set (segment, square, cube), Hausdorff dimension is equal to the topological dimension (i.e. 1, 2, or 3). This is not true for a more complex set, such as the Cantor set. Minkowski dimension d_M (and $d_H(F) \leq d_M(F)$) is defined in a similar way using Minkowski measure.

By definition, we have:

$$M^{d_M} = \lim_{\delta \to 0} \delta^{d_M} N_B(\delta) \tag{8.34}$$

When $\delta \to 0$, we have $d_M \ln M = d_M \ln \delta + \ln N_N(\delta)$. If M is finite, the Minkowski dimension, also called box-counting, can be defined by

$$d_M = \lim_{\delta \to 0} \frac{\ln N_B(\delta)}{-\ln \delta} \tag{8.35}$$

In the case of the Cantor set, at iteration n, we have 2^n segments of size 3^{-n} ($\delta = 3^{-n}$). When $n \to \infty$, we have

$$d_M(Cantor) = \frac{\ln 2^n}{-\ln 3^{-n}} = \frac{\ln 2}{\ln 3} \tag{8.36}$$

8.5.4 Multifractality

The multifractal picture is a refinement and generalization of the fractal properties that arise naturally in the case of self-similar distributions. The singularity spectrum $f(\alpha)$ can be introduced as a quantity which characterizes the degree of regularity and homogeneity of a fractal measure.

Hölder Exponent. A multifractal measure describes a non-homogeneous set A. Such a measure is called multifractal if it is everywhere self-similar, i.e. if the measure varies locally as a power law, at any point of A. Denoting μ a measure, we call the Hölder exponent or singularity exponent at x_0 the limit

$$\alpha(x_0) = \lim_{\delta \to 0} \frac{\ln \mu(B_{x_0}(\delta))}{\ln \delta} \tag{8.37}$$

where B_{r_0} is a box centered at r_0 of size δ. We have:

$$\mu(B_{x_0}(\delta)) \propto \delta^{\alpha(x_0)} \tag{8.38}$$

The smaller the value $\alpha(x_0)$, the less the measure is regular around x_0. For example, if μ corresponds to a Dirac distribution centered at 0, then $\alpha(0) = 0$, and if μ corresponds to a Gaussian distribution, then $\alpha(0) = -1$.

Singularity Spectrum. The singularity spectrum, associated with a measure μ, is the function which associates with α the fractal dimension of any point x_0 such that $\alpha(x_0) = \alpha$:

$$f(\alpha) = d_F(\{x_0 \in A \mid \alpha(x_0) = \alpha\}) \tag{8.39}$$

The function $f(\alpha)$ is usually (Paladin and Vulpiani, 1987) a single-humped function with the maximum at $\max_\alpha f(\alpha) = D$, where D is the dimension of the support. In the case of a single fractal, the function $f(\alpha)$ is reduced to a single point: $f(\alpha) = \alpha = D$.

The singularity spectrum describes statistically the α exponent distribution on the measure support. For example, if we split the support into boxes of size δ, then the number of boxes with a measure varying as δ^α for a given α is

$$N_\alpha(\delta) \propto \delta^{-f(\alpha)} \tag{8.40}$$

$f(\alpha)$ describes the histogram of $N_\alpha(\delta)$ when δ is small. A measure is homogeneous if its singularity spectrum is concentrated in a single point. If $f(\alpha)$ is large, the measure is multifractal.

Multifractal Quantities. From a practical point of view one does not determine directly the spectrum of exponents $[f(\alpha), \alpha]$; it is more convenient to compute its Legendre transformation $[\tau(q), q]$ given by

$$\begin{cases} f(\alpha) = q \cdot \alpha - \tau(q) \\ \alpha = \frac{d\tau(q)}{dq} \end{cases} \tag{8.41}$$

In the case of a simple fractal one has $\alpha = f(\alpha) = D$. In terms of the Legendre transformation this corresponds to

$$\tau(q) = D(q-1) \tag{8.42}$$

i.e. the behavior of $\tau(q)$ versus q is a straight line with coefficient given by the fractal dimension.

8.5.5 Generalized Fractal Dimension

Definition. The generalized fractal dimension, also called Rényi dimension of order q, is given by:

$$D_q = \frac{\tau(q)}{q-1} \qquad (8.43)$$

D_0 is also called capacity dimension, and coincides with the Hausdorff dimension. Dimensions D_1, and D_2 are respectively called information and correlation dimension.

Partition Function. The partition function Z is defined by:

$$Z(q,\delta) = \sum_{i=1}^{N(\delta)} \mu_i^q(\delta) \qquad (8.44)$$

where we denote $\mu_i(\delta) = \mu(B_i(\delta))$. If the measure μ is multifractal, Z follows a power law in the limit $\delta \to 0$.

$$Z(q,\delta) \propto \delta^{\tau(q)} \qquad (8.45)$$

The box-counting method consists of calculating the partition function, to derive $\tau(q)$ from Z, and to obtain the multifractal spectrum by a Legendre transform.

8.5.6 Wavelets and Multifractality

Singularity Analysis. Let $f(x)$ be the input signal, x_0 the singularity location, $\alpha(x_0)$ the Hölder exponent at the singularity point x_0 and n the degree of Taylor development such that $n \leq \alpha(x_0) < n+1$. We have

$$\begin{aligned} f(x) &= f(x_0) + (x-x_0)f^{(1)}(x_0) + ... + \\ &\quad \frac{(x-x_0)^n}{n!} f^{(n)}(x_0) + C\,|x-x_0|^{\alpha(x_0)} \end{aligned} \qquad (8.46)$$

Letting ψ be the wavelet with $n_\psi > n$ vanishing moments, then we have for the wavelet transform of $f(x)$ at x_0 when the scale goes to 0 (ψ is orthogonal to polynomials up to order n):

$$\lim_{\text{scale} s \to 0^+} T_\psi[f](x_0, s) \sim a^{\alpha(x_0)} \qquad (8.47)$$

One can prove that if f is C^∞, then we have

$$\lim_{\text{scale} s \to 0^+} T_\psi[f](x_0, s) \sim a^{n_\psi} \qquad (8.48)$$

Thus, we have

$$\begin{cases} T_\psi[f] \sim s^{n_\psi} & \text{where the signal } f \text{ is regular} \\ T_\psi[f] \sim s^\alpha (\gg s^{n_\psi}) & \text{around a singular zone} \end{cases} \quad (8.49)$$

For a fixed scale s, $T_\psi[f](.,s)$ will be greater when the signal is singular. This local maximum is organized in maxima lines (function of s) which converges, when s goes to 0, to a singularity of the signal. Mallat and Hwang (1992) demonstrate that to recover the Hölder exponent $\alpha(x_0)$ at x_0, one need only study the wavelet transform along these lines of maxima which converge (when the scale goes to 0) towards the singularity point x_0.

Along this maxima line l we have

$$T_\psi[f](b,s) \sim a^{\alpha(x_0)}, \ (b,a) \in l, s \to 0^+ \quad (8.50)$$

Fig. 8.10 displays the function $f(x) = K(x - x_0)^{0.4}$ with a singular point at x_0. The Hölder exponent at x_0 is equal to 0.4. In Fig. 8.11, we display the wavelet transform of $f(x)$ with a wavelet ψ which is the first derivative of a Gaussian. In Fig. 8.12, we display $log_2|T_\psi[f](x,s)|$ as a function of $log_2(s)$.

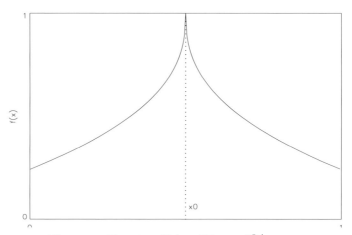

Fig. 8.10. Function $f(x) = K(x - x_0)^{0.4}$.

When we compute the slope of the curve $log_2|T_\psi[f](x,s)|$ versus $log_2(s)$ along a maxima line which converges at x_0, we obtain an estimation of the Hölder exponent (in this case $\alpha(x_0) \approx 0.4$ which corresponds to the theoretical value).

Wavelet Transform of Multifractal Signals. The estimation of the Hölder exponent by this method becomes inaccurate in the case of multifractal signals (Arneodo et al., 1995). We need to use a more global method. One can define the wavelet-based partition function by

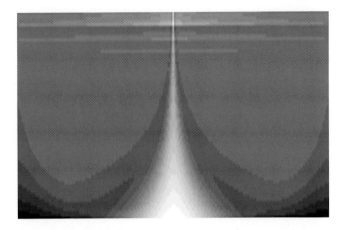

Fig. 8.11. Wavelet transform of $f(x)$ with a wavelet ψ which is the first derivative of a Gaussian. The small scales are at the top. The maxima line converges to the singularity point at x_0.

$$Z(q,s) = \sum_{b_i} |T_\psi[\mu](b_i,s)|^q \tag{8.51}$$

where $(b_i, s)_i$ are all local maxima at scale s.

Let $\tau(q)$ be the scaling exponent. We can prove that we have

$$Z(q,s) \sim a^{\tau(q)} \tag{8.52}$$

We can then calculate the singularity spectrum $D(\alpha)$ by its Legendre transformation

$$D(\alpha) = \min_q (q\alpha - \tau(q)) \tag{8.53}$$

This method is called the *Wavelet Transform Modulus Maxima* (WTMM).

Numerical Applications of WWTM Method. The calculation of the singularity spectrum of signal f proceeds as follows:

- compute the wavelet transform and the modulus maxima $T_\psi[f]$ for all (s,q). We chain all maxima across scale lines of maxima,
- compute the partition function $Z(q,s) = \sum_{b_i} |T_\psi[f](b_i,s)|^q$,
- compute $\tau(q)$ with $\log_2 Z(q,s) \approx \tau(q) \log_2(s) + C(q)$,
- compute $D(\alpha) = \min_q (q\alpha - \tau(q))$.

The Triadic Cantor Set. Definition: The measure associated with the triadic Cantor set is defined by $f(x) = \int_0^x d\mu$ where μ is the uniform measure lying on the triadic Cantor set described in section 8.5.1. To compute $f(x)$, we used the next recursive function called the Devil's Staircase function (which

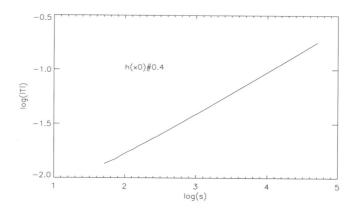

Fig. 8.12. Estimation of the Hölder exponent $\alpha(x_0)$ at x_0 by computing the slope of $\log_2 |T_\psi[f](x,s)|$ versus $\log_2(s)$ along a maxima line which converges to x_0.

looks like a staircase whose steps are uncountable and infinitely small: see Fig. 8.13).

$$f(x) = \begin{cases} p_1 f(3x) & \text{if } x \in [0, \tfrac{1}{3}] \\ p_1 & \text{if } x \in [\tfrac{1}{3}, \tfrac{2}{3}] \\ p_1 + p_2 f(3x - 2) & \text{if } x \in [\tfrac{2}{3}, 1] \end{cases} \tag{8.54}$$

This is a continuous function that increases from 0 to 1 on [0,1]. The recursive construction of $f(x)$ implies that $f(x)$ is self-similar.

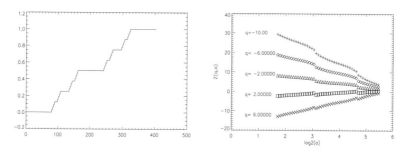

Fig. 8.13. Classical Devil's Staircase function (associated with triadic Cantor set) with $p_1 = 0.5$ and $p_2 = 0.5$ (*left*) and partition function $Z(q, s)$ for several values of q (*right*). The wavelet transform is calculated with ψ equal to the first derivative of a Gaussian.

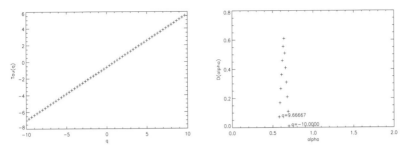

Fig. 8.14. Scaling Exponent estimation $\tau(q)$ (*left*) and Singularity Spectrum $D(\alpha)$ (*right*). The Scaling Exponent curve corresponds to the theoretical curve $\tau(q) = (q-1)\log_2(2)/\log_2(3)$. Points of the Singularity Spectrum where $q \neq \infty$(max q) and $q \neq -\infty$(min q) are reduced to a single point ($\alpha = \log_2(2)/\log(3), D(\alpha) = \log_2(2)/\log_2(3)$). This point corresponds to the Hausdorff dimension of the triadic Cantor Set.

The Generalized Devil's Staircase with $p_1 = 0.4$ and $p_2 = 0.6$. One can prove (Arneodo et al., 1995) that the theoretical singularity spectrum $D(\alpha)$ of the generalized Devil's Staircase function $f(x) = \int_0^x d\mu$ verifies the following:

- The singular spectrum is a convex curve with a maximum value α_{max} which corresponds to the fractal dimension of the support of the measure (μ).
- The theoretical support of $D(\alpha)$ is reduced at the interval $[\alpha_{min}, \alpha_{max}]$:

$$\begin{cases} \alpha_{min} = \min(\frac{\ln p_1}{\ln(1/3)}, \frac{\ln p_2}{\ln(1/3)}) \\ \alpha_{min} = \min(\frac{\ln p_1}{\ln(1/3)}, \frac{\ln p_2}{\ln(1/3)}) \end{cases} \quad (8.55)$$

Fig. 8.15 displays the generalized Devil's Staircase and its partition function $Z(q,s)$. In Fig. 8.16, we can see the Scaling Exponent and the Singularity Spectrum. This one is in perfect "accord" with the theoretical values: bell curve, $D(\alpha)_{max} = \log_2(2)/\log_2(3)$, $\alpha_{min} \approx 0.47$ and $\alpha_{max} \approx 0.83$.

8.6 Spanning Trees and Graph Clustering

The best match or nearest neighbor problem is important in many disciplines. In the database and more particularly data mining field, nearest neighbor searching is called similarity query, or similarity join (Bennett et al., 1999). A database record or tuple may be taken as a point in a space of dimensionality m, the latter being the associated number of fields or attributes. Fast methods for nearest neighbor finding, and their use as algorithmic building blocks in hierarchical clustering, can be found in (Murtagh, 1985). Applications of best

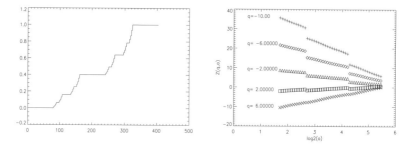

Fig. 8.15. Devil's Staircase function with $p_1 = 0.4$ and $p_2 = 0.6$ (*left*) and partition function $Z(q, s)$ for several values of q (*right*). The wavelet transform is calculated with ψ equal to the first derivative of a Gaussian.

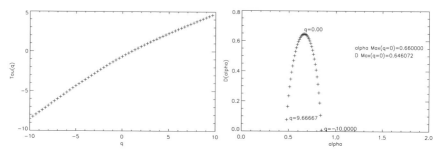

Fig. 8.16. Scaling Exponent estimation $\tau(q)$ (*left*) and Singularity Spectrum $D(\alpha)$ (*right*). The theoretical maximum value of $D(\alpha)$ is obtained for $\alpha_{maxD} = \log_2(2)/\log_2(3)$ and $D(\alpha_{maxD}) = \log_2(2)/\log_2(3)$.

match finding in solar astronomy are to be found in (Csillaghy et al., 2000; Csillaghy and Benz, 1999). Data preprocessing may be used to attempt to break the $O(n)$ barrier for determining the nearest neighbor of a data object, and many algorithms have been proposed over the years which remain linear time algorithms but with low constants of proportionality.

Data mining is often carried out in sparse information spaces, i.e., relationships between the data objects are few and far between. Under these circumstances, a graph is a good (and storage-efficient) data model.

Single linkage clustering is a very commonly used clustering method. Rohlf (1982) reviews algorithms for the single link method with complexities ranging from $O(n \log n)$ to $O(n^5)$. The criterion used by the single link method for cluster formation is weak, meaning that noisy data in particular give rise to results which are not robust.

The minimal spanning tree (MST) and the single link agglomerative clustering method are closely related: the MST can be transformed irreversibly into the single link hierarchy (Rohlf, 1973). The MST is defined as of minimal total weight, it spans all nodes (vertices) and is an unrooted tree. The MST has been a method of choice for at least four decades now either in its own right for data analysis (Zahn, 1971), as a data structure to be approximated, e.g. using shortest spanning paths (Murtagh, 1985), or as a basis for clustering. MST has been used for many years for large-scale galaxy distribution analysis (Bhavsar and Splinter, 1996; Krzewina and Saslaw, 1996; Doroshkevich et al., 2001). We will look at some fast algorithms for the MST in the remainder of this section.

Perhaps the most basic MST algorithm, due to Prim and Dijkstra (Horowitz and Sahni, 1978), grows a single fragment through $n-1$ steps. We find the closest vertex to an arbitrary vertex, calling these a fragment of the MST. We determine the closest vertex, not in the fragment, to any vertex in the fragment, and add this new vertex into the fragment. While there are fewer than n vertices in the fragment, we continue to grow it.

This algorithm leads to a unique solution. A default $O(n^3)$ implementation is clear, and $O(n^2)$ computational cost is possible (Murtagh, 1985).

Sollin's algorithm constructs the fragments in parallel. For each fragment in turn, at any stage of the construction of the MST, determine its closest fragment. Merge these fragments, and update the list of fragments. A tree can be guaranteed in this algorithm (although care must be taken in cases of equal similarity) and our other requirements (all vertices included, minimal total edge weight) are very straightforward. Given the potential for roughly halving the data remaining to be processed at each step, the computational cost reduces from $O(n^3)$ to $O(n^2 \log n)$.

The real interest of Sollin's algorithm arises when we are clustering on a graph and do not have all $n(n-1)/2$ edges present. Sollin's algorithm can be shown to have computational cost $m \log n$ where m is the number of edges. When $m \ll n(n-1)/2$ then we have the potential for appreciable gains.

The MST in feature spaces can of course make use of the fast nearest neighbor finding methods. See (Murtagh, 1985) for various examples.

Other graph data structures which have been proposed for data analysis are related to the MST. We know, for example, that the following subset relationship holds:

$$\text{MST} \subseteq \text{RNG} \subseteq \text{GG} \subseteq \text{DT}$$

where RNG is the relative neighborhood graph, GG is the Gabriel graph, and DT is the Delaunay triangulation. The latter, in the form of its dual, the Voronoi tessellation, has been used for analyzing the clustering of galaxy locations. References to these and related methods can be found in (Murtagh, 1993).

8.7 Voronoi Tessellation and Percolation

A Voronoi tessellation constructs a convex Voronoi cell around each occupied pixel and assigns fluxes to them based on the number of photons in the pixel, the cell area, and the effective exposure time at the pixel location. This of course is an imaging perspective, and *mutatis mutandis* we may consider continuous spaces, and unit regions. Thus background photons have large cells and low fluxes associated with them, whereas source photons are characterized by small cells and high fluxes.

The flux at which the observed cumulative distribution of fluxes for the selected region begins to deviate from the one expected for random Poisson noise is used as a threshold to discriminate between background events and pixels that may belong to sources. Pixels with associated fluxes exceeding the threshold are passed to a non-parametric percolation algorithm that groups adjacent high-flux pixels into sources. The advantage of Voronoi tessellation and percolation is that no assumptions are made about the geometrical properties of the sources and that very extended sources are detected as significant without a single pixel in the image being required to feature a photon count significantly different from the background.

Therefore, Voronoi tessellation and percolation is particularly well-suited for resolved sources of low surface brightness and potentially irregular shape. The main disadvantage of the approach is that it tends to produce blends when run with a low flux threshold on crowded fields. Some examples of the use of Voronoi tessellations for astronomical catalog analysis can be found in (van de Weygaert, 1994; González et al., 2000)

We will return to Voronoi tessellations later in this chapter when dealing with noise filtering.

8.8 Model-Based Clustering

Motivation for considering a model for our data is that it may help us answer important ancillary questions. A prime example is the question of how many inherent clusters or groups of data objects we have. We review recent progress in this section.

8.8.1 Modeling of Signal and Noise

A simple and widely applicable model is a distribution mixture, with the signal modeled by Gaussians, in the presence of Poisson background noise.

Consider data which are generated by a mixture of $(G-1)$ bivariate Gaussian densities, $f_k(x;\theta) \sim \mathcal{N}(\mu_k, \Sigma_k)$, for clusters $k = 2, \ldots, G$, and with Poisson background noise corresponding to $k = 1$. The overall population thus has the mixture density

$$f(x;\theta) = \sum_{k=1}^{G} \pi_k f_k(x;\theta)$$

8.8 Model-Based Clustering

where the mixing or prior probabilities, π_k, sum to 1, and $f_1(x;\theta) = \mathcal{A}^{-1}$, where $\mathcal{A}$ is the area of the data region. This is referred to as mixture modeling When constraints are set on the model parameters, θ, this leads to *model-based clustering* (Banfield and Raftery, 1993; Dasgupta and Raftery, 1998; Murtagh and Raftery, 1984; Banerjee and Rosenfeld, 1993).

The parameters, θ and π, can be estimated efficiently by maximizing the mixture likelihood

$$L(\theta, \pi) = \prod_{i=1}^{n} f(x_i; \theta),$$

with respect to θ and π, where x_i is the ith observation.

Now let us assume the presence of two clusters, one of which is Poisson noise, the other Gaussian. This yields the mixture likelihood

$$L(\theta, \pi) = \prod_{i=1}^{n} \left[\pi_1 \mathcal{A}^{-1} + \pi_2 \frac{1}{2\pi \sqrt{|\Sigma|}} \exp\left\{ -\frac{1}{2}(x_i - \mu)^T \Sigma^{-1}(x_i - \mu) \right\} \right],$$

where $\pi_1 + \pi_2 = 1$.

An iterative solution is provided by the expectation-maximization (EM) algorithm (Dempster et al., 1977). Let the "complete" (or "clean" or "output") data be $y_i = (x_i, z_i)$ with indicator set $z_i = (z_{i1}, z_{i2})$ given by $(1,0)$ or $(0,1)$. Vector z_i has a multinomial distribution with parameters $(1; \pi_1, \pi_2)$. This leads to the *complete data log-likelihood*:

$$l(y, z; \theta, \pi) = \Sigma_{i=1}^{n} \Sigma_{k=1}^{2} z_{ik} [\log \pi_k + \log f_k(x_k; \theta)]$$

The E-step then computes $\hat{z}_{ik} = E(z_{ik} \mid x_1, \ldots, x_n, \theta)$, i.e. the posterior probability that x_i is in cluster k. The M-step involves maximization of the *expected complete data log-likelihood*:

$$l^*(y; \theta, \pi) = \Sigma_{i=1}^{n} \Sigma_{k=1}^{2} \hat{z}_{ik} [\log \pi_k + \log f_k(x_i; \theta)].$$

The E- and M-steps are iterated until convergence.

For the 2-class case (Poisson noise and a Gaussian cluster), the complete-data likelihood is

$$L(y, z; \theta, \pi) = \prod_{i=1}^{n} \left[\frac{\pi_1}{\mathcal{A}} \right]^{z_{i1}} \left[\frac{\pi_2}{2\pi \sqrt{|\Sigma|}} \exp\left\{ -\frac{1}{2}(x_i - \mu)^T \Sigma^{-1}(x_i - \mu) \right\} \right]^{z_{i2}}$$

The corresponding expected log-likelihood is then used in the EM algorithm. This formulation of the problem generalizes to the case of G clusters, of arbitrary distributions and dimensions.

Fraley (1999) discusses implementation of model-based clustering, including publicly available software.

In order to assess the evidence for the presence of a signal-cluster, we use the *Bayes factor* for the mixture model, M_2, that includes a Gaussian density as well as background noise, against the "null" model, M_1, that contains only background noise. The Bayes factor is the posterior odds for the mixture model against the pure noise model, when neither is favored a priori. It is defined as $B = p(x|M_2)/p(x|M_1)$, where $p(x|M_2)$ is the *integrated likelihood* of the mixture model M_2, obtained by integrating over the parameter space. For a general review of Bayes factors, their use in applied statistics, and how to approximate and compute them, see (Kass and Raftery, 1995).

We approximate the Bayes factor using the *Bayesian Information Criterion* or BIC (Schwarz, 1978). For a Gaussian cluster and Poisson noise, this takes the form:

$$2 \log B \approx BIC = 2 \log L(\hat{\theta}, \hat{\pi}) + 2n \log \mathcal{A} - 6 \log n,$$

where $\hat{\theta}$ and $\hat{\pi}$ are the maximum likelihood estimates of θ and π, and $L(\hat{\theta}, \hat{\pi})$ is the maximized mixture likelihood.

A review of the use of the BIC criterion for model selection – and more specifically for choosing the number of clusters in a data set – can be found in (Fraley and Raftery, 1999).

Mixture modeling and the BIC criterion were applied to gamma-ray burst data (Mukherjee et al., 1998). Around 800 observations were assessed. A recently published paper (Hakkila et al., 2000) confirms that in our earlier work (Mukherjee et al., 1998) we were statistically correct, while also providing an instrumental rather than astrophysical origin of clustering.

8.8.2 Application to Thresholding

Consider an image or a planar or 3-dimensional set of object positions. For simplicity we consider the case of setting a single threshold in the image intensities, or the point set's spatial density.

We deal with a combined mixture density of two *univariate* Gaussian distributions $f_k(x; \theta) \sim \mathcal{N}(\mu_k, \sigma_k)$. The overall population thus has the mixture density

$$f(x; \theta) = \sum_{k=1}^{2} \pi_k f_k(x; \theta)$$

where the mixing or prior probabilities, π_k, sum to 1.

When the mixing proportions are assumed equal, the log-likelihood takes the form

$$l(\theta) = \sum_{i=1}^{n} \ln \left[\sum_{k=1}^{2} \frac{1}{2\pi \sqrt{|\sigma_k|}} \exp \left\{ -\frac{1}{2\sigma_k}(x_i - \mu_k)^2 \right\} \right]$$

The EM algorithm is then used to iteratively solve this (Celeux and Govaert, 1995). The Sloan Digital Sky Survey (SDSS, 2000) is producing a sky map of more than 100 million objects, together with 3-dimensional information (redshifts) for a million galaxies. Pelleg and Moore (1999) describe mixture modeling, using a k-D tree (also referred to as a multidimensional binary search tree) preprocessing to expedite the finding of the class (mixture) parameters, e.g. means, covariances.

8.9 Wavelet Analysis

Wavelets can be used for analyzing the projected 2D or the 3D galaxy distribution (Escalera et al., 1992; Slezak et al., 1993; Pagliaro et al., 1999; Kurokawa et al., 2001). For the noise model, given that this relates to point pattern clustering, we have to consider the Poisson noise case described in section 2.3.3 (Poisson noise with few counts).

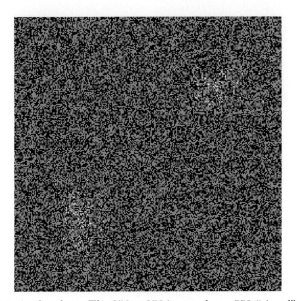

Fig. 8.17. Data in the plane. The 256 × 256 image shows 550 "signal" points – two Gaussian-shaped clusters in the lower left and in the upper right – with in addition 40,000 Poisson noise points added. Details of recovery of the clusters is discussed in Murtagh and Starck (1998).

Fig. 8.17 shows an example of where point pattern clusters – density bumps in this case – are sought, with a great amount of background clutter. Murtagh and Starck (1998) refer to the fact that there is no computational dependence on the number of points (signal or noise) in such a problem, when using a wavelet transform with noise modeling. Hence the dominant part of the processing is of constant computational cost, $O(1)$, which is quite an achievement.

264   8. Astronomical Catalog Analysis

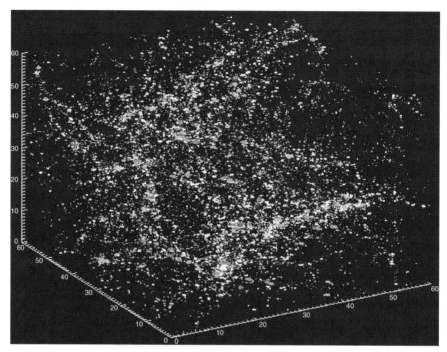

Fig. 8.18. Simulated data.

Fig. 8.18 shows a simulation of a $60h^{-1}$ Mpc box of universe, made by A. Klypin (simulated data at http://astro.nmsu.edu/~aklypin/PM/pmcode). It represents the distribution of dark matter in the present-day universe and each point is a dark matter halo where visible galaxies are expected to be located, i.e. the distribution of dark matter haloes can be compared with the distribution of galaxies in catalogs of galaxies.

Fig. 8.19 shows the same data set filtered by the 3D wavelet transform, using the algorithm described in section 2.4.5.

It has been shown that the genus curve derived from a density obtained by wavelet smoothing presents more information about the true topology of the distribution of galaxies than using the one derived from standard Gaussian smoothing (Martínez et al., 2005).

If wavelets are perfectly suited for detecting clusters in a 3D data set, they are not optimal for representing filaments and walls. For this reason, it has been proposed to combine the wavelet decomposition with two other multiscale transforms, the 3D ridgelet and the 3D beamlet, which represent well, respectively, walls and filaments (Starck et al., 2005).

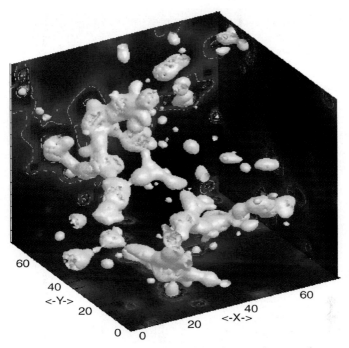

Fig. 8.19. Simulated data filtered by the wavelet transform.

8.10 Nearest Neighbor Clutter Removal

The wavelet approach is certainly appropriate when the wavelet function reflects the type of object sought (e.g. isotropic), and when superimposed point patterns are to be analyzed. However, non-superimposed point patterns of complex shape are very well treated by the approach described in (Byers and Raftery, 1998). Using a homogeneous Poisson noise model, Byers and Raftery derive the distribution of the distance of a point to its kth nearest neighbor.

Next, these authors consider the case of a Poisson process which is signal, superimposed on a Poisson process which is clutter. The kth nearest neighbor distances are modeled as a mixture distribution: a histogram of these, for given k, will yield a bimodal distribution if our assumption is correct. This mixture distribution problem is solved using the EM algorithm. Generalization to higher dimensions, e.g. 10, is also discussed.

Similar data was analyzed by noise modeling and a Voronoi tessellation preprocessing of the data in (Allard and Fraley, 1997). It is pointed out in this work how Voronoi tessellations can be very useful whenever the Voronoi tiles have meaning in relation to the morphology of the point patterns. However, this approach does not scale well to higher dimensions. Ebeling and Wiedenmann (1993), reproduced in (Dobrzycki et al., 1999), propose the use of a Voronoi tessellation for astronomical X-ray object detection and characterization.

8.11 Chapter Summary

Some old problems have been given new answers in this chapter, such as how we choose the inherent number of clusters in a data set. We have noted how a Voronoi graph structure is a good one when the morphology investigated is diverse.

Most of all in this chapter we have been concerned with the analysis of cosmology catalogs. Correlation and clustering were foremost in our minds. Scale was implicitly at issue thoughout the chapter, and the characterization of fractal or self-similar behavior was looked at in some depth.

9. Multiple Resolution in Data Storage and Retrieval

9.1 Introduction

Earlier chapters have covered important themes in data storage and access. Compression, in particular, is at the heart of technologies needed for compact storage and fast data access. Filtering, too, serves to allow the more essential elements of information to be retained. Deconvolution is available for refining our signals and images.

Recent innovative use of the Haar wavelet transform is to be found in the area of database querying. For very large databases, an approximate query result may be acceptable. For data confidentiality, an approximate query result may be required. The progressive refinement which is possible with the wavelet transform fits in well with support for approximate database querying.

When best match retrievals to our query are wanted, performing the search in wavelet space can have advantages. We will show below why this is so.

Data resolution and scale can be embodied in clustering schemes and more particularly in hierarchical clustering. A review of past and current work is presented, with particular reference to hyperlinked data repositories.

When data clustering is "active" – the user can interact with the structures imposed on the data – we have an open path towards innovative human-machine interaction. Visual user interfaces result from doing this.

We subsequently look closer at how we take ideas initially developed for signal and image processing towards application in more general information spaces. An enabling tool is array permutation. A number of examples are used to illustrate these innovative results.

9.2 Wavelets in Database Management

Catalog matching in astronomy gives rise to the problem of fuzzy joins, i.e. merging two tables based on object coordinates, but with different precisions associated with measurement of these coordinates (Read and Hapgood, 1992; Page, 1996). Here we will briefly review a different type of problem, namely that of approximate query processing. This arises when data must be kept confidential so that only aggregate or macro-level data can be divulged. Approximate query processing also provides a solution to access of information from massive data tables.

One approach to approximate database querying through aggregates is sampling. However a join operation applied to two uniform random samples results in a non-uniform result, which furthermore is sparse (Chakrabarti et al., 2000). A second approach is to keep histograms on the coordinates. For a multidimensional feature space, one is faced with a "curse of dimensionality" as the dimensionality grows. A third approach is wavelet-based. This will be briefly reviewed.

A form of progressive access to the data is sought, such that aggregated data can be obtained first, followed by greater refinement of the data. The Haar wavelet transform is a favored transform for such purposes, given that reconstructed data at a given resolution level is simply the mean of data values. Vitter et al. (1999; 1998) consider the combinatorial aspects of data access using a Haar wavelet transform, and based on a d-dimensional data hypercube. Such data, containing scores or frequencies, is often found in the commercial data mining context. The application area is referred to as OLAP, On-Line Analytical Processing.

As pointed out in Chakrabarti et al. (2000), one can treat multidimensional feature hypercubes as a type of high dimensional image, taking the given order of feature dimensions as fixed. As an alternative a uniform "shift and distribute" randomization can be used (Chakrabarti et al., 2000).

What if, however, one organizes the data such that adjacency has a meaning? This implies that similarly-valued objects, and similarly-valued features are close together. Later in this chapter we will describe exactly how this can be done. We will describe the close linkage with singular value decomposition, and we will also include consideration of the case of very high dimensional, and perhaps very sparse, feature spaces.

One further topic will be touched on before leaving the area of database querying. The best match problem seeks a set of closest retrievals in a feature space, and can be counterposed to the database exact match or partial match problems. Parseval's theorem indicates that Euclidean distance in the space or time domain is directly related to Euclidean distance in the frequency domain. The Euclidean distance is invariant under any orthonormal transformation of the given feature space. This allows the querying to be carried out in the Haar or any other orthonormal wavelet transform space. One important reason for doing this includes data filtering prior to querying. Furthermore a filtered database may well be very compressible, which points to possible interest when supporting querying of very large databases. Further theoretical and experimental results are presented in (Murtagh, 1998; Chang and Fu, 1999).

9.3 Fast Cluster Analysis

The non-technical person more often than not understands clustering as a partition. K-means provides such a solution.

A mathematical definition of a partition implies no multiple assignments of observations to clusters, i.e. no overlapping clusters. Overlapping clusters may be faster to determine in practice, and a case in point is the one-pass algorithm described in (Salton and McGill, 1983). The general principle followed is: make one pass through the data, assigning each object to the first cluster which is close enough, and making a new cluster for objects that are not close enough to any existing cluster.

This algorithm was the basis for clustering the web in the context of the Altavista search engine (Broder et al., 1997). A feature vector is determined for each HTML document considered, based on sequences of words. Similarity between documents is based on an inverted list. The similarity graph is thresholded, and components sought.

Broder (1998) solves the same clustering objective using a thresholding and overlapping clustering method similar to the Salton and McGill one. The application described is that of clustering the Altavista repository in April 1996, consisting of 30 million HTML and text documents, comprising 150 GBytes of data. The number of serviceable clusters found was 1.5 million, containing 7 million documents. Processing time was about 10.5 days. An analysis of the clustering algorithm used is in (Borodin et al., 1999).

The threshold-based pass of the data, in its basic state, is susceptible to lack of robustness. A bad choice of threshold leads to too many clusters or two few. To remedy this, we can work on a well-defined data structure such as the minimal spanning tree. Or, alternatively, we can iteratively refine the clustering. Partitioning methods, such as k-means, use iterative improvement of an initial estimation of a targeted clustering.

A very widely used family of methods for inducing a partition on a data set is called k-means, c-means (in the fuzzy case), ISODATA, competitive learning, vector quantization and other more general names (non-overlapping non-hierarchical clustering) or more specific names (minimal distance or exchange algorithms).

The usual criterion to be optimized is:

$$\frac{1}{|I|} \sum_{q \in Q} \sum_{i \in q} \|i - q\|^2$$

where I is the object set, $|\,.\,|$ denotes cardinality, q is some cluster, Q is the partition, and q denotes a set in the summation, whereas $\boldsymbol{q}$ denotes some associated vector in the error term, or metric norm. This criterion ensures that clusters found are compact, and therefore assumed homogeneous. The optimization criterion, by a small abuse of terminology, is ofter referred to as a minimum variance one.

A necessary condition that this criterion be optimized is that vector q be a cluster mean, which for the Euclidean metric case is:

$$q = \frac{1}{|q|} \sum_{i \in q} i$$

A batch update algorithm, due to (Lloyd, 1957; Forgy, 1965) and others makes assignments to a set of initially randomly-chosen vectors, q, as step 1. Step 2 updates the cluster vectors, q. This is iterated. The distortion error, equation 1, is non-increasing, and a local minimum is achieved in a finite number of iterations.

An online update algorithm is due to MacQueen (1976). After each presentation of an observation vector, i, the closest cluster vector, q, is updated to take account of it. Such an approach is well-suited for a continuous input data stream (implying "online" learning of cluster vectors).

Both algorithms are gradient descent ones. In the online case, much attention has been devoted to best learning rate schedules in the neural network (competitive learning) literature: (Darken and Moody, 1991; Darken et al., 1992; Darken and Moody, 1992; Fritzke, 1997).

A difficulty, less controllable in the case of the batch algorithm, is that clusters may become (and stay) empty. This may be acceptable, but also may be in breach of our original problem formulation. An alternative to the batch update algorithm is Späth's (1985) exchange algorithm. Each observation is considered for possible assignment into any of the *other* clusters. Updating and "downdating" formulas are given by Späth. This exchange algorithm is stated to be faster to converge and to produce better (smaller) values of the objective function. We have also verified that it is usually a superior algorithm to the minimal distance one.

K-means is very closely related to Voronoi (Dirichlet) tessellations, to Kohonen self-organizing feature maps, and various other methods.

The batch learning algorithm above may be viewed as:

1. An assignment step which we will term the E (estimation) step: estimate the posteriors,

    $P(\text{observations} \mid \text{cluster centers})$

2. A cluster update step, the M (maximization) step, which maximizes a cluster center likelihood.

Neal and Hinton (1998) cast the k-means optimization problem in such a way that both E- and M-steps monotonically increase the maximand's values. The EM algorithm may, too, be enhanced to allow for online as well as batch learning (Sato and Ishii, 1999).

In (Thiesson et al., 1999), k-means is implemented (i) by traversing blocks of data, cyclically, and incrementally updating the sufficient statistics and parameters, and (ii) instead of cyclic traversal, sampling from subsets of the data. Such an approach is admirably suited for very large data sets, where in-memory storage is not feasible. Examples used by Thiesson et al. (1999) include the clustering of a half million 300-dimensional records.

9.4 Nearest Neighbor Finding on Graphs

Hypertext is modeled by a graph. Hence search in information spaces often becomes a search problem on graphs. In this section, we will discuss nearest neighbor searching on graphs.

Clustering on graphs may be required because we are working with (perhaps complex non-Euclidean) dissimilarities. In such cases where we must take into account an edge between each and every pair of vertices, we will generally have an $O(m)$ computational cost where m is the number of edges. In a metric space we have seen that we can look for various possible ways to expedite the nearest neighbor search. An approach based on visualization – turning our data into an image – will be looked at below. However there is another aspect of our similarity (or other) graph which we may be able to turn to our advantage. Efficient algorithms for sparse graphs are available. Sparsity can be arranged – we can threshold our edges if the sparsity does not suggest itself more naturally. A special type of sparse graph is a planar graph, i.e. a graph capable of being represented in the plane without any crossovers of edges.

For sparse graphs, minimal spanning tree (MST) algorithms with $O(m \log \log n)$ computational cost were described by (Yao, 1975) and (Cheriton and Tarjan, 1976). A short algorithmic description can be found in (Murtagh, 1985), and we refer in particular to the latter.

The basic idea is to preprocess the graph, in order to expedite the sorting of edge weights (why sorting? – simply because we must repeatedly find smallest links, and maintaining a sorted list of edges is a good basis for doing this). If we were to sort all edges, the computational requirement would be $O(m \log m)$. Instead of doing that, we take the edge set associated with each and every vertex. We divide each such edge set into groups of size k. (The fact that the last such group will usually be of size $< k$ is taken into account when programming.)

Let n_v be the number of incident edges at vertex v, such that $\sum_v n_v = 2m$.

The sorting operation for each vertex now takes $O(k \log k)$ operations for each group, and we have n_v/k groups. For all vertices the sorting requires a number of operations which is of the order of $\sum_v n_v \log k = 2m \log k$. This looks like a questionable – or small – improvement over $O(m \log m)$.

Determining the lightest edge incident on a vertex requires $O(n_v/k)$ comparisons since we have to check all groups. Therefore the lightest edges incident on all vertices are found with $O(m/k)$ operations.

When two vertices, and later fragments, are merged, their associated groups of edges are simply collected together, therefore keeping the total number of groups of edges which we started out with. We will bypass the issue of edges which, over time, are to be avoided because they connect vertices in the same fragment: given the fact that we are building an MST, the total number of such edges-to-be-avoided cannot surpass $2m$.

To find what to merge next, again $O(m/k)$ processing is required. Using Sollin's algorithm, the total processing required in finding what to merge next is $O(m/k \ \log n)$. The total processing required for grouping the edges, and sorting within the edge-groups, is $O(m \log k)$, i.e. it is one-off and accomplished at the start of the MST-building process.

The total time is $O(m/k \ \log n) + O(m \log k)$. Let us fix $k = \log n$. Then the second term dominates and gives overall computational complexity as $O(m \log \log n)$.

This result has been further improved to near linearity in m by Gabow et al. (1986), who develop an algorithm with complexity $O(m \log \log \log \ldots n)$ where the number of iterated log terms is bounded by m/n.

Motwani and Raghavan (1995), chapter 10, base a stochastic $O(m)$ algorithm for the MST on random sampling to identify and eliminate edges that are guaranteed not to belong to the MST.

Let us turn our attention now to the case of a planar graph. For a planar graph we know that $m \leq 3n-6$ for $m > 1$. For proof, see for example (Tucker, 1980), or any book on graph theory.

Referring to Sollin's algorithm, described above, $O(n)$ operations are needed to establish a least cost edge from each vertex, since there are only $O(n)$ edges present. On the next round, following fragment-creation, there will be at most ceil$(n/2)$ new vertices, implying of the order of $n/2$ processing to find the least cost edge (where ceil is the ceiling function, or smallest integer greater than the argument). The total computational cost is seen to be proportional to: $n + n/2 + n/4 + \cdots = O(n)$.

So determining the MST of a planar graph is linear in numbers of either vertices or edges.

Before ending this review of very efficient clustering algorithms for graphs, we note that algorithms discussed so far have assumed that the similarity graph was undirected. For modeling transport flows, or economic transfers, the graph could well be directed. Components can be defined, generalizing the clusters of the single link method, or the complete link method. Tarjan (1983) provides an algorithm for the latter agglomerative criterion which is of computational cost $O(m \log n)$.

9.5 Cluster-Based User Interfaces

Information retrieval by means of "semantic road maps" was first detailed in (Doyle, 1961). The spatial metaphor is a powerful one in human information processing. The spatial metaphor also lends itself well to modern distributed computing environments such as the web. The Kohonen self-organizing feature map (SOM) method is an effective means towards this end of a visual information retrieval user interface. We will also provide an illustration of web-based semantic maps based on hyperlink clustering.

The Kohonen map is, at heart, k-means clustering with the additional constraint that cluster centers be located on a regular grid (or some other topographic structure) and furthermore their location on the grid be monotonically related to pairwise proximity (Murtagh and Hernández-Pajares, 1995). The nice thing about a regular grid output representation space is that it lends itself well to being a visual user interface.

Fig. 9.1 shows a visual and interactive user interface map, using a Kohonen self-organizing feature map (SOM). Color is related to density of document clusters located at regularly-spaced nodes of the map, and some of these nodes/clusters are annotated. The map is installed as a clickable imagemap, with CGI programs accessing lists of documents and – through further links – in many cases, the full documents. In the example shown, the user has queried a node and results are seen in the right-hand panel. Such maps are maintained for (currently) 18000 articles from the *Astrophysical Journal*, 10000 from *Astronomy and Astrophysics*, over 3300 astronomical catalogs, and other data holdings. More information on the design of this visual interface and user assessment can be found in (Poinçot et al., 1998; Poinçot et al., 2000).

A Java-based visualization tool (Guillaume and Murtagh, 2000) was developed for hyperlink-based data, consisting of astronomers, astronomical object names, article titles, and with the possibility of other objects (images, tables, etc.). Through weighting, the various types of links could be prioritized. An iterative refinement algorithm was developed to map the nodes (objects) to a regular grid of cells which, as for the Kohonen SOM map, are clickable and provide access to the data represented by the cluster. Fig. 9.2 shows an example for an astronomer.

9.6 Images from Data

9.6.1 Matrix Sequencing

We take our input object-attribute data, e.g. document-term or hyperlink array, as a 2-dimensional image. In general, an array is a mapping from the Cartesian product of observation set, I, and attribute set, J, onto the reals, $f : I \times J \longrightarrow \mathbb{R}$, while an image (single frame) is generally defined for discrete spatial intervals X and Y, $f : X \times Y \longrightarrow \mathbb{R}$. A table or array differs from a 2-dimensional image, however, in one major respect. There is an order relation defined on the row- and column-dimensions in the case of the image. To achieve invariance we must induce an analogous ordering relation on the observation and variable dimensions of our data table.

A natural way to do this is to seek to optimize contiguous placement of large (or nonzero) data table entries. Note that array row and column permutation to achieve such an optimal or suboptimal result leaves intact

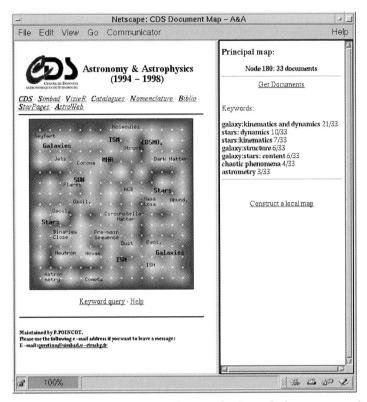

Fig. 9.1. Visual interactive user interface to the journal *Astronomy and Astrophysics*. Original in color.

each value x_{ij}. We simply have row and column, i and j, in different locations at output compared to input. Methods for achieving such block clustering of data arrays include combinatorial optimization (McCormick et al., 1972; Lenstra, 1974; Doyle, 1988) and iterative methods (Deutsch and Martin, 1971; Streng, 1991). In an information retrieval context, a simulated annealing approach was also used in (Packer, 1989). Further references and discussion of these methods can be found in (Murtagh, 1985; March, 1983; Arabie et al., 1988). Treating the results of such methods as an image for visualization purposes is a very common practice, e.g. (Gale et al., 1984).

We now describe briefly two algorithms which work well in practice.

Moments Method (Deutsch and Martin, 1971): Given a matrix, $a(i,j)$, for $i = 1, 2, \ldots, n$, and $j = 1, 2, \ldots, m$. Define row moments as $m(i) = (\sum_j j a(i,j))/(\sum_j a(i,j))$. Permute rows in order of nondecreasing row moments. Define column moments analogously. Permute columns in order of nondecreasing column moments. Reiterate until convergence.

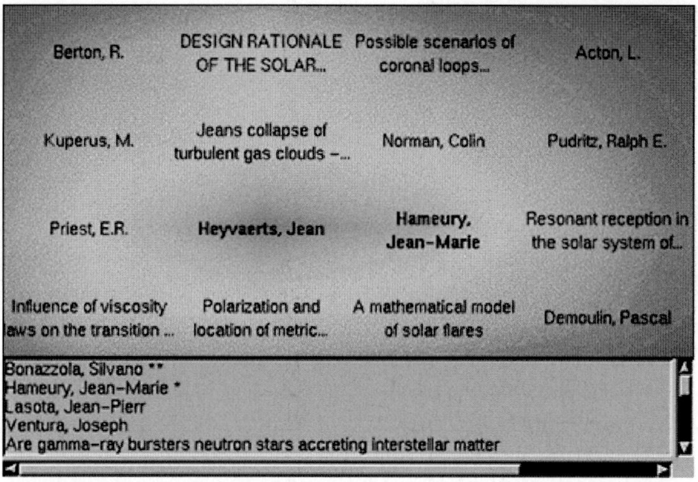

Fig. 9.2. Visual interactive user interfaces, based on graph edges. Map for astronomer Jean Heyvaerts. Original in color.

This algorithm results (usually) in large matrix entries being repositioned close to the diagonal. An optimal result cannot be guaranteed.

Bond Energy Algorithm (McCormick et al., 1972): Permute matrix rows and columns such that a criterion, BEA $= \sum_{i,j} a(i,j)(a(i-1,j) + a(i+1,j) + a(i,j-1) + a(i,j+1))$ is maximized.

An algorithm to implement the BEA is as follows: Position a row arbitrarily. Place the next row such that the contribution to the BEA criterion is maximized. Place the row following that such that the new contribution to the BEA is maximized. Continue until all rows have been positioned. Then do analogously for columns. No further convergence is required in this case. This algorithm is a particular use of the traveling salesperson problem, TSP, which is widely used in scheduling. In view of the arbitrary initial choice of row or column, and more particularly in view of the greedy algorithm solution, this is a suboptimal algorithm.

Matrix reordering rests on (i) permuting the rows and columns of an incidence array to some standard form, and then data analysis for us in this context involves (ii) treating the permuted array as an image, analyzed subsequently by some appropriate analysis method.

Dimensionality reduction methods, including principal components analysis (suitable for quantitative data), correspondence analysis (suitable for qualitative data), classical multidimensional scaling, and others, is based on singular value decomposition. It holds:

$$AU = \Lambda U$$

where we have the following. A is derived from the given data – in the case of principal components analysis, this is a correlation matrix, or a variance/covariance matrix, or a sums of squares and cross-products matrix.

Zha et al. (2001) formalize the reordering problem as the constructing of a sparse rectangular matrix

$$W = \begin{pmatrix} W_{11} & W_{12} \\ W_{21} & W_{22} \end{pmatrix}$$

so that W_{11} and W_{22} are relatively denser than W_{12} and W_{21}. Permuting rows and columns according to projections onto principal axes achieves this pattern for W. Proceeding recursively (subject to a feasibility cut-off), we can further increase near-diagonal density at the expense of off-diagonal sparseness.

A few comments on the computational aspects of array permuting methods when the array is very large and very sparse follow (Berry et al., 1996). Gathering larger (or nonzero) array elements to the diagonal can be viewed in terms of minimizing the envelope of nonzero values relative to the diagonal. This can be formulated and solved in purely symbolic terms by reordering vertices in a suitable graph representation of the matrix. A widely-used method for symmetric sparse matrices is the Reverse Cuthill-McKee (RCM) method.

The complexity of the RCM method for ordering rows or columns is proportional to the product of the maximum degree of any vertex in the graph represented by the array and the total number of edges (nonzeros in the matrix). For hypertext matrices with small maximum degree, the method would be extremely fast. The strength of the method is its low time complexity but it does suffer from certain drawbacks. The heuristic for finding the starting vertex is influenced by the initial numbering of vertices and so the quality of the reordering can vary slightly for the same problem for different initial numberings. Next, the overall method does not accommodate dense rows (e.g., a common link used in every document), and if a row has a significantly large number of nonzeros it might be best to process it separately; i.e., extract the dense rows, reorder the remaining matrix and augment it by the dense rows (or common links) numbered last.

One alternative approach is based on linear algebra, making use of the extremely sparse incidence data which one is usually dealing with. The execution time required by RCM may well require at least two orders of magnitude (i.e., 100 times) less execution time compared to such methods. However such methods, including for example sparse array implementations of correspondence analysis, appear to be more competitive with respect to bandwidth (and envelope) reduction at the increased computational cost.

Elapsed CPU times for a range of arrays (Berry et al., 1996), show performances between 0.025 to 3.18 seconds for permuting a 4000×400 array.

9.6.2 Filtering Hypertext

It is quite impressive how 2D (or 3D) image signals can handle with ease the scalability limitations of clustering and many other data processing operations. The contiguity imposed on adjacent pixels bypasses the need for nearest neighbor finding. It is very interesting therefore to consider the feasibility of taking problems of clustering massive data sets into the 2D image domain. We will look at a few recent examples of work in this direction.

Church and Helfman (1993) address the problem of visualizing possibly millions of lines of computer program code, or text. They consider an approach borrowed from DNA sequence analysis. The data sequence is tokenized by splitting it into its atoms (line, word, character, etc.) and then placing a dot at position i, j if the ith input token is the same as the jth. The resulting dotplot, it is argued, is not limited by the available display screen space, and can lead to discovery of large-scale structure in the data.

When data do not have a sequence we have an invariance problem which can be resolved by finding some row and column permutation which pulls large array values together, and perhaps furthermore into proximity to an array diagonal. Sparse data is a special case: a review of public domain software for carrying out SVD and other linear algebra operations on large sparse data sets can be found in (Berry et al., 1999).

Once we have a sequence-respecting array, we can immediately apply efficient visualization techniques from image analysis. Murtagh et al. (2000) investigate the use of noise filtering (i.e. to remove less useful array entries) using a multiscale wavelet transform approach.

An example follows. From the Concise Columbia Encyclopedia (1989 2nd ed., online version) a set of data relating to 12025 encyclopedia entries and to 9778 cross-references or links was used. Fig. 9.3 shows a 500 × 450 subarray, based on a correspondence analysis (i.e. ordering of projections on the first factor).

This part of the encyclopedia data was filtered using the wavelet and noise-modeling methodology described in (Murtagh et al., 2000) and the outcome is shown in Fig. 9.4. Overall the recovery of the more apparent alignments, and hence visually stronger clusters, is excellent. The first relatively long "horizontal bar" was selected – it corresponds to column index (link) 1733 = `geological era`. The corresponding row indices (articles) are, in sequence:

```
SILURIAN PERIOD
PLEISTOCENE EPOCH
HOLOCENE EPOCH
PRECAMBRIAN TIME
CARBONIFEROUS PERIOD
OLIGOCENE EPOCH
ORDOVICIAN PERIOD
```

278   9. Multiple Resolution in Data Storage and Retrieval

Fig. 9.3. Part (500 × 450) of original encyclopedia incidence data array.

```
TRIASSIC PERIOD
CENOZOIC ERA
PALEOCENE EPOCH
MIOCENE EPOCH
DEVONIAN PERIOD
PALEOZOIC ERA
JURASSIC PERIOD
MESOZOIC ERA
CAMBRIAN PERIOD
PLIOCENE EPOCH
CRETACEOUS PERIOD
```

The approach described here is based on a number of technologies: (i) data visualization techniques; (ii) the wavelet transform for data analysis – in particular data filtering; and (iii) data matrix permuting techniques. The wavelet transform has linear computational cost in terms of image row and column dimensions, and is not dependent on the pixel values.

9.6.3 Clustering Document-Term Data

Experiments were carried out on a set of bibliographical data – documents in the literature crossed by user-assigned index terms. This bibliographic data is from the journal *Astronomy and Astrophysics*. We looked at a set of such bibliography relating to 6885 articles published in *Astronomy and*

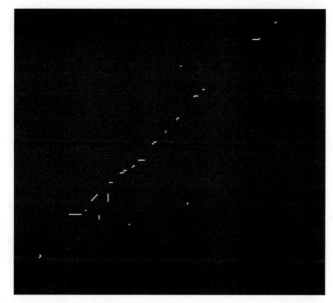

Fig. 9.4. End-product of the filtering of the array shown in the previous figure.

Astrophysics between 1994 and early 1999. A sample of the first 10 records is as follows.

```
1994A&A...284L...1I 102 167
1994A&A...284L...5W 4 5 14 16 52 69
1994A&A...284L...9M 29
1994A&A...284L..16F 15 64
1994A&A...284....1B 32 49 71
1994A&A...284...12A 36 153 202
1994A&A...284...17H 3 10 74 82 103
1994A&A...284...28M 17 42 102
1994A&A...284...33D 58
1994A&A...284...44S 111
```

A 19-character unique identifier (the *bibcode*) is followed by the sequence numbers of the index terms. There are 269 of the latter. They are specified by the author(s) and examples will be seen below. The experiments to follow were based on the first 512 documents in order to facilitate presentation of results. We investigated the row and column permuting of the 512 × 269 incidence array, based on the ordering of projections on the principal component, but limited clustering was brought about. This was due to the paucity of index term "overlap" properties in this dataset, i.e. the relatively limited numbers of index terms shared by any given pair of documents. For this reason, we elected to base subsequent work on the contingency table.

280   9. Multiple Resolution in Data Storage and Retrieval

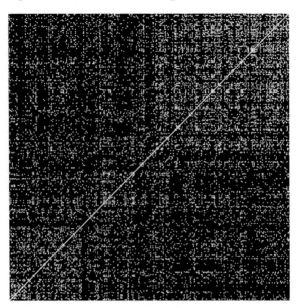

Fig. 9.5. Row/column-permuted contingency table of 512 documents, based on projections onto the first principal component.

A principal components analysis of the 512×269 dataset is dominated by the $O(m^3)$, $m = 269$, diagonalization requirement. Calculating the principal component projections for the rows takes linear (in document space) time. We used the order of principal component projections to provide a standard permutation of rows and columns of the document contingency table. The resulting permuted contingency table is shown in Fig. 9.5.

Figs. 9.6 and 9.7 show, respectively, the results of a wavelet transform (the redundant à trous transform is used) at wavelet resolution level 3 and the final smoothed version of the data. The latter is a background or continuum. In both of these figures, more especially in the continuum one, we have visual evidence for a cluster at the bottom left, another smaller one about one-third of the way up the diagonal, and a large one centered on the upper right-hand side of the image.

We are simply using the wavelet transform in this instance to facilitate analysis of a large, permuted data array. We wish to find contiguous clusters. Such clusters will for the most part be close to the diagonal. We recall that the contingency array used is symmetric, which explains the symmetry relative to the diagonal in what we see.

We can interpret the clusters on the basis of their most highly associated index terms. This in turn relates to the ordering of index terms on the first principal component axis in this case. Applying an arbitrary cut-off (± 0.2) to principal component projections, we find the index terms most associated with the two ends of the first principal component as follows:

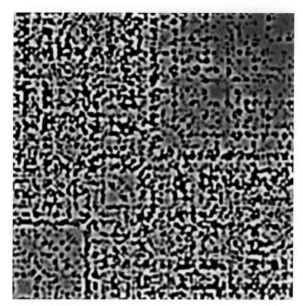

Fig. 9.6. Resolution level 3 from a wavelet transform of the data shown in Fig. 9.5.

```
stars:circumstellar matter
X-rays:stars
stars:abundances
stars:evolution
stars:mass loss
stars:binaries:close
stars:late type
stars:activity
stars:magnetic fields
stars:coronae
stars:flare
radio continuum:stars
stars:chromospheres
stars:binaries
```

The other extremity of the first principal component axis is associated with the following index terms:

```
ISM:molecules
galaxies:ISM
galaxies:kinematics and dynamics
galaxies:evolution
galaxies:spiral
galaxies:interactions
galaxies:structure
```

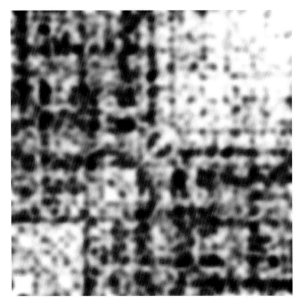

Fig. 9.7. The final smoothed version of the data resulting from a wavelet transform of the data shown in Fig. 9.5.

```
galaxies:abundances
galaxies:redshifts
galaxies:luminosity function,mass function
galaxies:compact
```

The distinction is clear – between stars, and stellar topics of inquiry, on the one hand, and interstellar matter (ISM) and galaxies, i.e. topics in cosmology, on the other hand. This distinction explains the two clusters clearly visible at the opposite ends of the diagonal in Fig. 9.7 (and less so in the original permuted data, Fig. 9.5). The distinction between stellar and cosmological foci of inquiry in the astronomical literature is a well-known one, which is linked directly and indirectly to astronomical instrumentation and even to shifts of professional interest over the past few decades.

9.7 Chapter Summary

The wavelet transform has been used in a novel and exciting way in various aspects of the work discussed in this chapter. Both database management, implying structured data, and best match information search, implying partially or weakly structured data, were covered.

Another separate issue is how the user interacts with data holdings. Interactive and responsive user interfaces can be created, based on data clustering and progressive refinement of views on data.

Finally we looked at how image and signal processing methods could be applied to multidimensional feature spaces.

The methods discussed in this chapter provide a broad range of tools and techniques for human navigation of data and information spaces.

10. Towards the Virtual Observatory

10.1 Data and Information

The Mosaic browser was released in early 1993, and fundamentally changed the way that, first, science, and later almost all other areas of social activity, are carried out. The drivers for the web, as is well-known, were scientific activities related to high energy physics – particularly Berners-Lee who developed the http protocol at CERN; and high performance computing and networking at NCSA – and particularly Andreesen who developed the Mosaic browser. Another comparable sea-change is currently being sought by many projects nationally and internationally relating to the Grid. Closely related is the concept of the virtual observatory in astronomy – viz., statistical properties of data in distributed databases as counterposed to traditional concern with single specified objects – and to e-science. In the latter area, the increasingly online scientific literature has led to enormous productivity growth in science.

To illustrate the quantitative and qualitative sea-change which our computing infrastructure is undergoing let us look at the issue of scalability. Our algorithms and software systems must be stable, robust over a wide range of scales in (storage) space and (computational) time. The well-known Internet search engine, Google, is reputed to use some 5000 CPUs, and Hotmail 8000. Massive data sets, and clever signal filtering and information fusion algorithms, are needed for example to find very distant objects which tell the story of the early Universe (the record in June 2001 is a quasar at redshift 6.2 derived from the processed data of the Sloan Digital Sky Survey – an appropriate illustration of the "virtual observatory" at work). Industrial tendencies are exemplified by the growth in web server farms, with real estate rented by the square meter, and caching and geolocation infrastructure to speed up data delivery. In networking the OSI (International Organisation for Standardization) 7-layer model is becoming a 2-layer model, services/content on top and transport/distribution on the bottom, with implications for protocols (a point made by J. Lawrence, Intel, in a keynote at FPL2001, European Conference on Field Programmable Logic, Queen's University Belfast, August 2001). Closer to research is the computational grid (driven by particle physics, and scientific visualization of simulation and modeling), the access grid (with a focus on videoconferencing and collaborative work), and the data grid (for fields such as bioinformatics/functional genomics and space-borne and Earth-based observational astronomy).

We will next give a short historical overview, first of cluster analysis, and then of multiresolution transforms.

Cluster analysis and related data analysis methods received significant boosts from different directions, and at different times. In the 1960s a range of very beautiful algorithmic methods were developed in the general area of multidimensional scaling, and have since become mainstream in the analysis of educational, political, and sociological data. Modeling of data through graph theoretical methods has a far longer history, with contributions by Euler, Hamilton, and later combinatoricists such as Erdős and Rényi. Nowadays graph theory is called upon to help with our understanding of the web cyberspace which is the infrastructure of modern society. Even old questions in clustering and data analysis can be given new answers, as exemplified by the issue of the number of clusters in a data set: Bayes factors (Statistics, University of Washington group), minimum description length (Wallace, and Computer Science, Monash group), Kolmogorov complexity (Computer Science, Royal Holloway group), are important contributions to this. New areas of clustering include chemical databases, multimedia data streams, astronomy, and other fields (see the special issue of The Computer Journal on this theme, Vol. 41, No. 8, 1998). Visualization and interactive analysis, from expensive beginnings in the work of Friedman and Stützle at SLAC (Stanford Linear Accelerator Center) in the 1960s, are now represented by such widely-used packages like S-Plus and R, Ggobi, and have strongly influenced image processing packages like IDL (and PV-Wave), and others. In the foreword to the special issue of The Computer Journal referred to above, we ask why all problems have not yet been solved in this field. We answer that there is continual turnover and renewal of applications, with requirements that are analogous to what went before but with important new elements, and that many theoretical and practical results have long ago passed into mainstream and widespread practice. As particular branches of the extended clustering and data analysis family, we can include machine learning, pattern recognition, and neural networks.

A short sketch of wavelet and multiresolution transforms follows. Fourier analysis has long occupied central ground in signal processing. Wavelets essentially do what Fourier does – frequency analysis – and incorporate also space/time analysis. The Fast Fourier Transform, the great invention of Tukey (who also gave us the terms "bit" and "software", and who died in July 2000), is bettered by the wavelet transform. Fundamental theories of wavelets were developed by Daubechies and others, Mallat linked wavelets to image processing, and Strang furthered comprehensive linkage with signal processing. In graphics, DeRose and others developed digital cinema technologies based on wavelets, with DeRose leaving for Pixar, to play a central role in the creation of such films as "Toy Story" (Disney/Pixar) and others.

Our work, (Starck et al., 1998a) and many papers, is characterized by the dominance of application drivers. Our concerns are to develop innovative solutions for problems in image, signal and data processing which include: compression, filtering, deconvolution, feature/object detection, fusion and registering, visualization, modeling and prediction, and quantitative characterization through information and noise models. Insofar as signal and noise modeling are fundamental to our work, we can claim with justice that for selected problems and classes of problem we provide optimal analysis of data and information.

10.2 The Information Handling Challenges Facing Us

Some of the grand challenges now opening up in astronomical data analysis are the following.

- What is to be understood by data mining in science (and areas of engineering), and why is this radically different from commercial and business data mining?
- How do we provide very long term semantic description of scientific data and information? Such challenges go far beyond writing a few DTDs (Document Type Definitions).
- Bayesian modeling can help with the all-important problem of model selection. What are the leading scientific decisions requiring such model-based decision making?
- Data and information storage and processing are integrally linked to delivery. What are the new paradigms emerging on the human-machine interface, and what are the implications for image and catalog archives?
- How will scientific "collaboratories" change given availability of broadband networks? What are the implications of 3G wireless communication, in conjunction with very high bandwidth backbone links? What are the implications – and the potential – of wide availability of inexpensive, very high quality display devices?

The topics discussed in this book are an important contribution to these, and other, challenges.

The future of astronomical data analysis harbors significant challenges. Our methodology is capable of meeting these challenges. The state of the art in data analysis described in this book is a contribution towards the objective of applying appropriate and necessary data analysis methodology to face these challenges.

Appendix

A. A Trous Wavelet Transform

A wavelet transform for discrete data is provided by the particular version known as à trous ("with holes", so called because of the interlaced convolution used in successive levels: see step 2 of the algorithm below) (Holschneider et al., 1989; Shensa, 1992; Starck et al., 1998a). This is a "stationary" or redundant transform, i.e. decimation is not carried out. One assumes that the sampled data $\{c_{0,l}\}$ are the scalar products, at pixels l, of the function $f(x)$ with a scaling function $\phi(x)$ which corresponds to a low-pass filter.

The wavelet function $\psi(x)$ obeys the dilation equation:

$$\frac{1}{2}\psi\left(\frac{x}{2}\right) = \sum_k g(k)\phi(x-k) \qquad (A.1)$$

The coefficients $g(x)$ are consistent with given ϕ and ψ. We compute the scalar products $\frac{1}{2^j}\langle f(x), \psi(\frac{x-l}{2^j})\rangle$, i.e. the discrete wavelet coefficients, with:

$$w_{j+1,l} = \sum_k g(k) c_{j,l+2^j k} \qquad (A.2)$$

The indexing is such that, here, $j = 1$ corresponds to the finest scale, implying high frequencies.

Generally, the wavelet resulting from the difference between two successive approximations is applied:

$$w_{j+1,l} = c_{j,l} - c_{j+1,l} \qquad (A.3)$$

The first filtering is then performed by a twice-magnified scale leading to the $\{c_{1,l}\}$ set. The signal difference $\{c_{0,l}\} - \{c_{1,l}\}$ contains the information between these two scales and is the discrete set associated with the wavelet transform corresponding to $\phi(x)$. The associated wavelet is $\psi(x)$.

$$\frac{1}{2}\psi\left(\frac{x}{2}\right) = \phi(x) - \frac{1}{2}\phi\left(\frac{x}{2}\right) \qquad (A.4)$$

The distance between samples increasing by a factor 2 (see Fig. A.1) from scale j to the next, $c_{j+1,l}$, is given by:

$$c_{j+1,l} = \sum_k h(k) c_{j,l+2^j k} \qquad (A.5)$$

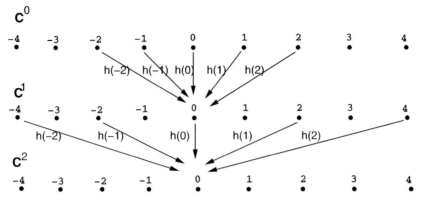

Fig. A.1. Passage from c_0 to c_1, and from c_1 to c_2.

The coefficients $\{h(k)\}$ derive from the scaling function $\phi(x)$:

$$\frac{1}{2}\phi\left(\frac{x}{2}\right) = \sum_l h(l)\phi(x-l) \tag{A.6}$$

The à trous wavelet transform algorithm is:

1. We initialize j to 0 and we start with the data $c_{j,k}$.
2. We carry out a discrete convolution of the data $c_{j,k}$ using the filter h. The distance between the central pixel and the adjacent ones is 2^j.
3. After this smoothing, we obtain the discrete wavelet transform from the difference $c_{j,k} - c_{j+1,k}$.
4. If j is less than the number J of resolutions we want to compute, we increment j and then go to step 2.
5. The set $\mathcal{W} = \{w_1, ..., w_J, c_J\}$ represents the wavelet transform of the data.

The algorithm allowing us to rebuild the data-frame is immediate: the last smoothed array c_J is added to all the differences, w_j.

$$c_{0,l} = c_{J,l} + \sum_{j=1}^{J} w_{j,l} \tag{A.7}$$

Triangle Function as the Scaling Function. Choosing the triangle function as the scaling function ϕ (see Fig. A.2, left) leads to piecewise linear interpolation:

$$\begin{aligned}\phi(x) &= 1- \mid x \mid \quad \text{if } x \in [-1,1] \\ \phi(x) &= 0 \quad \text{if } x \notin [-1,1]\end{aligned}$$

We have:

$$\frac{1}{2}\phi\left(\frac{x}{2}\right) = \frac{1}{4}\phi(x+1) + \frac{1}{2}\phi(x) + \frac{1}{4}\phi(x-1) \tag{A.8}$$

c_1 is obtained from:
$$c_{1,l} = \frac{1}{4}c_{0,l-1} + \frac{1}{2}c_{0,l} + \frac{1}{4}c_{0,l+1} \tag{A.9}$$
and c_{j+1} is obtained from c_j by:
$$c_{j+1,l} = \frac{1}{4}c_{j,l-2^j} + \frac{1}{2}c_{j,l} + \frac{1}{4}c_{j,l+2^j} \tag{A.10}$$

Figure A.2, right, shows the wavelet associated with the scaling function. The wavelet coefficients at scale j are:
$$w_{j+1,l} = -\frac{1}{4}c_{j,l-2^j} + \frac{1}{2}c_{j,l} - \frac{1}{4}c_{j,l+2^j} \tag{A.11}$$

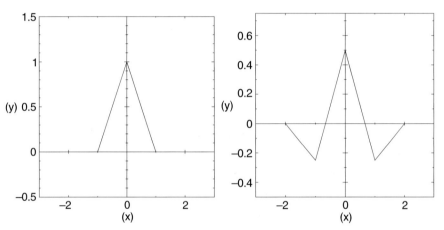

Fig. A.2. *Left:* triangle function ϕ; *right:* the wavelet ψ.

The above à trous algorithm is easily extended to two-dimensional space. This leads to a convolution with a mask of 3×3 pixels for the wavelet associated with linear interpolation. The coefficients of the mask are:

$$\begin{pmatrix} 1/4 & 1/2 & 1/4 \end{pmatrix} \begin{pmatrix} 1/4 \\ 1/2 \\ 1/4 \end{pmatrix}$$

At each scale j, we obtain a set $\{w_{j,l}\}$ which we will call a wavelet band or wavelet scale in the following. A wavelet scale has the same number of pixels as the input data.

B_3-spline Scaling Function. If we choose a B_3-spline for the scaling function:
$$\phi(x) = B_3(x) =$$

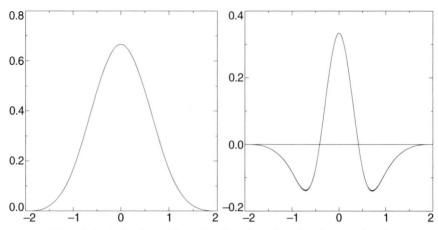

Fig. A.3. *Left:* the cubic spline function ϕ; *right:* the wavelet ψ.

$$\frac{1}{12}(\mid x-2\mid^3 -4\mid x-1\mid^3 +6\mid x\mid^3 -4\mid x+1\mid^3 +\mid x+2\mid^3) \qquad (A.12)$$

the coefficients of the convolution mask in one dimension are $(\frac{1}{16},\frac{1}{4},\frac{3}{8},\frac{1}{4},\frac{1}{16})$, and in two dimensions:

$$\begin{pmatrix} 1/16 & 1/4 & 3/8 & 1/4 & 1/16 \end{pmatrix} \begin{pmatrix} \frac{1}{16} \\ \frac{1}{4} \\ \frac{3}{8} \\ \frac{1}{4} \\ \frac{1}{16} \end{pmatrix} = \begin{pmatrix} \frac{1}{256} & \frac{1}{64} & \frac{3}{128} & \frac{1}{64} & \frac{1}{256} \\ \frac{1}{64} & \frac{1}{16} & \frac{3}{32} & \frac{1}{16} & \frac{1}{64} \\ \frac{3}{128} & \frac{3}{32} & \frac{9}{64} & \frac{3}{32} & \frac{3}{128} \\ \frac{1}{64} & \frac{1}{16} & \frac{3}{32} & \frac{1}{16} & \frac{1}{64} \\ \frac{1}{256} & \frac{1}{64} & \frac{3}{128} & \frac{1}{64} & \frac{1}{256} \end{pmatrix}$$

To facilitate computation, a simplification of this wavelet is to assume separability in the two-dimensional case. In the case of the B_3-spline, this leads to a row-by-row convolution with $(\frac{1}{16},\frac{1}{4},\frac{3}{8},\frac{1}{4},\frac{1}{16})$; followed by column-by-column convolution. Fig. A.3 shows the scaling function and the wavelet function when a cubic spline function is chosen as the scaling function ϕ.

The most general way to handle the boundaries is to consider that $c(k+N) = c(N-k)$ ("mirror"). But other methods can be used such as periodicity $(c(k+N) = c(N))$, or continuity $(c(k+N) = c(k))$.

Examples. In order to demonstrate the basic features of the wavelet transform, we created a simulated spectrum composed of a continuum, superimposed Gaussian noise, and three emission features with Gaussian line profiles. The features have widths of standard deviation equal to $15d$, $2d$, and $3d$ ($d = 9.76562e^{-4}\mu$m) and are located at 3.49 μm, 3.50 μm, and 3.60 μm, respectively. This spectrum is shown in Fig. A.4. Fig. A.5 contains its wavelet transform. Figs. A.5a-g plot the wavelet scales 1–7, and Fig. A.5h represents the smoothed array c_J. Note that the individual wavelet scales have zero mean. The original spectrum is obtained by an addition of the eight signals, i.e. the seven wavelet scales and the smoothed array. It can be seen that the

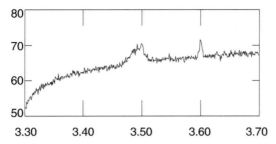

Fig. A.4. Simulated spectrum composed of a continuum, superimposed Gaussian noise, and three emission features with Gaussian line profiles (see text).

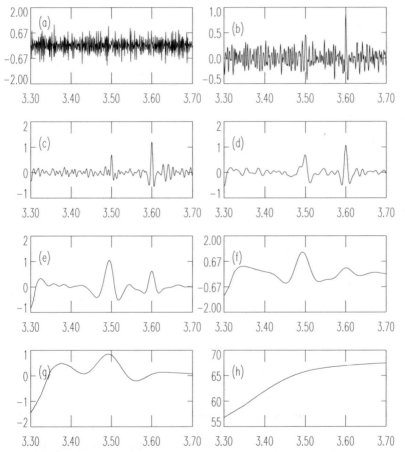

Fig. A.5. (a–g) The wavelet scales 1 to 7 and (h) the smoothed array c_J of the simulated spectrum shown in Fig. A.4. The addition of the eight signals gives exactly the original spectrum of Fig. A.4

narrow features near 3.50 μm and 3.60μm are detected at wavelet scales 2, 3, and 4, i.e. Fig. A.5 b-d, respectively. The broad band near 3.49 μm is detected at scales 5 and 6 (Figs. A.5 e,f, respectively). Because the two bands near 3.49 μm and 3.50 μm have a significantly different width, they are detected at different wavelet scales.

B. Picard Iteration

Suppose that an equation is given,

$$L(x) = 0 \tag{B.1}$$

where x is a vector, and L a function, and that it is possible to rewrite the equation into the form:

$$x = F(x) \tag{B.2}$$

where F is function obtained by rearrangement of the function L. Then the solution can be obtained from the sequence (Issacson and Keller, 1966; Hunt, 1994):

$$x^{n+1} = F(x^n) \tag{B.3}$$

The sequence converges if there exists a neighborhood such that, for any x and $x + \Delta$ in the neighborhood,

$$\| F(x + \Delta) - F(x) \| \leq C \| \Delta \| \tag{B.4}$$

for a constant $C < 1$. For example, the object-image relation is:

$$I - P * O = 0 \tag{B.5}$$

By convolving by P^* and adding O to both sides, we have

$$O = O + P^* * (I - P * O) \tag{B.6}$$

Picard iteration gives:

$$O^{n+1} = O^n + \lambda P^* * (I - P * O^n) \tag{B.7}$$

where λ is a parameter which controls the convergence.

C. Wavelet Transform Using the Fourier Transform

We start with the set of scalar products $c_0(k) = <f(x), \phi(x-k)>$. If $\phi(x)$ has a cut-off frequency $\nu_c \leq \frac{1}{2}$ (Starck et al., 1994; Starck and Bijaoui, 1994; Starck et al., 1998a), the data are correctly sampled. The data at resolution $j=1$ are:

$$c_1(k) = <f(x), \frac{1}{2}\phi(\frac{x}{2}-k)> \qquad (C.1)$$

and we can compute the set $c_1(k)$ from $c_0(k)$ with a discrete filter $\hat{h}(\nu)$:

$$\hat{h}(\nu) = \begin{cases} \frac{\hat{\phi}(2\nu)}{\hat{\phi}(\nu)} & \text{if } |\nu| < \nu_c \\ 0 & \text{if } \nu_c \leq |\nu| < \frac{1}{2} \end{cases} \qquad (C.2)$$

and

$$\forall \nu, \forall n \quad \hat{h}(\nu + n) = \hat{h}(\nu) \qquad (C.3)$$

where n is an integer. So:

$$\hat{c}_{j+1}(\nu) = \hat{c}_j(\nu)\hat{h}(2^j\nu) \qquad (C.4)$$

The cut-off frequency is reduced by a factor 2 at each step, allowing a reduction of the number of samples by this factor.

The wavelet coefficients at scale $j+1$ are:

$$w_{j+1}(k) = <f(x), 2^{-(j+1)}\psi(2^{-(j+1)}x - k)> \qquad (C.5)$$

and they can be computed directly from $c_j(k)$ by:

$$\hat{w}_{j+1}(\nu) = \hat{c}_j(\nu)\hat{g}(2^j\nu) \qquad (C.6)$$

where g is the following discrete filter:

$$\hat{g}(\nu) = \begin{cases} \frac{\hat{\psi}(2\nu)}{\hat{\phi}(\nu)} & \text{if } |\nu| < \nu_c \\ 1 & \text{if } \nu_c \leq |\nu| < \frac{1}{2} \end{cases} \qquad (C.7)$$

and

$$\forall \nu, \forall n \quad \hat{g}(\nu + n) = \hat{g}(\nu) \qquad (C.8)$$

The frequency band is also reduced by a factor 2 at each step. Applying the sampling theorem, we can build a pyramid of $N + \frac{N}{2} + \ldots + 1 = 2N$ elements.

For an image analysis the number of elements is $\frac{4}{3}N^2$. The overdetermination is not very high.

The B-spline functions are compact in direct space. They correspond to the autoconvolution of a square function. In Fourier space we have:

$$\hat{B}_l(\nu) = \left(\frac{\sin \pi \nu}{\pi \nu}\right)^{l+1} \tag{C.9}$$

$B_3(x)$ is a set of 4 polynomials of degree 3. We choose the scaling function $\phi(\nu)$ which has a $B_3(x)$ profile in Fourier space:

$$\hat{\phi}(\nu) = \frac{3}{2} B_3(4\nu) \tag{C.10}$$

In direct space we get:

$$\phi(x) = \frac{3}{8} \left[\frac{\sin \frac{\pi x}{4}}{\frac{\pi x}{4}}\right]^4 \tag{C.11}$$

This function is quite similar to a Gaussian and converges rapidly to 0. For 2-dimensions the scaling function is defined by $\hat{\phi}(u,v) = \frac{3}{2} B_3(4r)$, with $r = \sqrt{(u^2+v^2)}$. This is an isotropic function.

The wavelet transform algorithm with J scales is the following:

1. Start with a B_3-spline scaling function and derive ψ, h and g numerically.
2. Compute the corresponding FFT image. Name the resulting complex array T_0.
3. Set j to 0. Iterate:
4. Multiply T_j by $\hat{g}(2^j u, 2^j v)$. We get the complex array W_{j+1}. The inverse FFT gives the wavelet coefficients at scale 2^j;
5. Multiply T_j by $\hat{h}(2^j u, 2^j v)$. We get the array T_{j+1}. Its inverse FFT gives the image at scale 2^{j+1}. The frequency band is reduced by a factor 2.
6. Increment j.
7. If $j \leq J$, go back to 4.
8. The set $\{w_1, w_2, \ldots, w_J, c_J\}$ describes the wavelet transform.

If the wavelet is the difference between two resolutions, i.e.

$$\hat{\psi}(2\nu) = \hat{\phi}(\nu) - \hat{\phi}(2\nu) \tag{C.12}$$

and:

$$\hat{g}(\nu) = 1 - \hat{h}(\nu) \tag{C.13}$$

then the wavelet coefficients $\hat{w}_j(\nu)$ can be computed by $\hat{c}_{j-1}(\nu) - \hat{c}_j(\nu)$.

Reconstruction. If the wavelet is the difference between two resolutions, an evident reconstruction for a wavelet transform $\mathcal{W} = \{w_1, \ldots, w_J, c_J\}$ is:

$$\hat{c}_0(\nu) = \hat{c}_J(\nu) + \sum_j \hat{w}_j(\nu) \tag{C.14}$$

But this is a particular case, and other alternative wavelet functions can be chosen. The reconstruction can be made step-by-step, starting from the lowest resolution. At each scale, we have the relations:

$$\hat{c}_{j+1} = \hat{h}(2^j\nu)\hat{c}_j(\nu) \tag{C.15}$$
$$\hat{w}_{j+1} = \hat{g}(2^j\nu)\hat{c}_j(\nu) \tag{C.16}$$

We look for c_j knowing c_{j+1}, w_{j+1}, h and g. We restore $\hat{c}_j(\nu)$ based on a least mean square estimator:

$$\hat{p}_h(2^j\nu) \mid \hat{c}_{j+1}(\nu) - \hat{h}(2^j\nu)\hat{c}_j(\nu) \mid^2 +$$
$$\hat{p}_g(2^j\nu) \mid \hat{w}_{j+1}(\nu) - \hat{g}(2^j\nu)\hat{c}_j(\nu) \mid^2 \tag{C.17}$$

is to be minimum. $\hat{p}_h(\nu)$ and $\hat{p}_g(\nu)$ are weight functions which permit a general solution to the restoration of $\hat{c}_j(\nu)$. From the derivation of $\hat{c}_j(\nu)$ we get:

$$\hat{c}_j(\nu) = \hat{c}_{j+1}(\nu)\hat{\tilde{h}}(2^j\nu) + \hat{w}_{j+1}(\nu)\hat{\tilde{g}}(2^j\nu) \tag{C.18}$$

where the conjugate filters have the expression:

$$\hat{\tilde{h}}(\nu) = \frac{\hat{p}_h(\nu)\hat{h}^*(\nu)}{\hat{p}_h(\nu)|\hat{h}(\nu)|^2 + \hat{p}_g(\nu)|\hat{g}(\nu)|^2} \tag{C.19}$$
$$\hat{\tilde{g}}(\nu) = \frac{\hat{p}_g(\nu)\hat{g}^*(\nu)}{\hat{p}_h(\nu)|\hat{h}(\nu)|^2 + \hat{p}_g(\nu)|\hat{g}(\nu)|^2} \tag{C.20}$$

In this analysis, the Shannon sampling condition is always respected and no aliasing exists.

The denominator is reduced if we choose:

$$\hat{g}(\nu) = \sqrt{1 - \mid \hat{h}(\nu) \mid^2}$$

This corresponds to the case where the wavelet is the difference between the square of two resolutions:

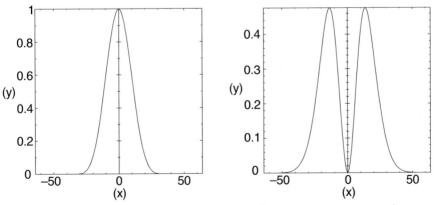

Fig. C.1. *Left:* the interpolation function $\hat{\phi}$ and *right:* the wavelet $\hat{\psi}$.

Appendix C

$$|\hat{\psi}(2\nu)|^2 = |\hat{\phi}(\nu)|^2 - |\hat{\phi}(2\nu)|^2 \qquad (C.21)$$

In Fig. C.1 the chosen scaling function derived from a B-spline of degree 3, and its resulting wavelet function, are plotted in frequency space.

The reconstruction algorithm is:

1. Compute the FFT of the image at the low resolution.
2. Set j to J. Iterate:
3. Compute the FFT of the wavelet coefficients at scale j.
4. Multiply the wavelet coefficients $\hat{w}_j$ by $\hat{\tilde{g}}$.
5. Multiply the image at the lower resolution $\hat{c}_j$ by $\hat{\tilde{h}}$.
6. The inverse Fourier transform of the addition of $\hat{w}_j\hat{\tilde{g}}$ and $\hat{c}_j\hat{\tilde{h}}$ gives the image c_{j-1}.
7. Set $j = j - 1$ and return to 3.

The use of a scaling function with a cut-off frequency allows a reduction of sampling at each scale, and limits the computing time and the memory size required.

D. Derivative Needed for the Minimization

The Error Function

The error function is written erf(x) and the complementary error function erfc(x). Their definitions are:

$$\text{erf}(x) = \frac{2}{\sqrt{\pi}} \int_0^x e^{-t^2} dt \tag{D.1}$$

$$\text{erfc}(x) = 1 - \text{erf}(x) = \frac{2}{\sqrt{\pi}} \int_x^\infty e^{-t^2} dt \tag{D.2}$$

These functions have the following limits and symmetries:

erf(0) = 0	erf(∞) = 1
erfc(0) = 1	erfc(∞) = 0
erf($-x$) = erf(x)	erfc($-x$) = 2$-$erfc(x)

(D.3)

N1-MSE

Now we compute the contribution of the wavelet coefficient x to the noise information:

$$h_n(x) = \frac{x^2}{2\sigma^2} \text{erfc}\left(\frac{x}{\sqrt{2}\sigma}\right) \tag{D.4}$$

$$\frac{dh_n(x)}{dx} = \frac{x}{\sigma^2} \text{erfc}\left(\frac{x}{\sqrt{2}\sigma}\right) + \frac{x^2}{2\sigma^2} \frac{\partial \text{erfc}\left(\frac{x}{\sqrt{2}\sigma}\right)}{\partial x} \tag{D.5}$$

We derive the function erfc:

$$\frac{\partial \text{erfc}(x)}{\partial x} = -\frac{2}{\sqrt{\pi}} e^{-x^2} \tag{D.6}$$

then

$$\frac{dh_n(x)}{dx} = \frac{x}{\sigma^2} \text{erfc}\left(\frac{x}{\sqrt{2}\sigma}\right) - \frac{x^2}{\sqrt{2\pi}\sigma^3} e^{-\frac{x^2}{2\sigma^2}} \tag{D.7}$$

In order to minimize the functional 7.44, we may want to calculate the derivative of $h_s(x-y)$, $h_s(x-y)$ measuring the amount of information contained in the residual (y being the data).

$$h_s(x-y) = \frac{(y-x)^2}{2\sigma^2}\text{erf}(\frac{y-x}{\sqrt{2}\sigma}) \tag{D.8}$$

Denoting $z = y - x$, we have

$$h_s(z) = \frac{z^2}{2\sigma^2}\text{erf}(\frac{z}{\sqrt{2}\sigma}) \tag{D.9}$$

$$\frac{dh_s(z)}{dz} = \frac{z}{\sigma^2}\text{erf}(\frac{z}{\sqrt{2}\sigma}) + \frac{z^2}{2\sigma^2}\frac{\partial \text{erf}(\frac{z}{\sqrt{2}\sigma})}{\partial z} \tag{D.10}$$

$$= \frac{z}{\sigma^2}\text{erf}(\frac{z}{\sqrt{2}\sigma}) + \frac{z^2}{\sqrt{2\pi}\sigma^3}e^{-\frac{z^2}{2\sigma^2}} \tag{D.11}$$

and

$$\frac{dh_s(x-y)}{dx} = -\frac{dh_s(z)}{dz} \tag{D.12}$$

N2-MSE

We compute the contribution of the wavelet coefficient x to the noise information:

$$h_n(x) = \frac{1}{\sigma^2}\int_0^x t\,\text{erfc}(\frac{x-t}{\sqrt{2}\sigma})dt \tag{D.13}$$

$$\frac{dh_n(x)}{dx} = h_n(x+dx) - h_n(x) \tag{D.14}$$

$$= \frac{1}{\sigma^2}\int_0^{x+dx}\text{erfc}(\frac{x+dx-t}{\sqrt{2}\sigma})dt - \frac{1}{\sigma^2}\int_0^x \text{erfc}(\frac{x-t}{\sqrt{2}\sigma})dt \tag{D.15}$$

$$\frac{dh_n(x)}{dx} = \frac{x}{\sigma^2}\text{erfc}(\frac{x-x}{\sqrt{2}\sigma}) + \frac{1}{\sigma^2}\int_0^x \frac{\partial\,t\,\text{erfc}(\frac{x-t}{\sqrt{2}\sigma})}{\partial x}dt \tag{D.16}$$

Now, because $\text{erfc}(0) = 1$ we have:

$$\frac{dH_n(x)}{dx} = \frac{x}{\sigma^2} + \frac{1}{\sigma^2}\int_0^x \frac{\partial\,t\,\text{erfc}(\frac{x-t}{\sqrt{2}\sigma})}{\partial x}dt \tag{D.17}$$

We derive the function erfc:

$$\frac{\partial\,\text{erfc}(x)}{\partial x} = -\frac{2}{\sqrt{\pi}}e^{-x^2} \tag{D.18}$$

$$\frac{\partial\,\mathrm{erfc}(\frac{(x-t)}{\sqrt{2}\sigma})}{\partial x} = -\frac{2}{\sqrt{\pi}}\frac{1}{\sqrt{2}\sigma}e^{-\frac{(x-t)^2}{2\sigma^2}} = -\sqrt{\frac{2}{\pi}}\frac{1}{\sigma}e^{-\frac{(x-t)^2}{2\sigma^2}} \qquad (D.19)$$

Now we deduce for the derivative of h_n:

$$\frac{dh_n(x)}{dx} = \frac{x}{\sigma^2} + \frac{1}{\sigma^2}\int_0^x -\sqrt{\frac{2}{\pi}}\frac{1}{\sigma}te^{-\frac{(x-t)^2}{2\sigma^2}}\,dt \qquad (D.20)$$

$$\frac{dh_n(x)}{dx} = \frac{x}{\sigma^2} + \frac{1}{\sigma^3}\sqrt{\frac{2}{\pi}}\int_0^x te^{-\frac{(x-t)^2}{2\sigma^2}}\,dt \qquad (D.21)$$

We create the variable J

$$J = \int_0^x te^{-\frac{(x-t)^2}{2\sigma^2}}\,dt \qquad (D.22)$$

We create the variable u

$$u = \frac{t-x}{\sqrt{2}\sigma} \qquad t = x + u\sqrt{2}\sigma$$
$$dt = \sqrt{2}\sigma\,du \qquad (D.23)$$
$$t = 0 \Rightarrow u = \frac{-x}{\sqrt{2}\sigma} \qquad t = x \Rightarrow u = 0$$

The variable J can be written with u

$$J = \int_{\frac{-x}{\sqrt{2}\sigma}}^{0} (x + u\sqrt{2}\sigma)e^{-u^2}\sqrt{2}\sigma\,du \qquad (D.24)$$

$$J = \sqrt{2}\sigma x \int_{\frac{-x}{\sqrt{2}\sigma}}^{0} e^{-u^2}\,du + 2\sigma^2 \int_{\frac{-x}{\sqrt{2}\sigma}}^{0} u e^{-u^2}\,du \qquad (D.25)$$

The first part of J can be rewritten as:

$$J_0 = \sqrt{2}\sigma x \int_0^{\frac{x}{\sqrt{2}\sigma}} e^{-u^2}\,du \qquad (D.26)$$

J_0 can be expressed with the error function.

$$J_0 = \sqrt{2}\sigma\frac{\sqrt{\pi}}{2}x\,\mathrm{erf}(\frac{x}{\sqrt{2}\sigma}) \qquad J_0 = \sigma\sqrt{\frac{\pi}{2}}x\,\mathrm{erf}(\frac{x}{\sqrt{2}\sigma}) \qquad (D.27)$$

Now the second part of J is obvious

$$J_1 = 2\sigma^2 \int_{\frac{-x}{\sqrt{2}\sigma}}^{0} u e^{-u^2}\,du \qquad (D.28)$$

or

$$\frac{de^{-u^2}}{du} = -2ue^{-u^2} \qquad (D.29)$$

We replace

$$J_1 = -\sigma^2 \int_{\frac{-x}{\sqrt{2}\sigma}}^{0} d(e^{-u^2}) \tag{D.30}$$

$$J_1 = \sigma^2 [e^{-\frac{x^2}{2\sigma^2}} - 1] \tag{D.31}$$

Now we can write J

$$J = J_0 + J_1 = \sigma\sqrt{\frac{\pi}{2}} \, x \, \text{erf}(\frac{x}{\sqrt{2}\sigma}) + \sigma^2[e^{-\frac{x^2}{2\sigma^2}} - 1] \tag{D.32}$$

We can write the derivative of h_n

$$\frac{dh_n(x)}{dx} = \frac{x}{\sigma^2} - \frac{1}{\sigma^3}\sqrt{\frac{2}{\pi}} J \tag{D.33}$$

$$= \frac{x}{\sigma^2} - \frac{x}{\sigma^2} x \, \text{erf}(\frac{x}{\sqrt{2}\sigma}) + \frac{1}{\sigma}\sqrt{\frac{2}{\pi}}[1 - e^{-\frac{x^2}{2\sigma^2}}] \tag{D.34}$$

$$= \frac{x}{\sigma^2}\text{erfc}(\frac{x}{\sqrt{2}\sigma}) + \frac{1}{\sigma}\sqrt{\frac{2}{\pi}}[1 - e^{-\frac{x^2}{2\sigma^2}}] \tag{D.35}$$

In order to minimize the functional 7.44, we may want to calculate the derivative of $h_s(x \mid y)$, $h_s(x \mid y)$ measuring the amount of information contained in the residual (y being the data).

$$h_s(x \mid y) = \frac{1}{\sigma^2} \int_0^{y-x} t \, \text{erf}(\frac{y-x-t}{\sqrt{2}\sigma}) dt \tag{D.36}$$

Denoting $z = y - x$, we have

$$h_s(z) = \frac{1}{\sigma^2}\int_0^z t \, \text{erf}(\frac{z-t}{\sqrt{2}\sigma})dt \tag{D.37}$$

$$= \frac{1}{\sigma^2}\int_0^z t dt - \frac{1}{\sigma^2}\int_0^z t \, \text{erfc}(\frac{z-t}{\sqrt{2}\sigma})dt \tag{D.38}$$

and

$$\frac{dh_s(x)}{dx} = \frac{\partial h_s(z)}{\partial z}\frac{\partial z}{\partial x} \tag{D.39}$$

$$\frac{\partial h_s(z)}{\partial z} = \frac{z}{\sigma^2} - \frac{\partial h_n(z)}{\partial z} \tag{D.40}$$

Then

$$\frac{dh_s(x)}{dx} = -\frac{y-x}{\sigma^2} + \frac{y-x}{\sigma^2}\text{erfc}(\frac{y-x}{\sqrt{2}\sigma}) + \sqrt{\frac{2}{\pi}}\frac{1}{\sigma}[1 - e^{-\frac{(y-x)^2}{2\sigma^2}}] \tag{D.41}$$

$$= -\frac{y-x}{\sigma^2}\text{erf}(\frac{y-x}{\sqrt{2}\sigma}) + \sqrt{\frac{2}{\pi}}\frac{1}{\sigma}[1 - e^{-\frac{(y-x)^2}{2\sigma^2}}] \tag{D.42}$$

E. Generalization of the Derivative Needed for the Minimization

The contribution of the wavelet coefficient x to the noise and signal information in the general case is

$$h_n(x) = \int_0^{|x|} p_n(u|x)(\frac{\partial h(x)}{\partial x})_{x=u} du \qquad (E.1)$$

$$h_s(x) = \int_0^{|x|} p_s(u|x)(\frac{\partial h(x)}{\partial x})_{x=u} du$$

Assuming $h(x) = \frac{1}{2}x^2$, we have

$$h_n(x) = \int_0^{|x|} p_n(x-u) u \, du \qquad (E.2)$$

$$h_s(x) = \int_0^{|x|} p_s(x-u) u \, du$$

$$\frac{dh_s(x)}{dx} = \int_0^x \frac{\partial P_s(x-u)}{\partial x}\Big|_{x=u} u \, du + \frac{1}{dx}\int_x^{x+dx} P_s(x-u) u \, du \qquad (E.3)$$

Since $P_s(0) = 0$, the second term tends to zero.
Denoting $\frac{\partial P_s(x-u)}{\partial x} = -\frac{\partial P_s(x-u)}{\partial u}$, we have

$$\frac{dh_s(x)}{dx} = -\int_0^x \frac{\partial P_s(x-u)}{\partial u} u \, du \qquad (E.4)$$

$$= -([u P_s(x-u)]_0^x - \int_0^w P_s(x-u) du) \qquad (E.5)$$

$$= \int_0^x P_s(x-u) du \qquad (E.6)$$

$$= \int_0^x P_s(u) du \qquad (E.7)$$

and from $h_n = h - h_s$ we get

$$\frac{dh_n(x)}{dx} = x - \int_0^x P_s(u) du \qquad (E.8)$$

and

$$\frac{dh_s(y-x)}{dx} = -\int_0^{y-x} P_s(u)du \qquad (E.9)$$

It is easy to verify that replacing $P_s(x) = \mathrm{erf}(x)$, and $P_n(x) = \mathrm{erfc}(x)$, (case of Gaussian noise) we find the same equation as in Appendix B.

F. Software and Related Developments

Software: Public Domain and Commercial

- An excellent first stop is at AstroWeb. Many aspects of astronomy are included, including software. Accessible from `cdsweb.u-strasbg.fr/astroweb.html`, there are nearly 200 links to astronomy software resources. Links to large software packages are included in these lists: ESO-MIDAS, AIPS, IRAF, the Starlink Software Collection, the IDL Astronomy User's Library, and information on the FITS storage standard.
- Statistical software is available at `www.astro.psu.edu/statcodes`. For statistics software in general, the point of reference is Statlib, `lib.stat.cmu.edu`.
- For clustering, an annotated resources list is at `astro.u-strasbg.fr/~fmurtagh/mda-sw/online-sw.html`. This includes independent component analysis, decision trees, Autoclass, etc. For data mining or knowledge discovery, KDNuggets is at `www.kdnuggets.com`.
- For wavelets, a list of resources, mostly in the public domain, is available at `www.amara.com/current/wavesoft.html`.
- A commercial package for mathematical morphology is at `malte.ensmp.fr/Micromorph`. Matlab (referred to again below) contains a great number of specialized toolboxes, including mathematical morphology, wavelets, and neural networks.
- Links to some neural network code is accessible from AstroWeb. The Stuttgart Neural Network Simulator, an all-round neural net modeling package, is at `www-ra.informatik.uni-tuebingen.de/SNNS`.
- Important commercial data analysis packages include IDL (with its twin, PV-Wave), S-Plus (and public domain version, R: see `www.r-project.org`), Mathematica, and Matlab (public domain versions include Octave and SciLab).
- The MR software package includes MR/1 which implements multiscale methods for 1D and 2D data, MR/2 which implements methods derived from multiscale entropy, and MR/3 which implements multiscale methods for 3D and multi-channel data. Nearly all figures in this book are based on this package. Information is available at `www.multiresolution.com`.

310   Appendix Software

- MRS, a package in IDL for multiresolution analysis on the sphere is available at `jstarck.free.fr/mrs.html`. MRL, also in IDL, is for weak lensing, and is available at `jstarck.free.fr/mrl.html`.
- MaxEnt software for maximum entropy signal analysis, including applications in chemical spectroscopy and imaging tomography, can be found at `www.maxent.co.uk`.
- The Pixon software for image analysis, again with wide-ranging application areas including nuclear medicine, is described at `www.pixon.com`.
- Finally, for other topics, including support vector machines, machine learning, rough sets, multivalued logic, and much more, see `homepage.ntlworld.com/jg.campbell/links`.
- There are many display programs available, which are certainly to be found in larger systems. Stand-alone FITS viewers for Windows are available at `http://www.eia.brad.ac.uk/rti/guide/fits.html`. A "gold standard" of astronomical imaging display programs is SAO DS9: `http://hea-www.harvard.edu/RD/ds9`. A stand-alone data visualization system is GGobi (a successor to XGobi): `www.ggobi.org`.
- For image compression, the following were cited in Chapter 5:
  - HCOMPRESS: `www.stsci.edu/software/hcompress.html`.
  - FITSPRESS: `www.eia.brad.ac.uk/rti/guide/fits.html`.
  - Wavelet: `www.geoffdavis.net/dartmouth/wavelet/wavelet.html`.
  - Fractal: `inls.ucsd.edu/y/Fractals/book.html`.
- For source detection: SExtractor, `terapix.iap.fr/soft/sextractor/index.html`.

New Developments

A number of important national and international projects are targeting a new generation of software systems. Some of them are listed here.

- Both new and old, the online astronomical literature is available at the NASA Astrophysics Data System, `adswww.harvard.edu`.
- Major virtual observatory projects include the Astrophysical Virtual Observatory, `www.eso.org/projects/avo`, and the National Virtual Observatory, `www.us-vo.org`. The Virtual Observatory Forum is at `www.voforum.org`.
- Opticon, the Optical Infrared Coordination Network, is at `www.astro-opticon.org`, and AstroWise, Astronomical Wide-Field Imaging System for Europe at `www.astro-wise.org`.
- For Grid-related work, AstroGrid (UK) is at `www.astrogrid.org`. The iAstro coordination initiative is at `www.iAstro.org`.

Bibliography

Ables, J.: 1974, *Astronomy and Astrophysics* **15**, 383
Acar, R. and Vogel, C.: 1994, *Physica D* **10**, 1217
Adler, R. J.: 1981, *The Geometry of Random Fields*, The Geometry of Random Fields, Chichester: Wiley, 1981
Adorf, H., Hook, R., and Lucy, L.: 1995, *International Journal of Imaging Systems and Technology* **6**, 339
Aghanim, N. and Forni, O.: 1999, *Astronomy and Astrophysics* **347**, 409
Aghanim, N., Forni, O., and Bouchet, F. R.: 2001, *Astronomy and Astrophysics* **365**, 341
Aghanim, N., Kunz, M., Castro, P. G., and Forni, O.: 2003, *Astronomy and Astrophysics* **406**, 797
Allard, D. and Fraley, C.: 1997, *Journal of the American Statistical Association* **92**, 1485
Alloin, D., Pantin, E., Lagage, P. O., and Granato, G. L.: 2000, *Astronomy and Astrophysics* **363**, 926
Alvarez, L., Lions, P.-L., and Morel, J.-M.: 1992, *SIAM Journal on Numerical Analysis* **29**, 845
Amato, U. and Vuza, D.: 1997, *Revue Roumaine de Mathématiques Pures et Appliquées* **42**, 481
Andersson, W. and Andersson, B.: 1993, *American Astronomical Society Meeting* **183**, 1604
Andrews, H. and Patterson, C.: 1976, *IEEE Transactions on Acoustics, Speech and Signal Processing* **24**, 26
Anscombe, F.: 1948, *Biometrika* **15**, 246
Appleton, P. N., Siqueira, P. R., and Basart, J. P.: 1993, *Astronomical Journal* **106**, 1664
Arabie, P., Schleutermann, S., Dawes, J., and Hubert, L.: 1988, in W. Gaul and M. Schader (eds.), *Data, Expert Knowledge and Decisions*, pp 215–224, Springer-Verlag
Arnaud, M., Maurogordato, S., Slezak, E., and Rho, J.: 2000, *Astronomy and Astrophysics* **355**, 461
Arneodo, A., Argoul, F., Bacry, E., Elezgaray, J., and Muzy, J. F.: 1995, *Ondelettes, Multifractales et Turbulences*, Diderot, Arts et Sciences, Paris
Aussel, H., Elbaz, D., Cesarsky, C., and Starck, J.-L.: 1999, *Astronomy and Astrophysics* **342**, 313
Averbuch, A., Coifman, R., Donoho, D., Israeli, M., and Waldén, J.: 2001, *SIAM J. Sci. Comput.*, To appear
Baccigalupi, C., Bedini, L., Burigana, C., De Zotti, G., Farusi, A., Maino, D., Maris, M., Perrotta, F., Salerno, E., Toffolatti, L., and Tonazzini, A.: 2000a, *Monthly Notices of the Royal Astronomical Society* **318**, 769

Baccigalupi, C., Bedini, L., Burigana, C., De Zotti, G., Farusi, A., Maino, D., Maris, M., Perrotta, F., Salerno, E., Toffolatti, L., and Tonazzini, A.: 2000b, *MNRAS* **318**, 769

Balastegui, A., Ruiz-Lapuente, P., and Canal, R.: 2001, *Monthly Notices of the Royal Astronomical Society* **328**, 283

Ballester, P.: 1994, *Astronomy and Astrophysics* **286**, 1011

Banday, A. J., Zaroubi, S., and Górski, K. M.: 2000, *Astrophysical Journal* **533**, 575

Banerjee, S. and Rosenfeld, A.: 1993, *Pattern Recognition* **26**, 963

Banfield, J. and Raftery, A.: 1993, *Biometrics* **49**, 803

Barreiro, R. B. and Hobson, M. P.: 2001, *Monthly Notices of the Royal Astronomical Society* **327**, 813

Bazell, D. and Aha, D.: 2001, *Astrophysical Journal* **548**, 219

Bendjoya, P.: 1993, *Astronomy and Astrophysics, Supplement Series* **102**, 25

Bendjoya, P., Petit, J.-M., and Spahn, F.: 1993, *Icarus* **105**, 385

Benjamini, Y. and Hochberg, Y.: 1995, *J. R. Stat. Soc. B* **57**, 289

Bennet, W.: 1948, *Bell Systems Technical Journal* **27**, 446

Bennett, C. L., Hill, R. S., Hinshaw, G., Nolta, M. R., Odegard, N., Page, L., Spergel, D. N., Weiland, J. L., Wright, E. L., Halpern, M., Jarosik, N., Kogut, A., Limon, M., Meyer, S. S., Tucker, G. S., and Wollack, E.: 2003, *Astrophysical Journal, Supplement Series* **148**, 97

Bennett, K., Fayyad, U., and Geiger, D.: 1999, *Density-based indexing for approximate nearest neighbor queries*, Technical Report MSR-TR-98-58, Microsoft Research

Benoît, A., et al.: 2003, *Astronomy and Astrophysics* **399**, L19

Bernardeau, F. and Uzan, J.: 2002, *Phys. Rev. D* **66**, 103506

Bernardeau, F., van Waerbeke, L., and Mellier, Y.: 2003, *Astronomy and Astrophysics* **397**, 405

Berry, M., Drmač, Z., and Jessup, E.: 1999, *SIAM Review* **41**, 335

Berry, M., Hendrickson, B., and Raghavan, P.: 1996, in J. Renegar, M. Shub, and S. Smale (eds.), *Lectures in Applied Mathematics Vol. 32: The Mathematics of Numerical Analysis*, pp 99–123, American Mathematical Society

Bertero, M. and Boccacci, P.: 1998, *Introduction to Inverse Problems in Imaging*, Institute of Physics

Bertin, E. and Arnouts, S.: 1996, *Astronomy and Astrophysics, Supplement Series* **117**, 393

Bhavsar, S. P. and Splinter, R. J.: 1996, *Monthly Notices of the Royal Astronomical Society* **282**, 1461

Bibring, J.-P. and OMEGA: 2004, *AAS/Division for Planetary Sciences Meeting Abstracts* **36**,

Bijaoui, A.: 1980, *Astronomy and Astrophysics* **84**, 81

Bijaoui, A.: 1984, *Introduction au Traitement Numérique des Images*, Masson

Bijaoui, A.: 1993, in Y. Meyer and S. Roques (eds.), *Progress in Wavelet Analysis and Applications*, pp 551–556, Editions Frontières

Bijaoui, A., Bobichon, Y., and Huang, L.: 1996, *Vistas in Astronomy* **40**, 587

Bijaoui, A., Bury, P., and Slezak, E.: 1994, *Catalog analysis with multiresolution insights. 1. Detection and characterization of significant structures*, Technical report, Observatoire de la Côte d'Azur

Bijaoui, A. and Jammal, G.: 2001, *Signal Processing* **81**, 1789

Bijaoui, A. and Rué, F.: 1995, *Signal Processing* **46**, 229

Birkinshaw, M.: 1999, *Physics Reports* **310**, 97

Bishop, C.: 1995, *Neural Networks and Pattern Recognition*, Oxford University Press

Blanc-Féraud, L. and Barlaud, M.: 1996, *Vistas in Astronomy* **40**, 531
Blanco, S., Bocchialini, K., Costa, A., Domenech, G., Rovira, M., and Vial, J. C.: 1999, *Solar Physical Journal* **186**, 281
Bobichon, Y. and Bijaoui, A.: 1997, *Experimental Astronomy* **7**, 239
Bonnarel, F., Fernique, P., Genova, F., Bartlett, J. G., Bienaymé, O., Egret, D., Florsch, J., Ziaeepour, H., and Louys, M.: 1999, in D. Mehringer, R. Plante, and D. Roberts (eds.), *Astronomical Data Analysis Software and Systems VIII*, pp 229–232, Astronomical Society of the Pacific
Bonnarel, F., Fernique, P., Genova, F., Bienaymé, O., and Egret, D.: 2001, in J. F.R. Harnden, F. Primini, and H. Payne (eds.), *Astronomical Data Analysis Software and Systems X*, p. 74, Astronomical Society of the Pacific
Bontekoe, T., Koper, E., and Kester, D.: 1994, *Astronomy and Astrophysics* **284**, 1037
Borodin, A., Ostrovsky, R., and Rabani, Y.: 1999, in *Proc. 31st ACM Symposium on Theory of Computing, STOC-99*
Bouchet, F. R., Bennett, D. P., and Stebbins, A.: 1988, *Nature* **335**, 410
Bouman, C. A. and Sauer, K.: 1993, *IEEE Transactions on Image Processing* **2(3)**, 296
Bovik, A.: 2000, *Handbook of Image and Video Processing*, Academic
Breen, E., Jones, R., and Talbot, H.: 2000, *Statistics and Computing* **10**, 105
Bridle, S. L., Hobson, M. P., Lasenby, A. N., and Saunders, R.: 1998, *Monthly Notices of the Royal Astronomical Society* **299**, 895
Broder, A.: 1998, in *Compression and Complexity of Sequences, SEQUENCES'97*, pp 21–29, IEEE Computer Society
Broder, A., Glassman, S., Manasse, M., and Zweig, G.: 1997, in *Proceedings Sixth International World Wide Web Conference*, pp 391–404
Bromley, B. C. and Tegmark, M.: 1999, *Astrophysical Journal Letters* **524**, L79
Brosch, N. and Hoffman, Y.: 1999, *Monthly Notices of the Royal Astronomical Society* **305**, 241
Buonanno, R., Buscema, G., Corsi, C., Ferraro, I., and Iannicola, G.: 1983, *Astronomy and Astrophysics* **126**, 278
Burg, J.: 1978, *Multichannel maximum entropy spectral analysis*, Annual Meeting International Society Exploratory Geophysics, Reprinted in Modern Spectral Analysis, D.G. Childers, ed., IEEE Press, 34–41
Burrus, C., Gopinath, R., and Guo, H.: 1998, *Introduction to Wavelets and Wavelet Transforms*, Prentice-Hall
Burt, P. and Adelson, A.: 1983, *IEEE Transactions on Communications* **31**, 532
Burud, I., Courbin, F., Magain, P., Lidman, C., Hutsemékers, D., Kneib, J.-P., Hjorth, J., Brewer, J., Pompei, E., Germany, L., Pritchard, J., Jaunsen, A. O., Letawe, G., and Meylan, G.: 2002, *Astronomy and Astrophysics* **383**, 71
Byers, S. and Raftery, A.: 1998, *Journal of the American Statistical Association* **93**, 577
Calderbank, R., Daubechies, I., Sweldens, W., and Yeo, B.-L.: 1998, *Applied and Computational Harmonic Analysis* **5**, 332
Cambrésy, L.: 1999, *Astronomy and Astrophysics* **345**, 965
Campbell, J. and Murtagh, F.: 2001, Image Processing and Pattern Recognition, online book, http://www.qub.ac.uk/ivs/resour.html
Candès, E. and Donoho, D.: 1999, *Philosophical Transactions of the Royal Society of London A* **357**, 2495
Candès, E. and Donoho, D.: 2000a, *J. Approx. Theory*. **113**, 59
Candès, E. and Donoho, D.: 2000b, in *SPIE conference on Signal and Image Processing: Wavelet Applications in Signal and Image Processing VIII*
Candès, E. and Guo, F.: 2002, *Signal Processing* **82(11)**, 1519

Candès, E. J., Demanet, L., Donoho, D. L., and Ying, L.: 2005, *Fast discrete curvelet transforms*, Technical report, Caltech
Cardoso, J.: 1998a, *Proceedings of the IEEE* **86**, 2009
Cardoso, J.-F.: 1998b, *Proceedings of the IEEE. Special Issue on Blind Identification and Estimation* **9(10)**, 2009
Cardoso, J.-F.: 1999, *Neural Computation* **11(1)**, 157
Cardoso, J.-F.: 2001, in *Proc. ICA 2001, San Diego*
Cardoso, J.-F.: 2003, *Journal of Machine Learning Research* **4**, 1177
Cardoso, J.-F. and et al., H. S.: 2002, in *Proc. EUSIPCO2002, Toulouse (France)*, pp 561–564
Carlsohn, M., Paillou, P., Louys, M., and Bonnarel, F.: 1993, in INRIA-ESA-Aérospatiale (ed.), *Computer Vision for Space Applications*, pp 389–395, Antibes
Carroll, T. and Staude, J.: 2001, *Astronomy and Astrophysics* **378**, 316
Castleman, K.: 1995, *Digital Image Processing*, Prentice-Hall
Castro, P. G.: 2003, *Phys. Rev. D* **67**, 123001
Caulet, A. and Freudling, W.: 1993, *ST-ECF Newsletter No. 20* pp 5–7
Cayón, L., Sanz, J. L., Barreiro, R. B., Martínez-González, E., Vielva, P., Toffolatti, L., Silk, J., Diego, J. M., and Argüeso, F.: 2000, *Monthly Notices of the Royal Astronomical Society* **315**, 757
Cayón, L., Sanz, J. L., Martínez-González, E., Banday, A. J., Argüeso, F., Gallegos, J. E., Górski, K. M., and Hinshaw, G.: 2001, *Monthly Notices of the Royal Astronomical Society* **326**, 1243
Celeux, G. and Govaert, G.: 1995, *Pattern Recognition* **28**, 781
Cesarsky, C., Abergel, A., Agnese, P., et al.: 1996, *Astronomy and Astrophysics* **315**, L32
Chakrabarti, K., Garofalakis, M., Rastogi, R., and Shim, K.: 2000, in *Proceedings of the 26th VLDB Conference*
Chambolle, A., DeVore, R., Lee, N., and Lucier, B.: 1998, *IEEE Transactions on Signal Processing* **7**, 319
Champagnat, F., Goussard, Y., and Idier, J.: 1996, *IEEE Transactions on Image Processing* **44**, 2988
Chang, K.-P. and Fu, A. W.-C.: 1999, in *Proceedings of the 15th International Conference on Data Engineering*, pp 126–133
Charbonnier, P., Blanc-Feraud, L., Aubert, G., and Barlaud, M.: 1997a, *IEEE Transactions on Image Processing* **6(2)**, 298
Charbonnier, P., Blanc-Féraud, L., Aubert, G., and Barlaud, M.: 1997b, *IEEE Transactions on Image Processing* **6**, 298
Charter, M.: 1990, in W. Linden and V. Dose (eds.), *Maximum Entropy and Bayesian Methods*, pp 325–339, Kluwer
Cheriton, D. and Tarjan, D.: 1976, *SIAM Journal on Computing* **5**, 724
Chipman, H., Kolaczyk, E., and McCulloch, R.: 1997, *Journal of the American Statistical Association* **92**, 1413
Choi, H. and Baraniuk, R.: 1998, in *Proceedings of the IEEE-SP International Symposium on Time-Frequency and Time-Scale Analysis*
Christou, J. C., Bonnacini, D., Ageorges, N., and Marchis, F.: 1999, *The Messenger* **97**, 14
Chui, C.: 1992, *Wavelet Analysis and Its Applications*, Academic Press
Church, K. and Helfman, J.: 1993, *Journal of Computational and Graphical Statistics* **2**, 153
Cittert, P. V.: 1931, *Zeitschrift für Physik* **69**, 298
Claret, A., Dzitko, H., Engelmann, J., and Starck, J.-L.: 2000, *Experimental Astronomy* **10**, 305

Claypoole, R., Davis, G., Sweldens, W., and Baraniuk, R.: 2000, *IEEE Transactions on Image Processing*, submitted
Cohen, A.: 2003, *Numerical Analysis of Wavelet Methods*, Elsevier
Cohen, A., Daubechies, I., and Feauveau, J.: 1992, *Communications in Pure and Applied Mathematics* **45**, 485
Cohen, A., DeVore, R., Petrushev, P., and Xu, H.: 1999, *Amer. J. Math.* **121**, 587
Cohen, L.: 1995, *Time-Frequency Analysis*, Prentice-Hall
Coifman, R. and Donoho, D.: 1995, in A. Antoniadis and G. Oppenheim (eds.), *Wavelets and Statistics*, pp 125–150, Springer-Verlag
Coles, P., Davies, A. G., and Pearson, R. C.: 1996, *Monthly Notices of the Royal Astronomical Society* **281**, 1375
Combettes, P. and Vajs, V.: 2005, *preprint*, submitted
Cooray, A.: 2001, *Phys. Rev. D* **64**, 3514
Corbard, T., Blanc-Féraud, L., Berthomieu, G., and Provost, J.: 1999, *Astronomy and Astrophysics* **344**, 696
Cornwell, T. J.: 1989, in *NATO ASIC Proc. 274: Diffraction-Limited Imaging with Very Large Telescopes*, pp 273–+
Cortiglioni, F., Mähönen, P., Hakala, P., and Frantti, T.: 2001, *Astrophysical Journal* **556**, 937
Costa, G. D.: 1992, in S. Howel (ed.), *ASP Conference Series 23, Astronomical CCD Observing and Reduction Techniques*, p. 90, Astronical Society of the Pacific
Courbin, F., Lidman, C., and Magain, P.: 1998, *Astronomy and Astrophysics* **330**, 57
Coustenis, A., Gendron, E., Lai, O., Véran, J., Woillez, J., Combes, M., Vapillon, L., Fusco, T., Mugnier, L., and Rannou, P.: 2001, *Icarus* **154**, 501
Couvidat, S.: 1999, DEA dissertation, Strasbourg Observatory
Crouse, M., Nowak, R., and Baraniuk, R.: 1998, *IEEE Transactions on Signal Processing* **46**, 886
Cruz, M., Martínez-González, E., Vielva, P., and Cayón, L.: 2005, *Monthly Notices of the Royal Astronomical Society* **356**, 29
Csillaghy, A. and Benz, A.: 1999, *Solar Physics* **188**, 203
Csillaghy, A., Hinterberger, H., and Benz, A.: 2000, *Information Retrieval* **3**, 229
Damiani, F., Sciortino, S., and Micela, G.: 1998, *Astronomische Nachrichten* **319**, 78
Darken, C., Chang, J., and Moody, J.: 1992, in *Neural Networks for Signal Processing 2, Proceedings of the 1992 IEEE Workshop*, IEEE Press
Darken, C. and Moody, J.: 1991, in M. Lippmann and Touretzky (eds.), *Advances in Neural Information Processing Systems 3*, Morgan Kaufman
Darken, C. and Moody, J.: 1992, in H. Moody and Lippmann (eds.), *Advances in Neural Information Processing Systems 4*, Morgan Kaufman
Dasgupta, A. and Raftery, A.: 1998, *Journal of the American Statistical Association* **93**, 294
Daubechies, I.: 1992, *Ten Lectures on Wavelets*, Society for Industrial and Applied Mathematics
Daubechies, I., Defrise, M., and Mol, C. D.: 2004, *Comm. Pure Appl. Math* **57**, 1413
Daubechies, I. and Sweldens, W.: 1998, *Journal of Fourier Analysis and Applications* **4**, 245
Davis, M. and Peebles, P.: 1983, *Astrophysical Journal* **267**, 465
Davoust, E. and Pence, W.: 1982, *Astronomy and Astrophysics, Supplement Series* **49**, 631
de Bernardis, P., et al.: 2000, *Nature* **404**, 955
Deans, S.: 1983, *The Radon Transform and Some of Its Applications*, Wiley

Debray, B., Llebaria, A., Dubout-Crillon, R., and Petit, M.: 1994, *Astronomy and Astrophysics* **281**, 613
Delabrouille, J., Cardoso, J.-F., and Patanchon, G.: 2003, *Monthly Notices of the Royal Astronomical Society* **346(4)**, 1089, http://arXiv.org/abs/astro-ph/0211504
Dempster, A., Laird, N., and Rubin, D.: 1977, *Journal of the Royal Statistical Society, Series B* **39**, 1
Deutsch, S. and Martin, J.: 1971, *Operations Research* **19**, 1350
Dixon, D., Johnson, W., Kurfess, J., Pina, R., Puetter, R., Purcell, W., Tuemer, T., Wheaton, W., and Zych, A.: 1996, *Astronomy and Astrophysics, Supplement Series* **120**, 683
Dixon, D. D., Hartmann, D. H., Kolaczyk, E. D., Samimi, J., Diehl, R., Kanbach, G., Mayer-Hasselwander, H., and Strong, A. W.: 1998, *New Astronomy* **3**, 539
Djorgovski, S.: 1983, *Journal of Astrophysics and Astronomy* **4**, 271
Dobrzycki, A., Ebeling, H., Glotfelty, K., Freeman, P., Damiani, F., Elvis, M., and Calderwood, T.: 1999, *Chandra Detect 1.0 User Guide, Version 0.9*, Technical report, Chandra X-Ray Center, Smithsonian Astrophysical Observatory
Dollet, C., Bijaoui, A., and Mignard, F.: 2004, *Astronomy and Astrophysics* **426**, 729
Donoho, D.: 1993, in A. M. Society (ed.), *Proceedings of Symposia in Applied Mathematics*, Vol. 47, pp 173–205
Donoho, D.: 1995, *Applied and Computational Harmonic Analysis* **2**, 101
Donoho, D. and Duncan, M.: 2000, in H. Szu, M. Vetterli, W. Campbell, and J. Buss (eds.), *Proc. Aerosense 2000, Wavelet Applications VII*, Vol. 4056, pp 12–29, SPIE
Donoho, D., Johnson, I., Hoch, J., and Stern, A.: 1992, *Journal of the Royal Statistical Society* **B54**, 41
Donoho, D. and Johnstone, I.: 1994, *Biometrika* **81**, 425
Doroshkevich, A. G., Tucker, D. L., Fong, R., Turchaninov, V., and Lin, H.: 2001, *Monthly Notices of the Royal Astronomical Society* **322**, 369
Doyle, J.: 1988, *Applied Mathematical Modelling* **12**, 86
Doyle, L.: 1961, *Journal of the ACM* **8**, 553
Dubaj, D.: 1994, DEA dissertation, Strasbourg Observatory
Durand, S. and Froment, J.: 2003, *SIAM Journal of Scientific Computing* **24(5)**, 1754
Durrer, R. and Labini, F. S.: 1998, *Astronomy and Astrophysics* **339**, L85
Ebeling, H. and Wiedenmann, G.: 1993, *Physical Review E* **47**, 704
Efron, B. and Tibshirani, R.: 1986, *Statistical Science* **1**, 54
Epstein, B., Hingorani, R., Shapiro, J., and Czigler, M.: 1992, in *Data Compression Conference*, pp 200–208, Snowbird, UT
Escalera, E., Slezak, E., and Mazure, A.: 1992, *Astronomy and Astrophysics* **264**, 379
ESO: 1995, *MIDAS, Munich Image Data Analysis System*, European Southern Observatory
Faure, C., Courbin, F., Kneib, J. P., Alloin, D., Bolzonella, M., and Burud, I.: 2002, *Astronomy and Astrophysics* **386**, 69
Feauveau, J.: 1990, *Ph.D. thesis*, Université Paris Sud
Ferraro, M., Boccignone, G., and Caelli, T.: 1999, *IEEE Transactions on Pattern Analysis and Machine Intelligence* **21**, 1199
Figueiredo, M. and Nowak, R.: 2003, *IEEE Transactions on Image Processing* **12(8)**, 906
Fisher, F.: 1994, *Fractal Image Compression: Theory and Applications*, Springer-Verlag

Fisher, K., Huchra, J., Strauss, M., Davis, M., Yahil, A., and Schlegel, D.: 1995, *Astrophysical Journal Supplement* **100**, 69
Fixsen, D. J., Cheng, E. S., Cottingham, D. A., Folz, W. C., Inman, C. A., Kowitt, M. S., Meyer, S. S., Page, L. A., Puchalla, J. L., Ruhl, J. E., and Silverberg, R. F.: 1996, *Astrophysical Journal* **470**, 63
Fligge, M., Solanki, S. K., and Beer, J.: 1999, *Astronomy and Astrophysics* **346**, 313
Forgy, E.: 1965, *Biometrics* **21**, 768
Forni, O. and Aghanim, N.: 1999, *Astronomy and Astrophysics, Supplement Series* **137**, 553
Forni, O., Poulet, F., Bibring, J.-P., Erard, S., Gomez, C., Langevin, Y., Gondet, B., and The Omega Science Team: 2005, in *36th Annual Lunar and Planetary Science Conference*, pp 1623–+
Fraley, C.: 1999, *SIAM Journal on Scientific Computing* **20**, 270
Fraley, C. and Raftery, A.: 1999, *Computer Journal* **41**, 578
Frei, W. and Chen, C.: 1977, *IEEE Transactions on Computers* **C-26**, 988
Freudling, W. and Caulet, A.: 1993, in P. Grosbøl (ed.), *Proceedings of the 5th ESO/ST-ECF Data Analysis Workshop*, pp 63–68, European Southern Observatory
Frieden, B.: 1978a, *Image Enhancement and Restoration*, Springer-Verlag
Frieden, B.: 1978b, *Probability, Statistical Optics, and Data Testing: A Problem Solving Approach*, Springer-Verlag
Fritzke, B.: 1997, *Some competitive learning methods*, Technical report, Ruhr-Universität Bochum
Fryźlewicz, P. and Nason, G. P.: 2004, *J. Comp. Graph. Stat.* **13**, 621
Furht, B.: 1995, *Real-Time Imaging* **1**, 49
Gabow, H., Galil, Z., Spencer, T., and Tarjan, R.: 1986, *Combinatorica* **6**, 109
Gailly, J.-L.: 1993, "gzip: The Data Compression Program", htpp://www.gzip.org
Gaite, J. and Manrubia, S. C.: 2002, *Monthly Notices of the Royal Astronomical Society* **335**, 977
Galatsanos, N. and Katsaggelos, A.: 1992, *IEEE Transactions on Image Processing* **1**, 322
Gale, N., Halperin, W., and Costanzo, C.: 1984, *Journal of Classification* **1**, 75
Gammaitoni, L., Hänggi, P., Jung, P., and Marchesoni, F.: 1998, *Reviews of Modern Physics* **70**, 223
Gastaud, R., Popoff, F., and Starck, J.-L.: 2001, in *Astronomical Data Analysis Software and Systems Conference XI*, Astronomical Society of the Pacific
Geman, S. and McClure, D.: 1985, in A. S. Assoc. (ed.), *Proc. Statist. Comput. Sect.*, Washington DC
Gerchberg, R.: 1974, *Optica Acta* **21**, 709
Ghael, S., Sayeed, A., and Baraniuk, R.: 1997, in *Proc. SPIE Int. Soc. Opt. Eng*
Gleisner, H. and Lundstedt, H.: 2001, *Journal of Geophysical Research* **106**, 8425
Golub, G., Heath, M., and Wahba, G.: 1979, *Technometrics* **21**, 215
González, J. A., Quevedo, H., Salgado, M., and Sudarsky, D.: 2000, *Astronomy and Astrophysics* **362**, 835
Gonzalez, R. and Woods, R.: 1992, *Digital Image Processing*, Addison-Wesley
Gorski, K.: 1998, in *Evolution of Large-Scale Structure: From Recombination to Garching*, p. E49
Gott, J. R., Dickinson, M., and Melott, A. L.: 1986a, *Astrophysical Journal* **306**, 341
Gott, J. R., Dickinson, M., and Melott, A. L.: 1986b, *Astrophysical Journal* **306**, 341

Goutsias, J. and Heijmans, H.: 2000, *IEEE Transactions on Image Processing* **9**, 1862
Green, P. J.: 1990, *IEEE Transactions on Medical Imaging* **9(1)**, 84
Greene, J., Norris, J., and Bonnell, J.: 1997, *American Astronomical Society Meeting* **191**, 4805
Grossmann, A., Kronland-Martinet, R., and Morlet, J.: 1989, in J. Combes, A. Grossmann, and P. Tchamitchian (eds.), *Wavelets: Time-Frequency Methods and Phase-Space*, pp 2–20, Springer-Verlag
Guibert, J.: 1992, in *Digitised Optical Sky Surveys*, p. 103
Guillaume, D. and Murtagh, F.: 2000, *Computer Physics Communications* **127**, 215
Gull, S. and Skilling, J.: 1991, *MEMSYS5 Quantified Maximum Entropy User's Manual*, Royston,England
Hakkila, J., Haglin, D., Pendleton, G., Mallozzi, R., Meegan, C., and Roiger, R.: 2000, *Astronomical Journal* **538**, 165
Halverson, N. W., Leitch, E. M., Pryke, C., Kovac, J., Carlstrom, J. E., Holzapfel, W. L., Dragovan, M., Cartwright, J. K., Mason, B. S., Padin, S., Pearson, T. J., Readhead, A. C. S., and Shepherd, M. C.: 2002, *Astrophysical Journal* **568**, 38
Hamilton, A.: 1993, *Astrophysical Journal* **417**, 19
Hamilton, A. J. S., Gott, J. R. I., and Weinberg, D.: 1986, *Astrophysical Journal* **309**, 1
Hanany, S., Ade, P., Balbi, A., Bock, J., Borrill, J., Boscaleri, A., de Bernardis, P., Ferreira, P. G., Hristov, V. V., Jaffe, A. H., Lange, A. E., Lee, A. T., Mauskopf, P. D., Netterfield, C. B., Oh, S., Pascale, E., Rabii, B., Richards, P. L., Smoot, G. F., Stompor, R., Winant, C. D., and Wu, J. H. P.: 2000, *Astrophysical Journal Letters* **545**, L5
Hanisch, R. J. and White, R. L. (eds.): 1994, *The restoration of HST images and spectra - II*, Space Telescope Science Institute, Baltimore
Haralick, R. and Shapiro, L.: 1985, *CVGIP: Graphical Models and Image Processing* **29**, 100
Hebert, T. and Leahy, R.: 1989, *IEEE Transactions on Medical Imaging* **8(2)**, 194
Hecquet, J., Augarde, R., Coupinot, G., and Aurière, M.: 1995, *Astronomy and Astrophysics* **298**, 726
Heijmans, H. and Goutsias, J.: 2000, *IEEE Transactions on Image Processing* **9**, 1897
Held, G. and Marshall, T.: 1987, *Data Compression*, Wiley
Hobson, M. P., Jones, A. W., and Lasenby, A. N.: 1999, *Monthly Notices of the Royal Astronomical Society* **309**, 125
Högbom, J.: 1974, *Astronomy and Astrophysics Supplement Series* **15**, 417
Holschneider, M., Kronland-Martinet, R., Morlet, J., and Tchamitchian, P.: 1989, in *Wavelets: Time-Frequency Methods and Phase-Space*, pp 286–297, Springer-Verlag
Hook, R.: 1999, *ST-ECF Newsletter No. 26* pp 3–5
Hook, R. and Fruchter, A.: 2000, in *ASP Conference Series 216: Astronomical Data Analysis Software and Systems IX*, p. 521, Astronomical Society of the Pacific
Hopkins, A. M., Miller, C. J., Connolly, A. J., Genovese, C., Nichol, R. C., and Wasserman, L.: 2002, *Astronomical Journal* **123**, 1086
Horowitz, E. and Sahni, S.: 1978, *Fundamentals of Computer Algorithms*, Pitman
Hotelling, H.: 1933, *Journal of Educational Psychology* **24**, 417
Huang, L. and Bijaoui, A.: 1991, *Experimental Astronomy* **1**, 311
Hunt, B.: 1994, *International Journal of Modern Physics C* **5**, 151
Hyvärinen, A., Karhunen, J., and Oja, E.: 2001, *Independent Component Analysis*, Wiley, New York

Irbah, A., Bouzaria, M., Lakhal, L., Moussaoui, R., Borgnino, J., Laclare, F., and Delmas, C.: 1999, *Solar Physical Journal* **185**, 255

Ireland, J., Walsh, R. W., Harrison, R. A., and Priest, E. R.: 1999, *Astronomy and Astrophysics* **347**, 355

Irwin, M. J.: 1985, *Monthly Notices of the Royal Astronomical Society* **214**, 575

Issacson, E. and Keller, H.: 1966, *Analysis of Numerical Methods*, Wiley

Jaffe, T., Bhattacharya, D., Dixon, D., and Zych, A.: 1997, *Astrophysical Journal Letters* **484**, L129

Jain, A. K.: 1990, *Fundamentals of Digital Image Processing*, Prentice-Hall

Jalobeanu, A.: 2001, Ph.D. thesis, Université de Nice Sophia Antipolis

Jammal, G. and Bijaoui, A.: 1999, in *44th Annual SPIE Meeting, Wavelet Applications in Signal and Image Processing VII*, Vol. 3813, pp 842–849, SPIE

Jansen, M. and Roose, D.: 1998, in *Proceedings of the Joint Statistical Meeting, Bayesian Statistical Science*, American Statistical Association

Jansson, P., Hunt, R., and Peyler, E.: 1970, *Journal of the Optical Society of America* **60**, 596

Jaynes, E.: 1957, *Physical Review* **106**, 171

Jewell, J.: 2001, *Astrophysical Journal* **557**, 700

Jin, J., Starck, J.-L., Donoho, D., Aghanim, N., and Forni, O.: 2005, *Eurasip Journal* **15**, 2470

Johnstone, I.: 2001, in *Statistical Challenges in Modern Astronomy III*, Penn State, July 18-21

Johnstone, I.: 2004, *JR Statist. Soc. B* **66(3)**, 547

Joyce, M., Montuori, M., and Labini, F. S.: 1999, *Astrophysical Journal Letters* **514**, L5

Joye, W. and Mandel, E.: 2000, in *ASP Conference Series 216: Astronomical Data Analysis Software and Systems IX*, p. 91, Astronomical Society of the Pacific

Kaaresen, K.: 1997, *IEEE Transactions on Image Processing* **45**, 1173

Kalifa, J.: 1999, Ph.D. thesis, Ecole Polytechnique

Kalifa, J., Mallat, S., and Rougé, B.: 2003, *IEEE Transactions on Image Processing* **12(4)**, 446

Kalinkov, M., Valtchanov, I., and Kuneva, I.: 1998, *Astronomy and Astrophysics* **331**, 838

Karhunen, K.: 1947, English translation by I. Selin, On Linear Methods in Probability Theory, The Rand Corporation, Doc. T-131, August 11, 1960

Kass, R. and Raftery, A.: 1995, *Journal of the American Statistical Association* **90**, 773

Katsaggelos, A.: 1993, *Digital Image Processing*, Springer-Verlag

Kauffmann, G., Colberg, J. M., Diaferio, A., and White, S. D. M.: 1999, *Monthly Notices of the Royal Astronomical Society* **303**, 188

Kazubek, M.: 2003, *IEEE Signal Processing Letters* **10(11)**, 324

Kempen, G. and van Vliet, L.: 2000, *Journal of the Optical Society of America A* **17**, 425

Kerscher, M.: 2000, in K. Mecke and D. Stoyan (eds.), *Statistical Physics and Spatial Satistics: The Art of Analyzing and Modeling Spatial Structures and Pattern Formation*, Lecture Notes in Physics 554

Kerscher, M., Pons-Bordería, M. J., Schmalzing, J., Trasarti-Battistoni, R., Buchert, T., Martínez, V. J., and Valdarnini, R.: 1999, *Astrophysical Journal* **513**, 543

Kerscher, M., Szapudi, I., and Szalay, A. S.: 2000, *Astrophysical Journal Letters* **535**, L13

Kolaczyk, E.: 1997, *Astrophysical Journal* **483**, 349

Kolaczyk, E. and Dixon, D.: 2000, *Astrophysical Journal* **534**, 490

Kolaczyk, E. and Nowak, R.: 2004, *Annals of Statistics* **32(11)**, 500
Kolaczyk, E. and Nowak, R.: 2005, *Biometrika* 92(1)
Komatsu, E., Kogut, A., Nolta, M. R., Bennett, C. L., Halpern, M., Hinshaw, G., Jarosik, N., Limon, M., Meyer, S. S., Page, L., Spergel, D. N., Tucker, G. S., Verde, L., Wollack, E., and Wright, E. L.: 2003, *Astrophysical Journal, Supplement Series* **148**, 119
Konstantinides, K., Natarajan, B., and Yovanov, G.: 1997, *IEEE Transactions on Image Processing* **6**, 479
Konstantinides, K. and Yao, K.: 1988, *IEEE Transactions on Acoustics, Speech and Signal Processing* **36**, 757
Kriessler, J. R., Han, E. H., Odewahn, S. C., and Beers, T. C.: 1998, *American Astronomical Society Meeting* **193**, 3820
Kron, R. G.: 1980, *Astrophysical Journal Supplement Series* **43**, 305
Krywult, J., MacGillivray, H. T., and Flin, P.: 1999, *Astronomy and Astrophysics* **351**, 883
Krzewina, L. G. and Saslaw, W. C.: 1996, *Monthly Notices of the Royal Astronomical Society* **278**, 869
Kunz, M., Banday, A. J., Castro, P. G., Ferreira, P. G., and Górski, K. M.: 2001, *Astrophysical Journal Letters* **563**, L99
Kurokawa, T., Morikawa, M., and Mouri, H.: 2001, *Astronomy and Astrophysics* **370**, 358
Kurtz, M.: 1983, in *Statistical Methods in Astronomy*, pp 47–58, European Space Agency Special Publication 201
Kuruoglu, E., Bedini, L., Paratore, M., Salerno, E., and Tonazzini, A.: 2003, *Neural Networks* **16(3-4)**, 479
López-Caniego, M., Herranz, D., Barreiro, R. B., and Sanz, J. L.: 2005, *Monthly Notices of the Royal Astronomical Society* p. 368
Lahav, O., Naim, A., Sodré, L., and Storrie-Lombardi, M.: 1996, *Monthly Notices of the Royal Astronomical Society* **283**, 207
Lalich-Petrich, V., Bhatia, G., and Davis, L.: 1995, in Holzapfel (ed.), *Proceedings of the Third International WWW Conference*, p. 159, Elsevier
Landweber, L.: 1951, *American Journal of Mathematics* **73**, 615
Landy, S. and Szalay, A.: 1993, *Astrophysical Journal* **412**, 64
Langevin, Y. and Forni, O.: 2000, in A. Tescher (ed.), *Proc. SPIE Vol. 4115, Applications of Digital Image Processing XXIII*, pp 364–373
Lannes, A. and Roques, S.: 1987, *Journal of the Optical Society of America* **4**, 189
Lauer, T.: 1999, *Publications of the Astronomical Society of the Pacific* **111**, 227
Lawrence, J., Cadavid, A., and Ruzmaikin, A.: 1999, *Astrophysical Journal* **513**, 506
Lee, H.: 1991, U.S. Patent 5 010 504, Apr. 1991, Digital image noise noise suppression method using CVD block transform
Lee, J.: 1999, *IEEE Transactions on Image Processing* **8**, 453
Lefèvre, O., Bijaoui, A., Mathez, G., Picat, J., and Lelièvre, G.: 1986, *Astronomy and Astrophysics* **154**, 92
Lega, E., Bijaoui, A., Alimi, J., and Scholl, H.: 1996, *Astronomy and Astrophysics* **309**, 23
Lenstra, J.: 1974, *Operations Research* **22**, 413
Levy, P. (ed.): 1948, *Fonctions Aléatoires de Second Ordre*, Paris, Hermann
Liang, Z.-P. and Lauterbur, P.: 2000, *Principles of Magnetic Resonance Imaging*, SPIE
Lieshout, M. V. and Baddeley, A.: 1996, *Statistica Neerlandica* **50**, 344
Llebaria, A. and Lamy, P.: 1999, in *ASP Conference Series 172: Astronomical Data Analysis Software and Systems VIII*, p. 46, Astronomical Society of the Pacific

Lloyd, P.: 1957, *Least squares quantization in PCM*, Technical report, Bell Laboratories
Louys, M., Starck, J.-L., Mei, S., Bonnarel, F., and Murtagh, F.: 1999, *Astronomy and Astrophysics, Supplement Series* **136**, 579
Lucy, L.: 1974, *Astronomical Journal* **79**, 745
Lucy, L.: 1994, in R. J. Hanisch and R. L. White (eds.), *The Restoration of HST Images and Spectra II*, p. 79, Space Telescope Science Institute
MacQueen, J.: 1976, in *Proceedings of the Fifth Berkeley Symposium on Mathematical Statistics and Probability, Vol. 1*, pp 281–297
Maddox, S. J., Efstathiou, G., and Sutherland, W. J.: 1990, *Monthly Notices of the Royal Astronomical Society* **246**, 433
Magain, P., Courbin, F., and Sohy, S.: 1998, *Astrophysical Journal* **494**, 472
Maino, D., Farusi, A., Baccigalupi, C., Perrotta, F., Banday, A. J., Bedini, L., Burigana, C., De Zotti, G., Górski, K. M., and Salerno, E.: 2002, *MNRAS* **334**, 53
Maisinger, K., Hobson, M. P., and Lasenby, A. N.: 2004, *Monthly Notices of the Royal Astronomical Society* **347**, 339
Malgouyres, F.: 2002, *Journal of information processes* **2(1)**, 1
Mallat, S.: 1989, *IEEE Transactions on Pattern Analysis and Machine Intelligence* **11**, 674
Mallat, S.: 1991, *IEEE Transactions on Information Theory* **37**, 1019
Mallat, S.: 1998, *A Wavelet Tour of Signal Processing*, Academic Press
Mallat, S. and Hwang, W. L.: 1992, *IEEE Transactions on Information Theory* **38**, 617
Mandelbrot, B.: 1983, *The Fractal Geometry of Nature*, Freeman, New York
March, S.: 1983, *Computing Surveys* **15**, 45
Marchis, R., Prangé, R., and Christou, J.: 2000, *Icarus* **148**, 384
Marr, D. and Hildreth, E.: 1980, *Proceedings of the Royal Society of London B* **207**, 187
Marshall, P. J., Hobson, M. P., Gull, S. F., and Bridle, S. L.: 2002, *Monthly Notices of the Royal Astronomical Society* **335**, 1037
Martínez, V., López-Martí, B., and Pons-Bordería, M.: 2001, *Astrophysical Journal Letters* **554**, L5
Martínez, V., Paredes, S., and Saar, E.: 1993a, *Monthly Notices of the Royal Astronomical Society* **260**, 365
Martínez, V., Starck, J.-L., Saar, E., Donoho, D., de la Cruz, P., Paredes, S., and Reynolds, S.: 2005, *Astrophysical Journal* **634**, 744
Martínez, V. J., Jones, B. J. T., Domínguez-Tenreiro, R., and van de Weygaert, R.: 1990, *Astrophysical Journal* **357**, 50
Martínez, V. J., Paredes, S., and Saar, E.: 1993b, *Monthly Notices of the Royal Astronomical Society* **260**, 365
Martínez, V. J. and Saar, E.: 2002, *Statistics of the Galaxy Distribution*, Chapman and Hall/CRC press, Boca Raton
Matheron, G.: 1967, *Elements pour une Théorie des Milieux Poreux*, Masson
Matheron, G.: 1975, *Random Sets and Integral Geometry*, Wiley
Maurogordato, S. and Lachieze-Rey, M.: 1987, *Astrophysical Journal* **320**, 13
McCormick, W., Schweitzer, P., and White, T.: 1972, *Operations Research* **20**, 993
Mecke, K. R., Buchert, T., and Wagner, H.: 1994, *Astronomy and Astrophysics* **288**, 697
Michtchenko, T. A. and Nesvorny, D.: 1996, *Astronomy and Astrophysics* **313**, 674
Miller, A. D., Caldwell, R., Devlin, M. J., Dorwart, W. B., Herbig, T., Nolta, M. R., Page, L. A., Puchalla, J., Torbet, E., and Tran, H. T.: 1999, *Astrophysical Journal Letters* **524**, L1

Miller, C. J., Genovese, C., Nichol, R. C., Wasserman, L., Connolly, A., Reichart, D., Hopkins, A., Schneider, J., and Moore, A.: 2001, *Astronomical Journal* **122**, 3492

Moffat, A.: 1969, *Astronomy and Astrophysics* **3**, 455

Mohammad-Djafari, A.: 1994, *Traitement du Signal* **11**, 87

Mohammad-Djafari, A.: 1998, *Traitement du Signal* **15**, 545

Molina, R. and Cortijo, F.: 1992, in *Proc. International Conference on Pattern Recognition, ICPR'92*, Vol. 3, pp 147–150

Molina, R., Katsaggelos, A., Mateos, J., and Abad, J.: 1996, *Vistas in Astronomy* **40**, 539

Molina., R., Katsaggelos, A., Mateos., J., Hermoso, A., and Segall, A.: 2000, *Pattern Recognition* **33**, 555

Molina, R., Núñez, J., Cortijo, F., and Mateos, J.: 2001, *IEEE Signal Processing Magazine* **18**, 11

Molina, R., Ripley, B., Molina, A., Moreno, F., and Ortiz, J.: 1992, *Astrophysical Journal* **104**, 1662

Motwani, R. and Raghavan, P.: 1995, *Randomized Algorithms*, Cambridge University Press

Moudden, Y., Cardoso, J.-F., Starck, J.-L., and Delabrouille, J.: 2005, *Eurasip Journal on Applied Signal Processing*, to appear

Moulin, P. and Liu, J.: 1999, *IEEE Transactions on Information Theory* **45**, 909

MR/1: 2001, Multiresolution Image and Data Analysis Software Package, Version 3.0, Multi Resolutions Ltd., http://www.multiresolution.com

Mukherjee, P. and Wang, Y.: 2004, *Astrophysical Journal* **613**, 51

Mukherjee, S., Feigelson, E., Babu, G., Murtagh, F., Fraley, C., and Raftery, A.: 1998, *Astronomical Journal* **508**, 314

Murtagh, F.: 1985, *Multidimensional Clustering Algorithms*, Physica-Verlag

Murtagh, F.: 1993, in A. Sandqvist and T. Ray (eds.), *Central Activity in Galaxies: From Observational Data to Astrophysical Diagnostics*, pp 209–235, Springer-Verlag

Murtagh, F.: 1998, *Journal of Classification* **15**, 161

Murtagh, F. and Hernández-Pajares, M.: 1995, *Journal of Classification* **12**, 165

Murtagh, F. and Raftery, A.: 1984, *Pattern Recognition* **17**, 479

Murtagh, F. and Starck, J.: 1998, *Pattern Recognition* **31**, 847

Murtagh, F., Starck, J., and Berry, M.: 2000, *Computer Journal* **43**, 107

Murtagh, F. and Starck, J.-L.: 1994, *ST-ECF Newsletter No. 21* pp 19–20

Murtagh, F., Starck, J.-L., and Bijaoui, A.: 1995, *Astronomy and Astrophysics, Supplement Series* **112**, 179

Murtagh, F., Starck, J.-L., and Louys, M.: 1998, *International Journal of Imaging Systems and Technology* **9**, 38

Narayan, R. and Nityananda, R.: 1986, *Annual Review of Astronomy and Astrophysics* **24**, 127

Nason, G.: 1996, *Journal of the Royal Statistical Society B* **58**, 463

Nason, G. P.: 1994, *Wavelet regression by cross-validation*, Technical report, Department of Mathematics, University of Bristol

Natarajan, B.: 1995, *IEEE Transactions on Image Processing* **43**, 2595

Naylor, T.: 1998, *Monthly Notices of the Royal Astronomical Society* **296**, 339

Neal, R. and Hinton, G.: 1998, in M. Jordan (ed.), *Learning in Graphical Models*, pp 355–371, Kluwer

Neelamani, R.: 1999, *Wavelet-based deconvolution for ill-conditionned systems*, MS thesis, Deptment of ECE, Rice University

Neelamani, R., Choi, H., and Baraniuk, R. G.: 2004, *IEEE Transactions on Signal Processing* **52(2)**, 418

Norris, J., Nemiroff, R., Scargle, J., Kouveliotou, C., Fishman, G. J., Meegan, C. A., Paciesas, W. S., and Bonnel, J. T.: 1994, *Astrophysical Journal* **424**, 540

Novikov, D., Schmalzing, J., and Mukhanov, V. F.: 2000, *Astronomy and Astrophysics* **364**, 17

Nowak, R. and Baraniuk, R.: 1999, *IEEE Transactions on Image Processing* **8**, 666

Núñez, J. and Llacer, J.: 1998, *Astronomy and Astrophysics, Supplement Series* **131**, 167

Nuzillard, D. and Bijaoui, A.: 2000, *Astronomy and Astrophysics, Supplement Series* **147**, 129

Oberhumer, M.: 1998, "lzop: Compress or Expand Files", htpp://www.oberhumer.com/opensource/lzop

Ojha, D., Bienaymé, O., Robin, A., and Mohan, V.: 1994, *Astronomy and Astrophysics* **290**, 771

Okamura, S.: 1985, in *ESO Workshop On The Virgo Cluster of Galaxies*, pp 201–215

Olsen, S.: 1993, *CVGIP: Graphical Models and Image Processing* **55**, 319

Packer, C.: 1989, *Information Processing and Management* **25**, 307

Page, C.: 1996, in *Proceedings of the 8th International Conference on Scientific and Statistical Database Management, SSDBM'96*

Pagliaro, A., Antonuccio-Delogu, V., Becciani, U., and Gambera, M.: 1999, *Monthly Notices of the Royal Astronomical Society* **310**, 835

Paladin, G. and Vulpiani, A.: 1987, *Physics Reports* **156**, 147

Pando, J. and Fang, L.: 1998, *Astronomy and Astrophysics* **340**, 335

Pando, J., Lipa, P., Greiner, M., and Fang, L.: 1998a, *Astrophysical Journal* **496**, 9

Pando, J., Valls-Gabaud, D., and Fang, L.: 1998b, *American Astronomical Society Meeting* **193**, 3901

Pantin, E. and Starck, J.-L.: 1996, *Astronomy and Astrophysics, Supplement Series* **315**, 575

Parker, J.: 1996, *Algorithms for Image Processing and Computer Vision*, Wiley

Patanchon, G., Cardoso, J.-F., Delabrouille, J., and Vielva, P.: 2004a, *CMB and foregrounds in WMAP first year data, astro-ph/0410280*

Patanchon, G., Delabrouille, J., and Cardoso, J.-F.: 2004b, in *Proc. ICA2004, Granada, Spain*

Peebles, P.: 1980, *The Large-Scale Structure of the Universe*, Princeton University Press

Peebles, P.: 2001, in V. Martínez, V. Trimble, and M. Pons-Bordería (eds.), *Historical Development of Modern Cosmology*, ASP Conference Series, Astronomical Society of the Pacific

Pelleg, D. and Moore, A.: 1999, in *KDD-99, Fifth ACM SIGKDD International Conference on Knowledge Discovery and Data Mining*

Pen, U.: 1999, *Astrophysical Journal Supplement* **120**, 49

Pence, W. and Davoust, E.: 1985, *Astronomy and Astrophysics, Supplement Series* **60**, 517

Penzias, A. A. and Wilson, R. W.: 1965, *Astrophysical Journal* **142**, 1149

Percival, J. and White, R.: 1996, in G. Jacoby and J. Barnes (eds.), *Astronomical Data Analysis Software and Systems V*, pp 108–111

Perona, P. and Malik, J.: 1990, *IEEE Transactions on Pattern Analysis and Machine Intelligence* **12**, 629

Petit, J. and Bendjoya, P.: 1996, in *ASP Conference Series 107: Completing the Inventory of the Solar System*, pp 137–146, Astronomical Society of the Pacific

Petrou, M. and Bosdogianni, P.: 1999, *Image Processing: The Fundamentals*, Wiley

Pham, D.-T.: 2001, *SIAM Journal on Matrix Analysis and Applications* **22(4)**, 1136
Phillips, N. G. and Kogut, A.: 2001, *Astrophysical Journal* **548**, 540
Pierre, M. and Starck, J.-L.: 1998, *Astronomy and Astrophysics* **128**, 801
Pierre, M., et al.: 2004, *Journal of Cosmology and Astro-Particle Physics* **9**, 11
Pietronero, L. and Sylos Labini, F.: 2001, in *Current Topics in Astrofundamental Physics: the Cosmic Microwave Background*, p. 391
Pijpers, F. P.: 1999, *Monthly Notices of the Royal Astronomical Society* **307**, 659
Pirzkal, N., Hook, R., and Lucy, L.: 2000, in N. Manset, C. Veillet, and D. Crabtree (eds.), *Astronomical Data Analysis Software and Systems IX*, p. 655, Astronomical Society of the Pacific
Poinçot, P., Lesteven, S., and Murtagh, F.: 1998, *Astronomy and Astrophysics, Supplement Series* **130**, 183
Poinçot, P., Lesteven, S., and Murtagh, F.: 2000, *Journal of the American Society for Information Science* **51**, 1081
Pons-Bordería, M., Martínez, V., Stoyan, D., Stoyan, H., and Saar, E.: 1999, *Astrophysical Journal* **523**, 480
Popa, L.: 1998, *New Astronomy* **3**, 563
Portilla, J., Strela, V., Wainwright, M., and Simoncelli, E. P.: 2003, *IEEE Trans Image Processing* 12(12), In press
Powell, K., Sapatinas, T., Bailey, T., and Krzanowski, W.: 1995, *Statistics and Computing* **5**, 265
Pratt, W.: 1991, *Digital Image Processing*, Wiley
Press, W.: 1992, in D. Worrall, C. Biemesderfer, and J. Barnes (eds.), *Astronomical Data Analysis Software and Systems I*, pp 3–16, Astronomical Society of the Pacific
Prewitt, J.: 1970, in B. Lipkin and A. Rosenfeld (eds.), *Picture Processing and Psychopictorics*, Academic Press
Proakis, J.: 1995, *Digital Communications*, McGraw-Hill
Puetter, R. and Yahil, A.: 1999, in *ASP Conference Series 172: Astronomical Data Analysis Software and Systems VIII*, p. 307, Astronomical Society of the Pacific
Radomski, J. T., Piña, R. K., Packham, C., Telesco, C. M., and Tadhunter, C. N.: 2002, *Astrophysical Journal* **566**, 675
Ragazzoni, R. and Barbieri, C.: 1994, *Publications of the Astronomical Society of the Pacific* **106**, 683
Rauzy, S., Lachieze-Rey, M., and Henriksen, R.: 1993, *Astronomy and Astrophysics* **273**, 357
Read, B. and Hapgood, M.: 1992, in *Proceedings of the Sixth International Working Conference on Scientific and Statistical Database Management*, pp 123–131
Riazuelo, A., Uzan, J.-P., Lehoucq, R., and Weeks, J.: 2002, *Simulating Cosmic Microwave Background maps in multi-connected spaces*, astro-ph/0212223
Ribeiro, M. B.: 2005, *Astronomy and Astrophysics* **429**, 65
Richardson, W.: 1972, *Journal of the Optical Society of America* **62**, 55
Richter, G. (ed.): 1998, *IAU Commission 9 Working Group on Sky Surveys Newsletter*, AIP Potsdam
Ripley, B.: 1981, *Spatial Statistics*, Wiley
Ripley, B.: 1995, *Pattern Recognition and Neural Networks*, Cambridge University Press
Roberts, L.: 1965, in J. Tippett (ed.), *Optical and Electro-Optical Information Processing*, MIT Press
Rohlf, F.: 1973, *Computer Journal* **16**, 93
Rohlf, F.: 1982, in P. Krishnaiah and L. Kanal (eds.), *Handbook of Statistics*, Vol. 2, pp 267–284, North-Holland

Romeo, A. B., Horellou, C., and Bergh, J.: 2003, *Monthly Notices of the Royal Astronomical Society* **342**, 337
Romeo, A. B., Horellou, C., and Bergh, J.: 2004, *Monthly Notices of the Royal Astronomical Society* **354**, 1208
Rudin, L., Osher, S., and Fatemi, E.: 1992, *Physica D* **60**, 259
Saghri, J., Tescher, A., and Reagan, J.: 1995, *IEEE Signal Processing Magazine* **12**, 32
Said, A. and Pearlman, W.: 1996, *IEEE Transactions on Circuits and Systems for Video Technology* **6**, 243
Salton, G. and McGill, M.: 1983, *Introduction to Modern Information Retrieval*, McGraw-Hill
Samet, H.: 1984, *ACM Computing Surveys* **16**, 187
Sanz, J. L., Herranz, D., and Martínez-Gónzalez, E.: 2001, *Astrophysical Journal* **552**, 484
Sato, M. and Ishii, S.: 1999, in M. Kearns, S. Solla, and D. Cohn (eds.), *Advances in Neural Information Processing Systems 11*, pp 1052–1058, MIT Press
Scargle, J.: 1997, in *Astronomical Time Series*, p. 1
Schild, R.: 1999, *Astrophysical Journal* **514**, 598
Schlegel, D. J., Finkbeiner, D. P., and Davis, M.: 1998, *Astrophysical Journal* **500**, 525
Schmalzing, J., Kerscher, M., and Buchert, T.: 1996, in *Dark Matter in the Universe, Italian Physical Society, Proceedings of the International School of Physics Course CXXXII, Varenna on Lake Como, Villa Monastero, 25 July - 4 August 1995, Edited by S. Bonometto, J.R. Primack, and A. Provenzale Oxford, GB: IOS Press, 1996, p.281*, pp 281–+
Schröder, P. and Sweldens, W.: 1995, *Computer Graphics Proceedings (SIGGRAPH 95)* pp 161–172
Schwarz, G.: 1978, *The Annals of Statistics* **6**, 461
SDSS: 2000, *Sloan Digital Sky Survey*
Seales, W., Yuan, C., and Cutts, M.: 1996, *Analysis of compressed video*, Technical Report 2, University of Kentucky Computer Science Department
Sendur, L. and Selesnik, I.: 2002, *IEEE Signal Processing Letters* **9(12)**, 438
Serra, J.: 1982, *Image Analysis and Mathematical Morphology*, Academic Press
Seshadri, T. R.: 2005, *Bulletin of the Astronomical Society of India* **33**, 1
Seward, J.: 1998, "bzip2 and libzip2: a Program and Library for Data Compression", htpp://sources.redhat.com/bzip2
Shamir, L. and Nemiroff, R. J.: 2005, *Astronomical Journal* **129**, 539
Shandarin, S. F.: 2002, *Monthly Notices of the Royal Astronomical Society* **331**, 865+
Shannon, C.: 1948, *Bell System Technical Journal* **27**, 379
Shensa, M.: 1992, *IEEE Transactions on Signal Processing* **40**, 2464
Shepp, L. and Vardi, Y.: 1982, *IEEE Transactions on Medical Imaging* **MI-2**, 113
Sheth, J. V. and Sahni, V.: 2005, *astro-ph/0502105*
Sheth, J. V., Sahni, V., Shandarin, S. F., and Sathyaprakash, B. S.: 2003, *Monthly Notices of the Royal Astronomical Society* **343**, 22
Siebenmorgen, R., Starck, J.-L., Cesarsky, D., Guest, S., and Sauvage, M.: 1996, *ISOCAM Data Users Manual Version 2.1*, Technical Report SAI/95-222/Dc, ESA
Simoncelli, E. P.: 1999, in P. Müller and B. Vidakovic (eds.), *Bayesian Inference in Wavelet Based Models*, Chapt. 18, pp 291–308, Springer-Verlag, New York, Lecture Notes in Statistics, vol. 141
Skilling, J.: 1989, in *Maximum Entropy and Bayesian Methods*, pp 45–52, Kluwer
Slezak, E., de Lapparent, V., and Bijaoui, A.: 1993, *Astrophysical Journal* **409**, 517

Smoot, G. F., et al.: 1992, *Astrophysical Journal Letters* **396**, L1
Snider, S., Allende Prieto, C., von Hippel, T., Beers, T., Sneden, C., Qu, Y., and Rossi, S.: 2001, *Astrophysical Journal* **562**, 528
Snoussi, H., Patanchon, G., Macias-Peres, J., Mohammad-Djaffari, A., and Delabrouille, J.: 2004, in A. I. Physics (ed.), *MaxEnt Workshops on Bayesian Inference and MaxEnt methods*, pp 125–140
Snyder, D., Hammound, A., and White, R.: 1993, *Journal of the Optical Society of America* **10**, 1014
Soille, P.: 2003, *Morphological Image Analysis*, Springer
Soon, W., Frick, P., and Baliunas, S.: 1999, *Astrophysical Journal Letters* **510**, L135
Späth, H.: 1985, *Cluster Dissection and Analysis: Theory, Fortran Programs, Examples*, Ellis Horwood
Sporring, J. and Weickert, J.: 1999, *IEEE Transactions on Information Theory* **45**, 1051
Starck, J. and Bijaoui, A.: 1991, in *High Resolution Imaging by Interferometry*, European Southern Observatory
Starck, J.-L.: 2000, in F. Casoli, J.-L. Lequeux, and F. Davids (eds.), *IR Space Astronomy, Today and Tomorrow*, pp 63–87, EDP Sciences – Springer-Verlag
Starck, J.-L., Abergel, A., Aussel, H., Sauvage, M., Gastaud, R., Claret, A., Desert, X., Delattre, C., and Pantin, E.: 1999a, *Astronomy and Astrophysics, Supplement Series* **134**, 135
Starck, J.-L., Aghanim, N., and Forni, O.: 2004, *Astronomy and Astrophysics* **416**, 9
Starck, J.-L., Aussel, H., Elbaz, D., and Cesarsky, C.: 1997a, in G. Mamon, T. X. Thuân, and J. T. T. Vân (eds.), *Extragalactic Astronomy in the Infrared*, pp 481–486, Editions Frontières
Starck, J.-L., Aussel, H., Elbaz, D., Fadda, D., and Cesarsky, C.: 1999b, *Astronomy and Astrophysics, Supplement Series* **138**, 365
Starck, J.-L. and Bijaoui, A.: 1994, *Signal Processing* **35**, 195
Starck, J.-L., Bijaoui, A., Lopez, B., and Perrier, C.: 1994, *Astronomy and Astrophysics* **283**, 349
Starck, J.-L., Bijaoui, A., and Murtagh, F.: 1995, *CVGIP: Graphical Models and Image Processing* **57**, 420
Starck, J.-L., Candès, E., and Donoho, D.: 2002, *IEEE Transactions on Image Processing* **11(6)**, 131
Starck, J.-L., Candès, E., and Donoho, D.: 2003a, *Astronomy and Astrophysics* **398**, 785
Starck, J.-L., Donoho, D. L., and Candès, E. J.: 2001, in *Proc. SPIE Vol. 4478, Wavelets: Applications in Signal and Image Processing IX*, pp 9–19
Starck, J.-L., Martinez, V., Donoho, D., Levi, O., Querre, P., and Saar, E.: 2005, *Eurasip Journal* **15**, 2455
Starck, J.-L. and Murtagh, F.: 1994, *Astronomy and Astrophysics* **288**, 343
Starck, J.-L. and Murtagh, F.: 1998, *Publications of the Astronomical Society of the Pacific* **110**, 193
Starck, J.-L. and Murtagh, F.: 1999, *Signal Processing* **76**, 147
Starck, J.-L. and Murtagh, F.: 2001, *IEEE Signal Processing Magazine* **18**, 30
Starck, J.-L., Murtagh, F., and Bijaoui, A.: 1998a, *Image Processing and Data Analysis: The Multiscale Approach*, Cambridge University Press
Starck, J.-L., Murtagh, F., Candès, E., and Donoho, D.: 2003b, *IEEE Transactions on Image Processing* **12(6)**, 706
Starck, J.-L., Murtagh, F., and Gastaud, R.: 1998b, *IEEE Transactions on Circuits and Systems II* **45**, 1118

Starck, J.-L., Murtagh, F., Pirenne, B., and Albrecht, M.: 1996, *Publications of the Astronomical Society of the Pacific* **108**, 446
Starck, J.-L., Murtagh, F., Querre, P., and Bonnarel, F.: 2001, *Astronomy and Astrophysics* **368**, 730
Starck, J.-L., Nguyen, M., and Murtagh, F.: 2003c, *Signal Processing* **83(10)**, 2279
Starck, J.-L. and Pierre, M.: 1998, *Astronomy and Astrophysics, Supplement Series* **128**
Starck, J.-L. and Querre, P.: 2001, *Signal Processing* **81**, 2449
Starck, J.-L., Siebenmorgen, R., and Gredel, R.: 1997b, *Astrophysical Journal* **482**, 1011
Steidl, G., Weickert, J., Brox, T., Mrzek, P., and Welk, M.: 2003, *On the equivalence of soft wavelet shrinkage, total variation diffusion, total variation regularization, and SIDEs*, Technical Report 26, Department of Mathematics, University of Bremen, Germany
Steiman-Cameron, T., Scargle, J., Imamura, J., and Middleditch, J.: 1997, *Astrophysical Journal* **487**, 396
Stoyan, D., Kendall, W., and Mecke, J.: 1995, *Stochastic Geometry and its Applications*, Wiley
Strang, G. and Nguyen, T.: 1996, *Wavelet and Filter Banks*, Wellesley-Cambridge Press
Streng, R.: 1991, in H.-H. Bock and P. Ihm (eds.), *Classification, Data Analysis and Knowledge Organization Models and Methods with Applications*, pp 121–130, Springer-Verlag
Sunyaev, R. A. and Zeldovich, I. B.: 1980, *Annual Review of Astronomy and Astrophysics* **18**, 537
Sweldens, W.: 1997, *SIAM Journal on Mathematical Analysis* **29**, 511
Sweldens, W. and Schröder, P.: 1996, in *Wavelets in Computer Graphics*, pp 15–87, ACM SIGGRAPH Course notes
Sylos Labini, F.: 1999, in *Dark matter in Astrophysics and Particle Physics*, p. 336
Szapudi, S. and Szalay, A. S.: 1998, *Astrophysical Journal Letters* **494**, L41
Szatmary, K., Gal, J., and Kiss, L.: 1996, *Astronomy and Astrophysics* **308**, 791
Tagliaferri, R., Longo, G., D'Argenio, B., and Incoronato, A.: 2003, *Neural Networks* **16**, 295
Takase, B., Kodaira, K., and Okamura, S.: 1984, *An Atlas of Selected Galaxies*, University of Tokyo Press
Tarjan, R.: 1983, *Information Processing Letters* **17**, 37
Taubman, D. and Zakhor, A.: 1994, *IEEE Transactions on Image Processing* **3**, 572
Tegmark, M., et al.: 2004, *Astrophysical Journal* **606**, 702
Tekalp, A. and Pavlović, G.: 1991, in A. Katsaggelos (ed.), *Digital Image Restoration*, pp 209–239, Springer-Verlag
Thiesson, B., Meek, C., and Heckerman, D.: 1999, *Accelerating EM for large databases*, Technical Report MST-TR-99-31, Microsoft Research
Thonnat, M.: 1985, INRIA Rapport de Recherche, Centre Sophia Antipolis, No. 387, *Automatic morphological description of galaxies and classification by an expert system*
Tikhonov, A., Goncharski, A., Stepanov, V., and Kochikov, I.: 1987, *Soviet Physics – Doklady* **32**, 456
Timmermann, K. E. and Nowak, R.: 1999, *IEEE Transactions on Signal Processing* **46**, 886
Titterington, D.: 1985, *Astronomy and Astrophysics* **144**, 381
Toft, P.: 1996, *Ph.D. thesis*, Department of Mathematical Modelling, Technical University of Denmark

Touma, D.: 2000, *Astrophysics and Space Science* **273**, 233
Tretter, D. and Bouman, C.: 1995, *IEEE Transactions on Image Processing* **4**, 308
Tucker, A.: 1980, *Applied Combinatorics*, Wiley
Valtchanov, I., Gastaud, R., Pierre, M., and Starck, J.-L.: 2000, in *ASP Conference Series 200: Clustering at High Redshift*, p. 460, Astronomical Society of the Pacific
Valtchanov, I., Pierre, M., and Gastaud, R.: 2001, *Astronomy and Astrophysics* **370**, 689
van de Weygaert, R.: 1994, *Astronomy and Astrophysics* **283**, 361
Vandame, B.: 2001, in A. Banday, S. Zaroubi, and M. Bartelmann (eds.), *Mining the Sky*, pp 595–597, Springer
Véran, J. and Wright, J.: 1994, in D. Worrall, C. Biemesderfer, and J. Barnes (eds.), *Astronomical Data Analysis Software and Systems III*, p. 40, Astronomical Society of the Pacific
Verde, L., Wang, L., Heavens, A. F., and Kamionkowski, M.: 2000, *Monthly Notices of the Royal Astronomical Society* **313**, 141
Vidakovic, B.: 1998, *Journal of the American Statistical Association* **93**, 173
Vielva, P., Martínez-González, E., Barreiro, R. B., Sanz, J. L., and Cayón, L.: 2004, *Astrophysical Journal* **609**, 22
Ville, J.: 1948, *Cables et Transmissions* **2A**, 61
Vio, R., Tenorio, L., and Wamsteker, W.: 2002, *Astronomy and Astrophysics* **391**, 789
Vitter, J. and Wang, M.: 1999, in *Proceedings of the 1999 ACM SIGMOD International Conference on Management of Data*
Vitter, J., Wang, M., and Iyer, B.: 1998, in *Proceedings of the 7th International Conference on Information and Knowledge Management*
Wakker, B. and Schwarz, U.: 1988, *Annual Reviews of Astronomy and Astrophysics* **200**, 312
Walker, K. C., Schaefer, B. E., and Fenimore, E. E.: 2000, *Astrophysical Journal* **537**, 264
Watanabe, M., Kodaira, K., and Okamura, S.: 1982, *Astronomy and Astrophysics, Supplement Series* **50**, 1
Watson, A. M.: 2002, *Revista Mexicana de Astronomía y Astrofísica* **38**, 233
Weir, N.: 1991, in *3rd ESO/ST-ECF Data Analysis Workshop*
Weir, N.: 1992, in D. Worral, C. Biemesderfer, and J. Barnes (eds.), *Astronomical Data Analysis Software and System 1*, pp 186–190, Astronomical Society of the Pacific
Weistrop, D., Harvey, V., Pitanzo, D., Cruzen, S., Rogers, P., Beaver, M., and Pierce, D.: 1996, *American Astronomical Society Meeting* **189**, 122.16
White, R., Postman, M., and Lattanzi, M.: 1992, in H. MacGillivray and E. Thompson (eds.), *Digitized Optical Sky Surveys*, pp 167–175, Kluwer
White, R. L. and Allen, R. J. (eds.): 1991, *The restoration of HST images and spectra*
Wigner, E.: 1932, *Physical Review* **40**, 749
Willet, R. and Nowak, R.: 2005, *IEEE Transactions on Information Theory*, Submitted
Willett, R. M., Jermyn, I., Nowak, R. D., and Zerubia, J.: 2004, in *ASP Conf. Ser. 314: Astronomical Data Analysis Software and Systems (ADASS) XIII*, pp 107–+
Yadav, J., Bharadwaj, S., Pandey, B., and Seshadri, T. R.: 2005
Yamada, I.: 2001, in D. Butnariu, Y. Censor, and S. Reich (eds.), *Inherently Parallel Algorithms in Feasibility and Optimization and Their Applications*, Elsevier
Yao, A.: 1975, *Information Processing Letters* **4**, 21

Zahn, C.: 1971, *IEEE Transactions on Computers* **C-20**, 68
Zavagno, A., Lagage, P. O., and Cabrit, S.: 1999, *Astronomy and Astrophysics* **344**, 499
Zha, H., Ding, C., Gu, M., He, X., and Simon, H.: 2001, in *Proc. ACM CIKM 2001, 10th International Conference on Information and Knowledge Management*, pp 25–31
Zhang, S., Fishman, G., Harmon, B., Paciesas, W., Rubin, B., Meegan, C., Wilson, C., and Finger, M.: 1993, in *American Astronomical Society Meeting*, Vol. 182, p. 7105
Zibulevsky, M. and Pearlmutter, B.: 2001, *Neural Computation* **13**, 863
Zibulevsky, M. and Zeevi, Y.: 2001, *Extraction of a single source from multichannel data using sparse decomposition*, Technical report, Faculty of Electrical Engineering, Technion – Israel Institute of Technology

Index

à trous wavelet transform, 11, 17, 30, 31, 90, 280, 291, 292

ADS, 4
Aladin project, 150, 151, 153, 155, 166, 228
Altavista, 269
anisotropic diffusion, 84
Anscombe transformation, 42, 180
arithmetic coding, 139
astrometry, 155

background
– estimation, 112
– gamma-ray, 52
– model, 52
– removal, 52
Bayes
– Bayesian approach, 78
– classifier, 26, 27
– factor, 262
– information criterion, 261
BIC, 261
blind source separation, 175, 184
boundaries, treatment of, 294

c-means, 269
Cantor set, 250, 251
CAR model, 84
catalog, 233
CCD detector, 29
CDS, 151
Center for Astrophysics, 146
CGMRF model, 84
cirrus, 16
classifier, 24, 26
CLEAN, 78, 93–96
– multiresolution, 93
– wavelet, 94
closing, 16
clustering, 233, 269
– minimal spanning tree, 258

– model-based clustering, 260
– Voronoi tessellation, 259
– wavelet transform, 263
CMB, 4, 5, 27, 130, 132
competitive learning, 270
compression, 3, 137
– hyperspectral, 170
– lossless image compression, 161
– progressive decompression, 167
compression packages, 145
Compton Gamma Ray Observatory, 12
computed tomography, 74
Concise Columbia Encyclopedia, 277
contrast enhancement, 32, 57
correlation length, 237
Cosmic Microwave Background, 4, 5, 27, 176
cosmological model, 233
cosmology, 5
Cox process, 238
CT, 74
curse of dimensionality, 268
curvelet transform, 14, 37, 59, 110
– contrast enhancement, 57

data
– 2MASS, 137
– 3C120, 176
– A2390, 55
– AFGL4029, 73
– AGB, 98
– APM, 137
– APS, 137
– BATSE, 12
– Beta Pictoris, 100, 101
– black hole, 4
– CAMIRAS, 73
– CFHT, 2, 73, 169
– CHH12K, 169
– Coma cluster, 150, 152
– COSMOS, 137
– dark, 52

Index

- Deep Impact, 80
- DENIS, 137
- ESO 3.6m, 98, 101
- flat, 52
- gamma-ray burst, 4
- GLAST, 52
- GRB, 4
- Hen 1379, 98, 99
- HST, 150
- IN Com, 4
- Integral, 52
- IRAS, 16, 240
- ISOCAM, 73
- light curves, 12
- M5, 166
- M5 cluster, 139
- MAMA, 137, 150, 151, 166, 228
- MegaCam, 137, 167
- NGC 1068, 72
- NGC 2997, 32–34
- NOT, 73
- OMEGA, 176, 186
- PDSs, 137
- Planck, 227, 228
- planetary nebula, 4
- PMM, 137
- QSO, 4
- RXTE, 8
- Saturn, 4, 20, 22, 209, 211
- Schmidt plate, 139, 151
- solar system, 4, 27
- SuperCOSMOS, 137
- TIMMI, 101
- Uranus, 4
- VLT, 2, 167
- XMM-LSS, 70

database query, 268
deblending, 113, 119
decimation, 291
deconvolution, 2, 71, 97
- CLEAN, 78
- CLEAN multiresolution, 93
- detection, 126
- Gerchberg-Saxon Papoulis, 106
- Landweber method, 79, 85
- Maximum Entropy Method, 81
- multiscale entropy, 99, 220
- pixon, 87, 93
- Richardson-Lucy, 80, 85, 90, 91
- super-resolution, 105
- wavelet, 86
- wavelet CLEAN, 95
- wavelet-vaguelette decomposition, 87

Delaunay triangulation, 259
derivative of Gaussian operator, 22
detection, 111
- extended sources, 116
- matched filter, 132
- non-Gaussianity, 132
- point sources, 132
Digital Sky Survey, 145
dilation, 15, 16
discrete cosine transform, 6, 33, 139, 146, 175
discrete sine transform, 6
dithering, 107
drizzling, 107
dynamic range compression, 32

edge, 3, 18
- compass operator, 19, 21
- derivative of Gaussian operator, 19
- detection, 18
- Frei-Chen operator, 19, 20
- Kirsch operator, 21
- Laplacian of Gaussian operator, 22
- Laplacian operator, 20, 22
- Prewitt operator, 19, 20
- Roberts operator, 19, 20
- Robinson operator, 21
- Sobel operator, 19, 20
EM method, 80, 261, 263, 270
energy packing, 33
entropy, 138, 201, 204
- multiscale entropy, 99
equatorial coordinates, 233
erosion, 15, 16
ESO, 2, 150

False Discovery Rate, FDR, 48
feature extraction, 24, 25
FFT, 5
filtering, 2, 29, 215, 233, 263
- curvelet, 59
Fisz transform, 63
Fourier, 5
- fast Fourier transform, 5
- Fourier transform, 3, 33, 94, 300, 302
- power spectrum, 6, 24
- short-term Fourier transform, 7
- spectrogram, 7
fractal, 233, 249, 252
- Cantor set, 249, 251, 255–257
- Devil's staircase function, 255–258
- dimension, 251
- Hölder exponent, 252–254, 256
- Hausdorff measure, 250

– Minkowski measure, 250
– modulus maxima, 255
– multifractality, 251, 252
– partition function, 253
– Rényi dimension, 253
– singularity spectrum, 251, 252, 255
– wavelet transform, 253, 254
fuzzy join, 267

Gabor transform, 7
Gabriel graph, 259
galactic coordinates, 233
galaxy, 4
– cluster, 5
– morphology, 115
– photometry, 115
gamma-ray, 52
gamma-ray burst, 27, 63
genus function, 233, 245
Gerchberg-Saxon Papoulis method, 106
Gibbs
– energy, 83
– phenomenon, 89
Gibbs phenomenon, 77
gradient, 18

Haar multichannel transform, 183
Haar transform, 3, 34, 36, 63, 65, 67, 70, 101, 138, 145, 162, 183, 268
– Fisz transform, 63
Hadamard transform, 3
Hausdorff, 250, 251
– dimension, 251, 257
– measure, 250
Hough transform, 3, 12
Hubble Space Telescope, 71
Hubble Space Telescope WFC, 92, 93
Huffman coding, 139, 143, 144, 146
hyperspectral imagery, 170

ICA
– FastICA, 186
– wavelet, 193
– WJADE, 193
– WSMICA, 194
ICF, 104
independent component analysis, 175, 184
information theory, 202
interstellar
– matter, 282
– medium, 4
intrinsic correlation function, 206, 207
IRAS, 240, 241

ISO, 36
ISODATA, 269
isotropy, 36

J function, 244
Jacobi method, 79
JADE, 185
– wavelet, 193
JPEG 2000, 4

k-means, 269, 271
Karhunen-Loève transform, 24, 175–181
Kohonen map, 270, 273, 274
Kullback-Leibler, 176

Landweber method, 79
large images, 167
Legendre transformation, 252, 255
lifting scheme, 161
lossless compression, 161
lossy compression, 138

MAD, 41
MAP, 107
Markov random field, 82, 83
mathematical morphology, 3, 15
– closing, 16
– dilation, 15, 16
– erosion, 15, 16
– opening, 16
– skeleton, 16
matrix reordering, 274, 275
maximum a posteriori, 79
Maximum Entropy Method, 81, 99, 205–207
maximum likelihood, 26, 78
medial axis, 16
median, 140
– median absolute deviation, 41
– multiscale median transform, 140
– pyramidal median transform, 36, 138, 141, 142
MEM, 206, 207
Mexican hat wavelet, 9
MIDAS, 155
minimal spanning tree, 258, 271, 272
Minkowski, 250, 251
– dimension, 251
– functionals, 244, 247
– measure, 250
mirror wavelet basis, 88
mixture modeling, 261, 262
MMI, 66

moments, 24
morphological filter, 17
MR/1, 228, 230
MST, 258, 271, 272
multichannel data, 175, 226
multifractality, 251, 252
multilayer perceptron, 27
multiresolution support, 37, 38, 51, 90
multiscale entropy, 210, 231
– deconvolution, 72, 73, 99, 101, 109, 220
– filtering, 215, 218
– multichannel data, 226
multiscale median transform, 140
multiscale vision model, 116, 117

nearest neighbor, 271
neural network, 27
noise, 38, 39, 42, 43, 63, 65, 70, 90
– Anscombe transformation, 42, 180
– correlated, 48
– Gaussian, 29, 39, 42, 70, 79, 180
– median absolute deviation, 41
– multiplicative, 46
– non-stationary, 46, 48
– Poisson, 29, 42, 43, 63, 65, 67, 70, 80, 180, 183
– root mean square map, 46
– sigma clipping, 40
– variance stabilization, 42

object classification, 114
OLAP, 268
opening, 16

parametric, 26
partial differential equation, 84
pattern recognition, 24
PET, 74
photometry, 114, 156
Picard iteration, 80, 297
pixon, 87, 92, 93
planar graph, 271
positron emission tomography, 74
POSS-I, 150–152
posteriors, 26
potential function, 83
power spectrum, 6, 24
PRESS-optimal filter, 66
principal components analysis, 33, 175, 176, 280
priors, 26
progressive decompression, 167
PSF, 74, 75

pyramid, 299
pyramidal median transform, 36, 138, 141, 142

quadtree, 142, 144, 146

Radon transform, 3, 12, 13
regularization, 75, 206, 215
– Richardson-Lucy, 91
– Tikhonov, 75, 85
– wavelet, 89
relaxation, 23
Reverse Cuthill-McKee, 276
Richardson-Lucy method, 80, 90, 91
ridgelet transform, 12, 37, 110
ROSAT, 55
Rosette/Monoceros Region, 4

SAO-DS9, 169, 170
SAR model, 82
scalogram, 9
segmentation, 3, 23
– region growing, 23
SExtractor, 111, 228, 230
shape, 24
sigma clipping, 40
signal shaping, 108
SIMBAD, 2
simulated annealing, 274
simulation
– ΛCDM model, 241–243
– Cantor set, 255
– Cox process, 238
– Devil's staircase function, 255
single link, 258
singular value decomposition, 275
sinogram, 12
skeleton, 16
Sloan Digital Sky Survey, 137, 263
SMICA, 185, 189
– wavelet, 194
software, 309
Space Telescope Science Institute, 145
sphere, 186
– ICA on the sphere, 186
– map on the sphere, 186
SPIHT, 172
stationary signal, 40
stochastic resonance, 108
streaming, 1
super-resolution, 105
supervised classification, 26
support, 37, 90

support, multiresolution, 38, 51
SURE, 50

thresholding, 50
– adaptive thresholding, 51
– bivariate shrinkage, 52
– False Discovery Rate, FDR, 48, 51
– hard, 50, 88
– local Wiener, 52
– MMI, 66
– PRESS-optimal filter, 66
– soft, 50, 88
– SURE, 50
– universal threshold, 50
total variation, 57, 84, 101
training, 26
transform
– à trous wavelet transform, 31
– Anscombe transformation, 42
– bi-orthogonal wavelet transform, 30, 146
– curvelet transform, 14, 37, 59, 110
– discrete cosine transform, 6, 33, 139, 146, 175
– discrete sine transform, 6
– Feauveau transform, 30
– Fisz transform, 63
– Fourier transform, 3
– Gabor transform, 7
– Haar transform, 3, 63, 138
– Hadamard transform, 3
– Hough transform, 3, 12
– Karhunen-Loève transform, 24, 176
– lifting scheme, 161
– multiscale median transform, 140
– pyramidal median transform, 138, 141, 142
– Radon transform, 3, 12
– ridgelet transform, 12, 37, 110
– short-term Fourier transform, 7
– wavelet, 9
– wavelet Karhunen-Loève transform, 179, 180
– Wigner-Ville transform, 6
transformation
– Legendre, 255
two-point correlation function, 234, 235, 237

universal threshold, 35, 50
unsupervised classification, 26, 269

vaguelette, 87
Van Cittert method, 85, 90, 91, 96, 98
variance stabilization, 42
visibilities, 106
visualization, 2
Voronoi tessellation, 233, 259, 266, 270

wavelet, 9
– à trous wavelet transform, 30, 31, 291
– bi-orthogonal, 10, 36
– CLEAN deconvolution, 95
– clustering, 263
– compression, 145
– continuous wavelet transform, 9
– contrast enhancement, 32
– Daubechies wavelet, 34, 36, 163
– deconvolution, 86
– Feauveau transform, 30
– filtering, 263
– fractal analysis, 253
– Haar, 3, 63, 101, 146, 183
– hard threshold, 50
– ICA, 193
– integer wavelet transform, 165
– lifting scheme, 161
– local Wiener, 52
– Mexican hat function, 9, 132
– modulus maxima, 255
– multichannel data, 177, 179, 180
– multiscale entropy, 99, 208
– orthogonal, 10
– scalogram, 9
– significant coefficient, 38, 91
– soft threshold, 50
– thresholding, 35, 48, 50
– transform, 194
– wavelet transform, 4, 31, 34, 36, 180, 300
– wavelet-vaguelette decomposition, 87, 92
Wiener filter, 79
Wigner-Ville transform, 6

zero crossings, 22, 218

ASTRONOMY AND ASTROPHYSICS LIBRARY

Series Editors: G. Börner · A. Burkert · W. B. Burton · M. A. Dopita
A. Eckart · T. Encrenaz · B. Leibundgut · J. Lequeux
A. Maeder · V. Trimble

The Stars By E. L. Schatzman and F. Praderie

Modern Astrometry 2nd Edition
By J. Kovalevsky

The Physics and Dynamics of Planetary Nebulae By G. A. Gurzadyan

Galaxies and Cosmology By F. Combes, P. Boissé, A. Mazure and A. Blanchard

Observational Astrophysics 2nd Edition
By P. Léna, F. Lebrun and F. Mignard

Physics of Planetary Rings Celestial Mechanics of Continuous Media
By A. M. Fridman and N. N. Gorkavyi

Tools of Radio Astronomy 4th Edition, Corr. 2nd printing
By K. Rohlfs and T. L. Wilson

Tools of Radio Astronomy Problems and Solutions 1st Edition, Corr. 2nd printing
By T. L. Wilson and S. Hüttemeister

Astrophysical Formulae 3rd Edition (2 volumes)
Volume I: Radiation, Gas Processes and High Energy Astrophysics
Volume II: Space, Time, Matter and Cosmology
By K. R. Lang

Galaxy Formation By M. S. Longair

Astrophysical Concepts 4th Edition
By M. Harwit

Astrometry of Fundamental Catalogues
The Evolution from Optical to Radio Reference Frames
By H. G. Walter and O. J. Sovers

Compact Stars. Nuclear Physics, Particle Physics and General Relativity 2nd Edition
By N. K. Glendenning

The Sun from Space By K. R. Lang

Stellar Physics (2 volumes)
Volume 1: Fundamental Concepts and Stellar Equilibrium
By G. S. Bisnovatyi-Kogan

Stellar Physics (2 volumes)
Volume 2: Stellar Evolution and Stability
By G. S. Bisnovatyi-Kogan

Theory of Orbits (2 volumes)
Volume 1: Integrable Systems and Non-perturbative Methods
Volume 2: Perturbative and Geometrical Methods
By D. Boccaletti and G. Pucacco

Black Hole Gravitohydromagnetics
By B. Punsly

Stellar Structure and Evolution
By R. Kippenhahn and A. Weigert

Gravitational Lenses By P. Schneider, J. Ehlers and E. E. Falco

Reflecting Telescope Optics (2 volumes)
Volume I: Basic Design Theory and its Historical Development. 2nd Edition
Volume II: Manufacture, Testing, Alignment, Modern Techniques
By R. N. Wilson

Interplanetary Dust
By E. Grün, B. Å. S. Gustafson, S. Dermott and H. Fechtig (Eds.)

The Universe in Gamma Rays
By V. Schönfelder

Astrophysics. A New Approach 2nd Edition
By W. Kundt

Cosmic Ray Astrophysics
By R. Schlickeiser

Astrophysics of the Diffuse Universe
By M. A. Dopita and R. S. Sutherland

The Sun An Introduction. 2nd Edition
By M. Stix

Order and Chaos in Dynamical Astronomy
By G. J. Contopoulos

Astronomical Image and Data Analysis
2nd Edition By J.-L. Starck and F. Murtagh

The Early Universe Facts and Fiction
4th Edition By G. Börner

ASTRONOMY AND ASTROPHYSICS LIBRARY

Series Editors: G. Börner · A. Burkert · W. B. Burton · M. A. Dopita
A. Eckart · T. Encrenaz · B. Leibundgut · J. Lequeux
A. Maeder · V. Trimble

The Early Universe Facts and Fiction
4th Edition By G. Börner

The Design and Construction of Large Optical Telescopes By P. Y. Bely

The Solar System 4th Edition
By T. Encrenaz, J.-P. Bibring, M. Blanc, M. A. Barucci, F. Roques, Ph. Zarka

General Relativity, Astrophysics, and Cosmology By A. K. Raychaudhuri, S. Banerji, and A. Banerjee

Stellar Interiors Physical Principles, Structure, and Evolution 2nd Edition
By C. J. Hansen, S. D. Kawaler, and V. Trimble

Asymptotic Giant Branch Stars
By H. J. Habing and H. Olofsson

The Interstellar Medium
By J. Lequeux

Methods of Celestial Mechanics (2 volumes)
Volume I: Physical, Mathematical, and Numerical Principles
Volume II: Application to Planetary System, Geodynamics and Satellite Geodesy
By G. Beutler

Solar-Type Activity in Main-Sequence Stars
By R. E. Gershberg

Relativistic Astrophysics and Cosmology
A Primer By P. Hoyng

Magneto-Fluid Dynamics
Fundamentals and Case Studies
By P. Lorrain

```
SCI QB 51.3 .I45 S73 2006

Starck, J.-L 1965-

Astronomical image and data

analysis
```

DATE DUE

GAYLORD   PRINTED IN U.S.A.